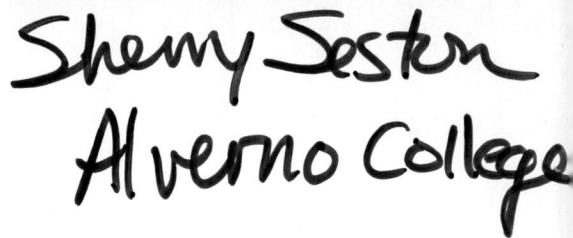

Unix and Perl to the Rescue!

Your research has generated gigabytes of data and now you need to analyze it. You hate using spreadsheets, but it's all you know, so what else can you do? This book will transform how you work with large and complex data sets, teaching you powerful programming tools for slicing and dicing data to suit your needs.

Written in a fun and accessible style, this step-by-step guide will inspire and inform non-programmers about the essential aspects of Unix and Perl. It shows how, with just a little programming knowledge, you can write programs that could save you hours, or even days. No prior experience is required, and new concepts are introduced using numerous code examples that you can try out for yourself. Going beyond the basics, the authors touch upon many broader topics that will help those new to programming, including debugging and how to write in a good programming style.

KEITH BRADNAM is a project scientist in the Genome Center at the University of California, Davis. He has extensive experience working with model organism databases and spent four years as a project leader at WormBase, helping to develop this important bioinformatics resource.

IAN KORF is an Associate Professor in Molecular and Cellular Biology at the University of California, Davis. His research seeks to understand structure and function in genomic DNA. He has developed new tools for gene prediction, co-authored the only book devoted to BLAST and helped in the development of BioPerl.

Unix and Perl to the Rescue!

A field guide for the life sciences (and other data-rich pursuits)

KEITH BRADNAM
University of California, Davis

IAN KORF
University of California, Davis

CAMBRIDGE
UNIVERSITY PRESS

University Printing House, Cambridge CB2 8BS, United Kingdom

Cambridge University Press is part of the University of Cambridge.

It furthers the University's mission by disseminating knowledge in the pursuit of
education, learning and research at the highest international levels of excellence.

www.cambridge.org
Information on this title: www.cambridge.org/9781107000681

© Keith Bradnam and Ian Korf 2012

First published 2012
Reprinted 2014

Printed in the United Kingdom by Clays, St Ives plc.

A catalogue record for this publication is available from the British Library

Library of Congress Cataloguing in Publication data
Bradnam, Keith.
  UNIX and Perl to the rescue! : a field guide for the life sciences (and other data-rich pursuits) / Keith
  Bradnam, Ian Korf.
    p.   cm.
  Includes bibliographical references and index.
  ISBN 978-1-107-00068-1 (hardback) – ISBN 978-0-521-16982-0 (paperback)
  1. Science – Data processing.   2. UNIX (Computer file)   3. Perl (Computer program language)
  I. Korf, Ian.   II. Title.
  Q183.9.B73   2012
  005.4'32–dc23   2011047927

ISBN 978-1-107-00068-1 Hardback
ISBN 978-0-521-16982-0 Paperback

Contents

Introduction and background 1

Why this book?

If this book had to have a mission statement, we would say that it is designed to help you make the transition from computer *user* to computer *programmer*.[1]

We wrote this book with life scientists in mind. But it is equally appropriate for anyone who needs to slice and dice large, diverse data sets. A few years ago, biologists did not need to know how to program. With the arrival of the Human Genome Project and other -omic technologies, biology has been transformed into an incredibly data-rich science. While the science is moving ahead at a staggering rate, most people have not changed themselves to match. Not everyone needs to know how to program, but for those that desire it, this book will help them catch up quickly.

We have both watched students struggle with trying to analyze mountains of data, and sometimes the struggle has not been because the students lack the ability to tackle the problem. Rather, it is because they frequently lack the *tools* to tackle the problem. For many people, data analysis means 'using a spreadsheet.' Sometimes this is all you need, but for many problems a programming solution will be faster, easier, and much more powerful.

This is not a book for dummies or idiots. Conversely, it's also not for super-geniuses. It's for ordinary educated people who haven't needed to program until now. Whether the topic is language, mathematics, or programming, some people learn faster than others. But we all learn to read, write, multiply, and divide. And we can all learn to program. Rest assured, you *can* program. We are happy to be your guides.

Learning to program is a journey. Like other journeys, it takes time and effort. But the rewards are worth every step. Not only will you be learning a new skill that you can apply to your work, you will be seeing the world of data from a completely different perspective. We guarantee you will find this personally enlightening, and we are not exaggerating when we say that your newfound knowledge will empower you more than you can imagine.

Why Unix?

The Unix OS has been around since 1969 and it's not likely to disappear any time soon. Back then there was no such thing as a graphical user interface (GUI). You typed everything. It may seem archaic to use a keyboard to issue commands today, but it's much easier to automate keyboard-driven tasks than mouse-driven tasks. There are several variants of Unix (including Linux), though the differences do not matter much. Although you may not have noticed it, Apple has been using Unix as the underlying OS on all of their computers since 2001.[2]

---

[1] Note that this doesn't mean you need to grow a beard, start reading science-fiction books, or wear T-shirts bearing unfathomable geeky slogans. Indeed, all of these clichés about programmers should be tossed aside. Programmers are real people ... well, most of us are anyway.

[2] If you haven't noticed it, that's probably because it is 'hidden' behind a very slick-looking GUI. But it's there nonetheless.

Increasingly, the raw output of biological research exists as *in silico* data, usually in the form of very large text files that can grow to several gigabytes in size. Unix is particularly suited to working with such files and has several powerful (and flexible) commands that can process your data for you. The real strength of learning Unix is that most of these commands can be combined in an almost unlimited fashion. If you can learn just five Unix commands, you will be able to do a lot more than just five things.

Why Perl?

Perl is one of the most popular programming languages, and has a particularly strong following in the bioinformatics community. People sometimes get argumentative about which language is best. There is no single best language for everything. Perl does most things very well, and is a fine programming language to learn. Other equally capable and easy to use languages include Python and Ruby. Once you learn how to program well in one language, adapting to other languages is trivial.

Originally developed in 1987, Perl remains under active development and there is therefore a *lot* of supporting material available to help you learn it.[3] You are very likely to find Perl pre-installed on just about every type of Unix/Linux-based OS, and it is also available for Windows.

Among programming languages, there is often a distinction between those that are *interpreted* (e.g., Perl, Python, Ruby) and those that are *compiled* (e.g., C, C++, Java). People often call interpreted programs *scripts*. It is generally easier to learn programming with a scripting language because you don't have to worry as much about variable types and memory allocation. The downside is that the interpreted programs often run much slower than compiled ones. But let's not get lost in petty details. Scripts are programs, scripting is programming, and computers can solve problems quickly regardless of the language.

About the authors

Keith Bradnam started out his academic career studying ecology. This involved lots of field trips and throwing quadrats around on windy hillsides. He was then lucky enough to be in the right place at the right time to do a Masters degree in bioinformatics (at a time when nobody was very sure what bioinformatics was). From that point onwards he has spent most of his waking life sat at a keyboard (often staring into a Unix terminal). A PhD studying eukaryotic genome evolution followed; this was made easier by the fact that only one genome had been completed at the time he started (this soon changed). After a brief stint working on an *Arabidopsis* genome database he moved to working on the excellent model organism database, WormBase, at the Wellcome Trust Sanger Institute. It was here that he first met Ian Korf and where they bonded over a shared love of Macs, neatly written code, and English puddings. Ian then tried to run away and hide in California at the UC Davis Genome Center, but Keith tracked him down and joined his lab. Apart from doing research, he also gets to look after all the computers in the lab

---

[3] A good 'first port of call' would be www.perl.org, the official web site of the Perl programming language.

and teach the occasional class or two. However, he would give it all up for the chance to be able to consistently beat Ian at foosball, but that seems unlikely to happen anytime soon. Keith still likes Macs and neatly written code, but now has a much harder job finding English puddings.

As a youth, Ian Korf's favorite classes were sciences and his favorite pastime was computer gaming. At the time, you wouldn't have thought that hacking and writing computer games would be very useful skills for a budding molecular biologist. Certainly nobody ever counseled Ian to do so, especially when he was doing it at 2 a.m.! But apparently the misspent hours of youth can sometimes turn out to be worthwhile investments. Ian's first experience with bioinformatics came as a post-doc at Washington University (St. Louis), where he was a member of the Human Genome Project. He then went 'across the pond' to the Sanger Institute for another post-doc. There he met Keith Bradnam, and found someone who truly understood the role of communication and presentation in science. Ian was somehow able to persuade Keith to join his new lab in Davis, California. This book is but one of their hopefully useful contributions.

Acknowledgments

This book evolved from a course that we both teach to graduate and undergraduate students at UC Davis. We are grateful to the students for their patience with us, as this course has evolved quite a bit since we started teaching it. Their feedback, and their enthusiasm for learning Unix and Perl, have made this book what it is. We also would like to thank Nancy Parmalee for helpful suggestions about the index.

Keith would also like to thank his wife Mel for her tireless support and understanding throughout the long book-writing process. He would also like to express profound gratitude to the wonders of caffeine, the relaxation afforded by his iTunes music library, and to the entire nation of Belgium.

Ian would like to thank all his students, past, present, and future. May your adventures take you to lands unimagined, and your skills see you safely home.

This book was written using Apple's excellent 'Pages' word-processing software, with extensive use of Dropbox software by Dropbox Inc. to make the process of collaborative writing a joy. Code examples were written using TextMate by MacroMates Ltd and TextWrangler by Bare Bones Software.

Or rather, how *not* to use this book

Organization

This book is divided into seven parts (you are currently reading Part 1). You may be impatient to start programming with Perl, but if you don't know any Unix we suggest that you start with Part 3, which will teach you the basics of Unix. When you finish that, you can optionally jump ahead to Part 5, which covers some advanced Unix topics. Or you might just want to proceed to Part 4, which covers all of the fundamentals of Perl. The choice is yours. Of course, if you don't yet have Unix and Perl installed on your computer, then you might want to start with Part 2, which covers how you can get Unix and Perl for your PC.

If you've never programmed, we hope that after learning the 'essential' Perl of Part 4, you will be able to write many fantastic and powerful scripts. More importantly, we hope that you will be able to write scripts that are actually *useful*. For this part of the book, we've tried, where possible, to only ever introduce one new concept at a time. Hopefully this will prevent you from being overloaded with too many new concepts at once. This also keeps chapters short and, mostly, self-contained. For a few topics that have increased complexity, we use two or more chapters to cover all aspects of that topic.

We have strived to make sure there are lots of examples. These are all scripts that we encourage you to copy and try yourself. However, you may still gain much understanding just from reading them. In addition to the examples, Part 4 of this book also features a number of problems at the end of most chapters.[4] You are strongly encouraged to tackle the problems. Ultimately, this it the best way to learn Perl (or any programming language). For each problem we provide a solution,[5] but be aware that one of the famous mottos associated with Perl is:

> *TMTOWTDI*[6] *– There's more than one way to do it*

We have hopefully provided solutions that are easily understandable, but if you want to solve each problem in a different way then that is great.

The topics covered in Part 4 might be all you ever need to know in order to solve many different problems. However, we go further into the more advanced aspects of Perl in Part 6. The distinction between 'essential' and 'advanced' is somewhat arbitrary. If you finish Part 4 then you should at least have a look at Part 6.

Part 7 covers many different subjects that are not unique to Perl. In general, this is the section that focuses on 'good programming practices.' Most subjects in this part are relevant to many programming languages, though we also include two sections on how to fix broken Perl scripts.

Finally, we should note that we do not cover every aspect of Unix and Perl. The world of Unix is especially vast, and several books would be needed in order to cover

---

[4] We include some problems in the Unix section too, but not as many.
[5] Included in an appendix.
[6] Some people pronounce this 'Tim Toady.'

the myriad number of Unix commands you could learn about. Likewise, we do not cover every feature or function available in Perl. However, we strongly feel that this book covers all of the basics (and much more besides). Readers are therefore encouraged to use this book as a launch pad for a journey into a much wider world of programming. If you develop a hunger for learning about new Unix commands, Perl functions, and even new programming languages, then dare to venture beyond the confines of this book. You will be rewarded!

Style conventions

Each chapter has a main heading and a subheading. The subheadings are one area where we have tried to inflict our pun-tastic sense of humor on you.[7] We will often include both Unix and Perl examples, which you should attempt to follow. The Unix examples will include simple instructions of Unix commands that you should type, whereas the Perl examples will contain complete scripts, accompanied by line numbering. E.g.:

Example 1.2.1

```
1.   #!/usr/bin/perl
2.   print "The shortest script in the world?\n";
```

The line numbers are just there so we can refer to them in the text. You are not meant to type the line numbers! Following just about every example will be a section that tries to explain what the point of the example was. For the Unix examples, this will be a section titled *Explanation*, but for the Perl scripts it will be a line-by-line breakdown of how the script is working. E.g.:

Understanding the script

Line 2 contains a simple `print` statement.

In addition to having worked-through examples, we will also set problems that you should try to solve. Where appropriate, answers will be provided, but we encourage you to try solving the problems without looking at the solution.

Hopefully you will have noticed that we use a fixed-width font for writing any Perl or Unix code. This will be done for complete scripts and even when we mention a single Unix command or Perl function within a sentence. E.g., we might mention that the Unix command `sed` shares similarities with Perl's substitution operator (`s//`).

Sometimes we will show fragments of Perl scripts, just to illustrate a point or to demonstrate the syntax of a command. We do not include line numbers for these examples, and they are not intended to be run as complete scripts. E.g.:

```
my @array_A = @array_B; # copying an array
```

---

[7] Footnotes, like this one, are another place where you might find occasional diversionary comments on matters which may not be entirely related to Unix and Perl. E.g., did you know that there are no words in the English language that rhyme with the word 'orange'?

Occasionally, we will want to shout something at you because it is *so* important, and the world will cease to exist if you fail to understand the critical point we are making. E.g.:

> *The world will cease to exist if you fail to understand the critical point we are making!*

Any time you see something written in this style, you should probably re-read it several times and remember that we will be not-inconsiderably displeased if you fail to remember our advice!

Installing Unix and Perl 2

What do I need in order to learn Unix and Perl?   2.1

In addition to time, patience, and a beverage of your choice…

This book is primarily intended to teach you how to program in Perl. But in order to do that we want you to first learn some Unix skills. You will need access to a computer that is capable of running the Perl programming language *and* a Unix or Linux Operating System (OS). You will also need to use a *code editor* program to write your Perl scripts. We'll talk about what specific software to choose in the next chapter, but first let's consider the bigger picture, including a brief overview of how Unix and Linux differ and whether that even matters.

What computers can run Perl?

As a programming language, Perl is platform agnostic. You can write (and run) Perl scripts on just about any modern computer.[1] We will assume that >99% of the people reading this use a PC (running Microsoft Windows and/or Linux), or an Apple Mac. A small proportion of you may be using some other type of dedicated Unix platform, such as a Sun or SGI machine. None of this really matters too much. All of the Perl examples that we demonstrate in this book should work on any machine that you can install Perl on.

What computers can run Unix?

While you can easily install Perl on most computers, the same is not quite so true for Unix. We decided that this book should include an introduction to Unix because a lot of *in silico* biological research happens on Unix/Linux platforms. So it makes sense to learn how to run your Perl scripts within the context of a Unix OS. As we mentioned in Chapter 1.1, all modern Mac computers use Unix as the underlying OS. This means that if you own (or have access to) a Mac computer that runs OS X, then you don't need to download or install anything in order to work through this book.[2] If you don't have access to a Mac and you own a PC running Windows, then we suggest you add Linux to your machine (an overview of how to do this is provided in the next chapter).

Linux

If you don't have access to Unix, then hopefully you can install Linux instead. Linux is often called a Unix-*like* OS, though this claim may offend any Linux purists reading this. It is unusual in that its core components are all based on free and open-source software. This doesn't make it any less powerful or less secure than other OSs, and you might feel more comfortable knowing that the majority of web servers around the world

---

[1] If your definition of 'modern' is running Windows 3.1 on a PC that features a 386DX processor, then you might be disappointed to discover that you will probably need to purchase a new computer.
[2] Except for installing a code editor.

run Linux. Because the underlying code used by all Linux systems is freely available, many companies have packaged together slightly different versions of the OS[3] that you can download for free. This very attractive price point, coupled with the fact that it is possible to run Linux without having to install anything on your computer's hard drive (see the next chapter) means that Linux is a great solution for PC owners wanting to work through this book.

Unix vs. Linux: part 1

When we say that we want to teach you Unix, we are not talking about learning the entire OS.[4] Instead, we want to teach you about a few of the Unix commands that you will have to type into a program known as a *terminal* (which we will explore in Chapter 3.2). There are probably a few thousand Unix commands, but we are only aiming to teach you 20 or so of the most essential ones. All of the Unix commands that we will teach you are also available on Linux. This means that all we need from a Unix or Linux OS is the ability to open a terminal program and run some Unix commands. All other differences between Unix and Linux are not important for this book.

Unix vs. Linux: part 2

You do know that there is an exception to every rule, right? While we are confident in saying that there are no important differences between Unix and Linux, you should be aware that there are still differences, both between Unix and Linux and between different versions of Unix/Linux. These differences can mean that some of the commands that you type *may* produce slightly different output to what we show in this book. Sometimes this is because one OS might have a newer version of a Unix command when compared to another OS.

This means it would be impossible to write this book in a way that makes it fully compatible with *all* possible Unix and Linux variants, so we have written this book using Apple's version of Unix.[5] We are 99.9% sure that every Unix command we mention will be available on whatever form of Unix or Linux you use, but bear in mind that the output of our Unix examples might sometimes look different to yours.

Learning Perl without learning Unix?

If you are using a Windows PC, then you might not want to install Linux and you might just want to learn Perl. That's fine, but bear in mind that our Perl examples are written from a Unix point-of-view, so we will show examples of how to run Perl scripts from the perspective of a Unix system.

---

[3] Such packages are known as *distributions* and usually contain a slightly different mix of software tools and sometimes a different GUI.
[4] Otherwise you could argue that learning to use a Mac is a way of learning how to use Unix.
[5] We have chosen Apple because we both use Apple computers for our work (including running Unix commands and writing Perl scripts).

There are virtually a dozen solutions

There are four main ways you can install a Linux OS on your Windows PC:

(1) Install Cygwin. This provides a Linux-like environment on your PC; it is free to download.
(2) Run Linux from a CD-ROM or install on a bootable flash drive. This is a good solution for people who don't want (or do not have permission) to modify the contents of their hard drive.
(3) Install a full Linux distribution on your computer, either as a replacement for Windows or as a dual-boot option.
(4) Run Linux by using *virtualization* software. There are many software packages that will allow you to effectively install an OS *inside* another one.

We will discuss each of these options in a little more detail in this chapter, though it is beyond the scope of this book to provide detailed installation instructions, not least because any instructions we could provide would quickly become out of date (the world of Linux moves very quickly). Bear in mind that the internet contains a lot of information about installing Linux on PCs.

Linux distributions

If you didn't already know, you should be aware that there are many different versions of Linux in existence. They differ in many respects, but the core functionality is very similar no matter which one you choose. The web site www.livecdlist.com lists most of the popular variants that are out there and provides a good starting point for choosing a distribution. Fashions come and go in computing, and it is likely that this list will look very different in a few years' time. One of the most popular full-featured Linux OSs out there at the time of writing is Ubuntu (available at ubuntu.com). For the purposes of working through this book, any of the popular distributions will be fine, but if you want to install Linux on a CD-ROM or flash drive you might want to choose a distribution that requires less space (see the relevant section below).

Installing Cygwin

It is important to note that Cygwin isn't a true form of Unix or Linux. It is software that will result in you having a terminal window through which you can use many Unix programs (including Perl). There are some differences between Cygwin and other types of Unix, which may mean that not every Unix example in this book will work exactly as described, but overall it should be sufficient for you to learn the basics of Unix. At the time of writing, Cygwin is free and is under active development. You can download it from www.cygwin.com.

Running from a CD-ROM or flash drive

Storage capacities of USB flash drives continue to increase (and prices decrease), which means it is possible to store entire OSs on them. It is also possible to boot your computer from a USB drive and run the OS that you have installed. Because not everyone has a large-capacity flash drive, and because some people would like to run Linux from a CD-ROM, there has been a demand for lightweight Linux distributions which omit some of the less essential parts in order to fit on a flash drive or CD-ROM. Linux distributions such as Damn Small Linux (www.damnsmalllinux.org) go so far as to fit an entire OS into less than 50 MB of space. Other popular versions of Linux which are also compact are Slax (available at www.slax.org) and Puppy Linux (www.puppylinux.com).

The obvious advantages to these methods are that you don't need to add anything to your computer and you can easily take your Linux OS anywhere you go. Some of these solutions will still require you to boot your computer from the flash drive or CD-ROM (meaning that you can't use Windows until you reboot). However, some of these solutions can also be run without restarting your computer, meaning that you can access Linux within a single window.

Install Linux

If you try out the method above and discover that you like Linux, you may want to make it your main OS, or at least have it available to run alongside your Windows OS. It is very common to find 'dual-boot' machines, which means you can choose which OS you want to run from a menu after turning the machine on. Setting up Linux in this manner will require sufficient free space on your hard drive and you will also need administrator privileges on your computer. Follow the instructions provided by your Linux distribution. Alternatively, rather than have a dual-boot system, you can make Linux your main OS and run Windows using virtualization software.

Virtualization software

By installing suitable software, virtualization effectively allows you to run one (or more) different OSs within your main OS. It is very commonly used to run Linux as a *guest* OS within a Windows OS, and vice versa.[6] Most modern computers will have sufficient hardware to do this. Some virtualization software is free to download. This is another fast-moving field within the software industry and it is therefore hard to make specific suggestions as to which software to use. Some popular virtualization solutions that are currently available include Microsoft Virtual PC, VirtualBox (free from Oracle), VMware Player, and Parallels Workstation. However, please note that there are also many other products available and we do not endorse any one of these products.

---

[6] It's also very common to see virtualization software used in such a way that one computer runs older versions of its own OS (e.g., running Windows XP from within Windows 7).

Installing a code editor

So I installed Perl and Linux, can I start writing code yet?

The majority of this book will teach you how to write Perl scripts, and you will therefore need something to write them with. Like other scripting languages, Perl scripts are just text files and can therefore be written with any software capable of producing a plain-text-format file. Note that *plain text* specifically means text that is devoid of formatting. All OSs come with basic text editors that are capable of producing plain-text files.[7] You should *not* use these editors. Nor should you use a fully fledged word-processor program such as Microsoft Word. This point bears repeating:

> *Do not use a word processor to write code, it will cause stress and grief!*

You should instead use a program that is specifically designed to write code. Such programs are known as *text editors* or *source code editors*, and include a number of features that will greatly help you as you learn to program in Perl. The most important reason for using a code editor is that they already know about the syntax of Perl (as well as many other programming languages) and will change the color of what you write in a process called *syntax highlighting*. A simple analogy would be to imagine that you could write a sentence in English and have all the verbs and nouns automatically change to red or blue. If you mistyped the name of a verb, then it wouldn't turn red, and this would give you instant visual feedback that there was a problem. As you will quickly learn, fixing bugs in code can be a troublesome task, so anything that helps you find bugs as you write them is to be welcomed.

Apart from syntax highlighting, code editors have many other useful features and it is essential to use them when writing any code. There are many free ones available and you should ideally try out several to find one that works for you. They all have slightly different combinations of features and some are cross-platform, whereas others will only be available for Macs, PCs, or Linux OSs. As a starting point, we would suggest Notepad++ for Windows, TextWrangler for Macs, and Gedit or jEdit for Linux. Once again, we are not trying to endorse any particular piece of software. As you start to type a lot of code, the relationship between you and your code editor becomes very important, and it is highly recommended that you 'test drive' other editors.[8]

---

[7] Typically, this would be TextEdit on a Mac and Notepad on a Windows PC.

[8] Wikipedia has a very detailed page that compares the features of various text editors: http://en.wikipedia.org/wiki/Comparison_of_text_editors.

Essential Unix 3

No mouse required!

By this point you should have a computer that runs a version of Unix or Linux. Everything we do in this part of the book will involve *typing* commands using a program known as the *terminal* (more on that in the next section). Unix contains many hundreds of commands, but we only need to learn a small number in order to achieve most of what we want to accomplish.

You are probably used to working with programs like the Apple Finder or the Windows File Explorer to navigate around the hard drive of your computer. Some people are habituated to using the mouse to move files, drag files to the trash, etc., and it can seem strange switching from this to typing commands instead. Be patient, and try – as much as possible – to stay within the world of the Unix terminal. We will teach you many basics of Unix, such as: renaming files, moving files, creating text files, etc. and you may sometimes be tempted to resort to doing this without using Unix. Initially it will feel wrong to do something as simple as moving a file from one folder to another by typing a command. Stick with it and it will start to become second nature. Learning to do things by typing commands also gives you a back-up plan if your mouse breaks!

Throughout this part of the book we will provide lots of Unix examples that you should also type yourself. Please make sure you complete and understand each task before moving on to the next one. We will sometimes show the output of running various commands. In some cases your output will look different to our output because it is very unlikely that any two filesystems will be identical (even on computers with the same OS). Hopefully, though, you will be able to follow all of these examples without getting too lost.

One final note: Throughout the remainder of this part of the book we will refer to Unix over and over again. Every time we mention the U-word, you can equally think of the L-word (Linux). From the perspective of what we are trying to teach you, the two are synonymous.[1]

---

[1] If this bothers you, then feel free to buy our special 'Linux-edition' of this book. It is identical, except we change all mentions of Unix to Linux.

Download puTTY.exe for terminal connection.

The Unix terminal

A window into a wider world

A 'terminal' is the common name[2] for a type of program that does two main things. It allows you to send typed instructions *to* the computer (i.e., run programs, move/ view files, etc.) and it allows you to see the output that results *from* those instructions. Historically, computers did not have any form of GUI, so the only way of interacting with them was by typing commands to do everything.[3] The keyboard was the only form of input and a single monitor screen was the only form of output. In modern-day OSs, the terminal will be run as just one of many different applications; some people refer to terminal applications as terminal *emulators*.

You can count on all Unix/Linux OSs to have a terminal program, and it is common for any such program to include 'term,' 'terminal,' or 'tty' as part of its name – e.g., on Apple computers the default terminal application is simply named Terminal. Bear in mind that all OSs will also allow you to download alternative terminal programs. These will offer you different degrees of customization as well as slightly different features. However, for the purposes of this book, the differences between any two terminal applications are trivial.

After launching the terminal program, you should see something that looks a bit like this:

This is the standard Apple terminal program. Yours might look very different,[4] but there should at least be some text inside the terminal window, and perhaps a blinking

---

[2] Also known as a 'term' or a 'tty.'

[3] Of course, there is the whole 'pre-keyboard' era of punch cards and punch tape as forms of data input. But enough with the history lesson.

[4] It is fairly common to see terminal programs use a two-color scheme of either: black on white, white on black, or the *Matrix*-esque green on black.

cursor. The text might just be a single character such as a $, %, or # symbol, or it might include other information such as the name of your computer or your login name.

Customizing your terminal

Before we go any further, you should note that your terminal program will very likely let you alter the appearance of the terminal window. If you explore its options/settings/ preferences menu you will probably be able to do things like change the default colors, style of font, and size of font. Initially it might be better to stick with the default settings until you are comfortable using the program, but at some point you should set up your terminal so it is to your liking.[5]

Note that you can resize terminal windows, or have multiple windows open side by side. Some terminal applications will also let you open multiple tabs within a single window. There will be many situations where it will be useful to have multiple terminals open and it will be a matter of preference as to whether you want to have multiple windows, or one window with multiple tabs (there will usually be keyboard shortcuts for switching between windows or tabs).

Before we go any further you might also want to check what keyboard commands are used to close windows or tabs, just so you don't *accidentally* do that.[6] For much of this part of the book you will *only* need to use the terminal, so feel free to resize it to its maximum window size. This might also help you avoid the temptation to start moving/ renaming files by using your OS's file browser. Before we proceed with learning our very first Unix command, let's reiterate that last point.

Do not use your mouse. In the land of the terminal, the keyboard is king!

---

[5] Terminal applications will use a *fixed-width* font for the default font. There are good reasons for this and you should probably *not* change it to any non-fixed-width font (e.g., Times, Arial, etc.).

[6] Closing a terminal window will often, but not always, stop the program or command that you were running. If your program had been running for a day and was just about to give you the answer to life, the universe, and everything else – then you will have to wait another day for the answer.

The Unix command prompt

We command you to read this section

Hopefully your terminal window already contains some text. As mentioned in the last chapter, this text might include your login name[7] or the name of the machine you are using. The text traditionally ends with a punctuation character of some kind (most commonly a $ or % sign). Collectively, this text is known as the *command prompt*.

Example 3.3.1 It's time for your first interaction with the world of Unix. After you make sure that your current terminal window is selected, take a deep breath and press the enter key on your keyboard.

Explanation

Congratulations! You have just interacted with your Unix terminal. Hopefully the world didn't end and your computer is still intact. Most importantly, you should have noticed that the text that was on the screen *before* you pressed enter has now been duplicated on a new line. Every time you type any Unix command and press enter, the computer will attempt to follow your instructions and then, when finished, return you to the command prompt. Sometimes you might have to wait a while for a program to finish before the command prompt returns, but once you see the prompt, then you know you are free to type your next command. Some forms of Unix provide a blinking cursor, which makes it a bit easier to focus your eye on where you can type. Usually, a new terminal window will have the current command prompt at the *top* of the window. But if you try pressing enter 20–30 times, the current command prompt will get moved to the bottom of the window.

Examples of command prompts

Depending on what version of Unix you are using, your command prompt might also be set up to include the name of the current directory.[8] This might change as you navigate to different directories on the computer (which we will be doing in a few chapters' time). Therefore, don't be surprised if the text that makes up the command prompt changes from time to time. Command prompts can also be customized to include a lot of other information. Here are a few examples of what some command prompts can look like:

---

[7] Also called the 'user name.' It's entirely possible to have one account name that you use to login to your computer (e.g., 'keith') and then have a different Unix login name (e.g., 'themaster').

[8] Directories are the same thing as what you might think of as 'folders' when using a graphical file manager.

Prompt	Description
$	A single-character prompt
%	Another common single-character prompt
bash-3.2$	The default prompt on Mac OS X. This prompt includes the name and version number of something called the *shell*. More of that in a later chapter.
nigel@stonehenge$	A prompt that includes both details of the user name (nigel) and the computer name (stonehenge). It's very common to see this type of information included in the prompt.
nigel:/home %	A prompt that contains the user name as well as the name of the current directory (/home). We'll explore the syntax of directory names later.

Because of this huge diversity in command prompts, we will stick with using a single dollar sign for all examples which include the command prompt. If we show you the following:

```
$ ls
```

This should be interpreted as *type the Unix command 'ls' at the command prompt*. And if you're wondering what the ls command does, then you only need to look at the next chapter!

Reminder, no mouse required!

Everything we do in this part of the book will involve *typing* commands at the command prompt. Sometimes, newcomers to Unix are tempted to click the mouse to try to reposition the cursor somewhere else in the terminal window. In nearly all cases, this will not achieve what you are trying to do. Clicking with a mouse will normally just grab the focus of the terminal window and will not move the cursor. In any case, we know that you will, of course, be following our advice *not to use the mouse for this part of the book!*

Your home directory

Unix, like most OSs, keeps files arranged in a hierarchical structure. From the 'top level' of the computer, there will be a number of directories, each of which can contain files and subdirectories, and each of those, in turn, can contain more files and directories and so on, *ad infinitum*. It's important to note that you will always be 'in' a directory when using the terminal. The default behavior is that when you open a new terminal you start in your own 'home' directory (containing files and directories that only you can modify).

The `ls` command

To see what files are in our home directory (or indeed in *any* directory), we need to use the `ls` command. This command 'lists' the contents of a directory. So why don't they call the command 'list' instead? Well, this is a good thing because typing long commands over and over again is tiring and time-consuming. There are many (frequently used) Unix commands that are just two or three letters long.[9]

Example 3.4.1 Let's execute the `ls` command, and hopefully we shall see some output:

```
$ ls
Desktop      Downloads    Movies    Pictures   Sites
Documents    Library      Music     Public
$
```

Explanation

There are five things that you should note from the output of this command:

(1) You will see different output to what is shown here.[10] This is expected. If you don't see any output at all that *might* still be okay (assuming that your current directory doesn't contain any files or other directories).

---

[9] Unix also features single-letter commands, such as w.

[10] Unless you also have an Apple computer and haven't added any other files or folders to your home directory, in which case your output might be identical.

(2) The $ character that you see before the ls command is the Unix command prompt (see the previous chapter).

(3) The output of the ls command lists nine different items. In this case, they are all directories,[11] but they could also be files. We'll learn how to tell them apart later. Note that these directories are listed alphabetically in *columns*.

(4) After the ls command finishes it produces a new command prompt, ready for you to type your next command.

(5) The ls command should be typed using lower-case letters. Unix usually (but not always) distinguishes between lower- and upper-case characters, i.e., it is a case-sensitive OS. Typing LS or Ls might not work.[12]

Example 3.4.2 The ls command is used to list the contents of *any* directory, not necessarily the one that you are currently in. Try the following:

```
$ ls /usr
X11      bin     include     libexec        local     share
X11R6    etc     lib         llvm-gcc-4.2   sbin      standalone
$
```

Explanation

We are now asking Unix to list the contents of the directory called 'usr'. More specifically, the forward-slash character implies that this directory must be located at the top level of the filesystem (we'll return to this point in the next chapter). Before we were simply running a Unix command, but we are now running a command and providing it with an *argument* (the '/usr' part). Many Unix commands can be run with or without arguments, though some commands will always require one or more arguments.

> *Always separate Unix commands from their arguments with a*
> *space character!*

The 'usr' directory is a common directory that is present on most (if not all) Unix systems. On many Unix systems this is where all of the user accounts will be located. However, we don't need to understand what all of those directories are doing.[13] One thing to note about this output is that the directories which start with upper-case letters (X11 and X11R6) are placed *before* directories that start with lower-case letters. This is one of the many things that can differ between different Unix systems. Your system might show very similar output, but might order all directories alphabetically, regardless of their case.

---

[11] These are the default directories that are created for all new user accounts with Mac OS X.

[12] This will work on Macs, which recognize commands typed in any combination of lower-and upper-case characters. This is not true of all Unix systems though.

[13] You might be wondering how we know that these are all directories and not files. Well, you'll have to wait a little longer before we explain this.

This will be a root-and-branch review

Looking at a list of directories from within a Unix terminal can often seem confusing. But bear in mind that these directories are exactly the same type of folders that you can see if you use your computer's graphical file management program. A tree analogy is often used when describing computer filesystems because of the branching nature of the directory structure. Like a tree, a Unix filesystem has roots, or more specifically, it has *a root*, which is represented by a single forward-slash character (/).[14] The analogy with the tree starts to break down a little as it is common to refer to the root level as the 'top' of the directory structure – think of it as an inverted tree if it helps. From the root level (/) there are usually many (10–20) top-level directories. A small number of these directories will be present on all Unix systems, but there will also be many directories that are specific to different types of Unix. Here is a fictional example of what *part* of a Unix tree might look like:

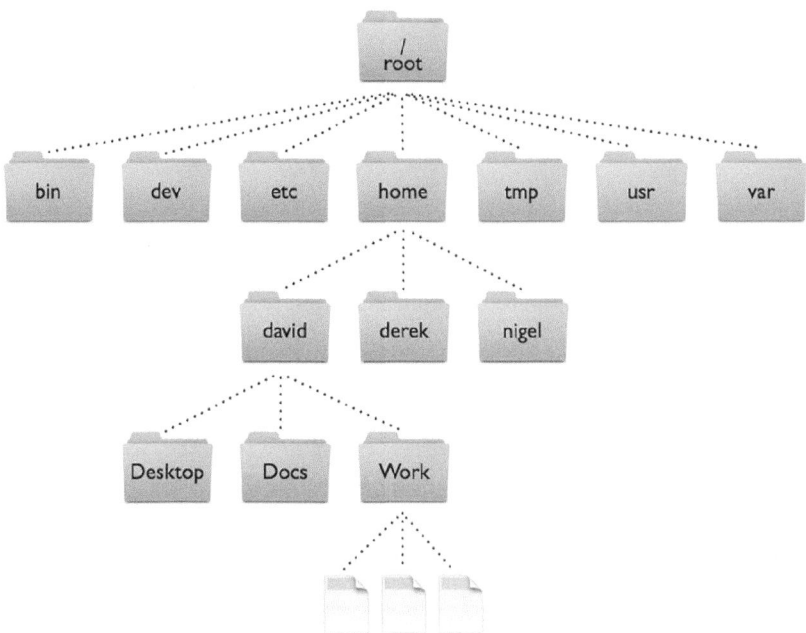

In this example we show seven top-level directories below the root level.[15] These directories may seem to have strange names, but you don't really need to know what they are for. The one thing to note from this schematic is that there is a 'home' directory which in turn contains the home directories of three users ('david,' 'derek,' and 'nigel'). We then show that David's home directory contains more directories, one of which ('Work') contains some files. This schematic is highly simplified – a full Unix

---

[14] On a standard Windows OS the equivalent to the root level would be C:\.
[15] These are the directories that occur on nearly all Unix systems.

filesystem may contain several hundred directories, and many thousands of files. Note that we shall return to this fictional filesystem in the next few chapters.

> *Directories that exist inside other directories, are often referred to as subdirectories.*

Example 3.5.1 If you want to see what your own root level looks like you can simply run the `ls` command and tell it that you want to list the contents of the root directory (represented by a single forward-slash character):

```
$ ls /
Applications    System      cores   mach_kernel    tmp
Developer       Users       dev     net            usr
Library         Volumes     etc     private        var
Network         bin         home    sbin
```

Explanation
This is the listing of the root directory on a computer running Mac OS X. Note that it has a directory called 'home' (like many Unix systems) but the actual home directories of users are stored as subdirectories within the 'Users' directory. There is often a lot of variation as to where different Unix systems keep home directories,[16] and in the next chapter we will learn how to find out where your own home directory is located.

Navigating the filesystem tree

If we briefly return to our fictional Unix filesystem, we should note that the directory called 'david' contains three subdirectories and is itself a subdirectory of the 'home' directory. If we wanted to copy or move some files that are in David's 'Work' directory to Nigel's home directory, we can trace a path that would navigate *up* three levels in the directory tree (→ Work → david → home), and then navigate down one level (→ nigel). If this concept of going up and down various branches seems intuitive to you, then great! If it doesn't, then it might help you to start thinking about your filesystem in this way. This means that if you wanted to copy a file from David's 'Work' directory to his 'Docs' directory, you actually have to go up one level and then down one level (rather than go directly across).[17]

---

[16] Though it is very common to name the actual home directory after your login/user name.

[17] Of course the concepts of 'up,' 'down,' or 'across' are somewhat misleading. All directories, programs, and other files exist as series of binary 1s and 0s on a magnetic disk or other storage medium. However, these concepts can sometimes make it a lot easier to understand how to use many different Unix commands.

Finding out where you are in the filesystem　3.6

You should learn to use pwd PDQ

As we have already mentioned, there may be many hundreds of directories on any Unix machine. So how do you know which one you are in? In Unix, the current directory you are in is known as the *working directory*. To find out where you are, you can run the Unix command pwd, which will *p*rint the *w*orking *d*irectory, and this is pretty much all this command does. If you have opened a new terminal window you will normally be placed in your home directory:[18]

```
$ pwd
/Users/nigel
```

In this example we are in the home directory of a user called 'nigel,' who has his home directory located as a subdirectory of the 'Users' directory. Conversely, 'Users' is 'above' the level of 'nigel' and Unix would refer to this directory as the *parent* directory of 'nigel'.

Slashes separate parts of the directory path

As we have just seen, Unix uses forward-slashes to separate out the various parts of a directory location. In the above example we can see that 'nigel' must be a subdirectory of 'Users' because the two are separated by a single forward-slash character. Collectively, the set of file and directory names that are combined with forward-slashes is known as a *path*. A single path will always specify some unique location in the filesystem.

If a path *starts* with a forward-slash character, then this is the same thing as the single forward-slash that represents the root level. In the above example, because it starts with a forward-slash, we can infer that 'Users' must be a directory one level beneath the root directory. We shall return to the issue of navigating paths in the next chapter.

Remember to use pwd!

As you become more familiar with Unix, you will find yourself trying to switch between different directories in the filesystem (which we will learn how to do in the next chapter). The more you move around, particularly as you start using multiple terminal windows, the more likely it is that you will get 'lost' in the filesystem. When you start running more complex commands (or Perl scripts) a common reason for them not appearing to work is that you are not in the correct directory. As you start to learn Unix, it is good to get into the habit of frequently running pwd to check that you always know where you are.

---

[18] You can configure Unix to start your terminal sessions somewhere other than your home directory and this is sometimes useful, but 99% of the time a new terminal will place you in your home directory.

It's time for a change

We have previously seen how the ls command allows us to 'look' at the contents of any directory in a Unix filesystem. It is entirely possible, and sometimes desirable, to perform actions with files or folders that are in different directories with respect to your current working directory. However, we frequently want to change directories so we are in the same directory as some file or program. We can do this using the cd command (change directory). Let's return to our fictional Unix filesystem and see how the user Nigel would change directory from his home directory to the temporary directory 'tmp'.[19] Feel free to repeat these steps from your own home directory:

```
$ pwd
/home/nigel
$ cd /tmp
$ pwd
/tmp
```

We start by confirming that we are in Nigel's home directory by running the pwd command. We then use the cd command to change to /tmp. Note that the cd command does not give you any output or feedback after you run it. If we knew what files we were expecting to see in /tmp then we could run the ls command to check that this is the intended location. However, it is always possible that two different directories could contain identical contents. That is why we run the pwd command again, to confirm that we are really in /tmp.

Changing directories in multiple steps

Let's imagine that we want to change directory to the 'home' directory and then change to David's 'Work' directory. We will omit the pwd confirmation step from now on:

```
$ cd /home
$ cd david/Work
```

There are two things to note here. First, notice how that second cd command does not start with a forward-slash. You only need to include a forward-slash at the start of a Unix path when you want to start navigating from the root level of the filesystem. Hopefully, you also noticed that performing this in two steps is a little pointless. If you just want to navigate from directory 'A' to directory 'B,' you can always do that in one step:

```
$ cd /home/david/Work
```

---

[19] The 'tmp' directory is used for storing various temporary output from programs. Be aware that it is automatically emptied at periodic intervals, so you should not use it to store important files.

Can't change directory?

Remember that many Unix systems are case-sensitive. This means that if we had typed 'David' instead of 'david' we might have seen an error message like the following:

```
$ cd /home/David/Work
cd /home/David/Work: No such file or directory
```

We would also see this type of error message if we had misspelled any part of the path, or missed out any of the forward-slashes. If we mistakenly typed something like:

```
$ cd home/nigel
```

instead of:

```
$ cd /home/nigel
```

without the leading forward-slash character, Unix will assume that we want to change directory to a subdirectory of our current directory called 'home'. It is important to be able to appreciate the difference between the two commands shown above. The second command will only ever specify a *single* location on the filesystem, as there can only be one directory called 'home' at the root level of the computer. In contrast, the first command could potentially work in multiple places as there could be another 'home' directory somewhere else in the filesystem.[20]

Changing to the parent directory

One of the most common uses of the cd command is to navigate to the directory above the one you're currently in – e.g., you want to change from the 'david' directory to 'home' or from 'home' to '/'. This is straightforward if you know what the directory above you is called. However, you may have forgotten where you are and you may not want to keep on running the pwd command to find out. Luckily, you can simply tell the cd command to go 'up' one level by using the following syntax:

```
$ cd ..
```

Two dots (without a space) are used by Unix to refer to the parent directory. You can also use this with the ls command to list the contents of the parent directory:

```
$ ls ..
```

If you wanted to navigate up *two* levels then you simply need to include a forward-slash between two sets of double dots:

```
$ cd ../..
```

The use of the forward-slash is consistent with what we have already seen about Unix paths, and the forward-slash acts as a delimiter that separates out different levels in the overall directory hierarchy.

---

[20] This is analogous to the fact that every US city can potentially have an address called '1600 Pennsylvania Avenue,' but within a *single* city it is likely that there is only going to be one address with that name.

When you run the previous command, Unix will try to go up two levels in the directory hierarchy. If your current working directory was /bin, then you only have one level above you, but Unix will not produce an error – it will simply take you up as far as it can go.

Problem 3.7.1 Assuming you are in a new terminal window which has put you in your home directory, navigate to the root level of your filesystem, and then try to navigate back to your home directory. Use the pwd and ls commands to keep on checking where you are and what files and directories are present as you navigate.

Absolute and relative paths   3.8

We're absolutely sure that this part is all relative

In the last chapter we saw how we could navigate up a level by using the .. command, which signifies the parent directory. What if we wanted to navigate up some levels and then back down again in one single operation? Can you think how we might navigate from inside David's 'Desktop' directory to David's 'Docs' directory in one go? Clearly we could do this with two separate cd commands:

```
$ pwd
/home/david/Desktop
$ cd ..
$ cd Docs
```

But rather than run two separate commands, we can simply combine both paths from each command:

```
$ cd ../Docs
```

You can read this as 'navigate up one level and then navigate down one level into the Docs directory.' In Unix this is known as a *relative* path – i.e., relative to our current location (/home/david/Desktop), we can get to the Docs directory by navigating up one level and then down one level. This has the net effect of traversing sideways in the directory structure. A slightly longer example of a relative path would be:

```
$ cd ../../../tmp
```

We are now going up three levels and then down one level into the 'tmp' directory. The third 'up' operation takes us to the root level. From here it becomes easier to navigate to the 'tmp' directly from the root level:

```
$ cd /tmp
```

This syntax is known as an *absolute* path. All absolute paths start from the root level of the OS and specify a file or directory in relation to this fixed starting point. It is important to understand the difference between these two types of path. When we start moving and copying files, we will often have to specify paths for both the source and destination directories, and using the wrong type of path will be a likely source of errors.[21]

Absolute vs. relative paths

Sometimes it will be quicker to change directories using the absolute path, and sometimes it will be quicker to use a relative path. However, you can only use an absolute path if you know all of the intermediate directories that separate the root level and the directory that you want to navigate to. With practice it will become obvious which type of path you should use. This partly depends on getting to know a little bit about the

---

[21] Being able to specify the correct path will be important for some other Unix commands, and also for when we want to start running our own Perl programs.

layout of your own filesystem. Changing to a new computer can sometimes be a little like moving to a new town. You have to learn where everything is all over again and it takes time to learn the fastest way of getting from 'A' to 'B.'

One final warning for this section. It's entirely possible to do pointless things with the cd command, such as:

```
$ pwd
/home/david/Desktop
$ cd ../../../home/david/Docs
```

or similarly:

```
$ pwd
/home/david/Desktop
$ cd /home/david/Docs
```

In these examples we needlessly navigated all the way to the top of the filesystem and then back down again in order to just change directory to a directory that is at the same level as where we started. If you want to do all of that unnecessary typing then just be aware that Unix isn't going to stop you.[22]

---

[22] By extension, you can even navigate up and down the filesystem and end up right back where you started. We bet you're going to try that right now, aren't you?

Working with your home directory   3.9

(The) home (directory) is where the heart is

What is a home directory?

Of all the directories on a Unix filesystem that you will work with, the home directory is probably the most important. This directory serves the same purpose as the 'My Documents' folder on a Windows computer, and it is where you will store various files that are owned by you.[23] Reflecting its special status, it has a few important properties that set it apart from other directories. It is common for your home directory to be named after your real name or your login name (of course, the two might also be the same). This means that if you have two or more users on your computer who have the same name, they will need to use different names for their home directories.

By default, new terminal windows should always place you inside your home directory. Of course, you can always confirm the location of your home directory (in a newly opened terminal window) with the pwd command:

```
$ pwd
/home/nigel
```

Finding your way home

One of the most common acts of 'directory navigation' that you will perform is to return to your home directory (from wherever you were). If you know where your home directory is, you can simply use the cd command to go there:

```
$ cd /home/nigel
```

Because you will want to perform this action over and over again and because you might not always remember where your home directory is,[24] Unix provides several useful shortcuts:

```
$ cd ~nigel
```

This command should be read as 'change directory to the home directory of the user called 'nigel.'' The ~ character (known as a tilde) is used by Unix to refer to a home directory. Note that you can use this syntax to navigate to directories beneath the level of your home directory – e.g., the path ~david/Docs would refer to the 'Docs' subdirectory of David's home directory. Additionally, you can use this syntax to navigate to, or list the contents of, *another* user's home directory:

```
$ cd ~derek
$ cd ~david
$ ls ~derek
$ ls ~david/Work
```

---

[23] These files will hopefully end up including the many Perl scripts you will be writing!

[24] On large, Unix-based networks, it's entirely possible that your home directory may change location occasionally as it might be moved from one disk to another by your system administrator.

If it is *your* home directory that you want to go to then you can save even more time by omitting the user name altogether:

```
$ cd ~
```

A tilde on its own will always be understood by Unix to refer to your home directory. If we want to save ourselves even more typing, we can take home directory navigation to its extreme:

```
$ cd
```

If you don't provide any other information to the cd command, then it will take you to your home directory. This is the format of the command that you will end up using the most. You shouldn't proceed any further without first trying to get lost somewhere in the labyrinth of your own Unix filesystem, before safely returning home with the cd command.

Problem 3.9.1 Try to get yourself 'lost' in your filesystem. Change directory to the root level and then start navigating to a different directory at the root level. Check that you can find your way home by simply typing cd and then confirming your location with pwd.

The Unix shell

It's time to come out of your shell

What is the shell?

The shell is a command-line interpreter that lets you interact with Unix. You might be thinking that this sounds an awful lot like the terminal, but the two are very distinct. A terminal is like a web browser. There are a lot of web browsers, and they all let you interact with the internet. Similarly, there are a lot of terminal programs, and they all give you a command-line prompt to issue commands and observe the output of those commands.

The shell takes what you type and decides what to do with it. Did you want to run a program? Assign a variable? Autocomplete the name of a file?[25] Pipe the output from one program to another? The shell is actually a scripting language somewhat like Perl. It is not as powerful as Perl, but for some simple tasks a shell script is sometimes more convenient and appropriate. In this book we only touch upon shell scripting, because we prefer to do our programming in a more fully featured language.

Your default shell

Unix is a very flexible OS, and it is therefore not surprising that there is more than one kind of shell.[26] Here is a list of the most common shells:

- *Bourne shell* – commonly known as sh. Named after its creator, Stephen Bourne, this shell has remained a popular default shell ever since its original development in 1977.
- *C shell* – known as csh. Developed shortly after the Bourne shell, it quickly gave rise to a related shell called the TENEX shell or tcsh. The latter shell contains everything in csh plus some other useful features such as *command-line completion*. This is something that we will introduce you to in a few chapters.
- *Korn shell* – known as ksh. Developed by David Korn in the early 1980s. It includes many features of csh and is backwards compatible with sh.
- *Bourne-again shell*[27] – known as bash. It is widely used and is currently the default shell on computers running Mac OS X. It was developed a decade after sh.
- *Z shell* – known as zsh. This is the newest of all the shells mentioned so far, and is gaining in popularity. Like most newer shells, it incorporates various elements of all of the other shells that have gone before it, but it also includes new features such as spelling correction (rare among Unix shells).

For the purposes of this book, we only use a small subset of a shell's capabilities, so the differences among shells are minor. For simplicity, use whichever shell is the

---

[25] We have not discussed auto-completing or tab-completing yet, but it is one of the more useful features of a shell.

[26] Programmers become evangelical about languages, OSs, editors, and to no surprise, shells also.

[27] Unix developers love nothing more than a good pun.

default on your system (we will show you how to find out what your default shell is in the next chapter). There are a few important differences among shells, and we shall cover these differences as and when they arise in the remaining chapters of this part of the book. Bear in mind, however, that it is always possible to change shells (either temporarily or permanently), so you shouldn't feel chained to whatever your default shell is.

Environment variables   3.11

How much it rains is also an environment variable

Unix keeps track of several special variables that are associated with your Unix account. These are specifically known as *environment variables*. They are always written in upper-case letters and always start with a dollar sign. We haven't really covered variables yet, but they are one of the fundamental ways by which computer programs store data.[28] A single variable will typically have a unique name and will contain some data that can vary (hence the name *variable*). The leading dollar sign is a common way of indicating that something is a variable. This helps distinguish variables from file/directory names, Unix commands, etc.

There are many different environment variables, and they are used to store useful pieces of information such as: which directories contain programs; your current directory; your favorite (Unix) text editor; and even your login name (in case you forget!). Explanations about environment variables often go hand in hand with explanations about the shell, because the two are closely coupled. It therefore makes sense that the first environment variable that we will introduce is $SHELL, which will contain the name of your default shell.

Example 3.11.1 We can display the contents of any environment variable by using a Unix command called echo. This command simply echoes back the contents of whatever text you type, but if you type the name of an environment variable, then it echoes back the *contents* of that variable. E.g.:

```
$ echo SHELL
SHELL
$ echo $SHELL
/bin/bash
```

Explanation
The first use of the echo command simply repeats the text 'SHELL' – indeed, it would repeat the text of any regular words that you typed.[29] However, if we add the $ sign to 'SHELL' then we are instead talking about the environment variable $SHELL, and we see that our user Nigel is using the bash shell on his computer. We also learn from this that bash is a file that lives in the '/bin' directory.

---

[28] Unix and all programming languages use variables extensively. They are also used by many data-heavy applications, such as database programs, spreadsheets, etc.

[29] Using the echo command to repeat regular text might seem a bit pointless, but it can be used to add some specified text to a file.

Example 3.11.2 Let's see some more environment variables for our user, Nigel:

```
$ echo $USER
nigel
$ echo $HOME
/home/nigel
$ echo $PWD
/tmp
$ echo $EDITOR
Emacs
```

Explanation

These commands reveal that: Nigel's user name is 'nigel', his home directory is '/home/nigel', his current working directory is '/tmp', and his preferred/default text editor is the program 'Emacs.'

The environment variables listed in these examples are part of the default set of environment variables that are part of all Unix systems. Any additional software that is installed on your computer may well add some new variables which will store data relating to that software. The Unix command `printenv` can also be used to inspect the content of any environment variable, but note that you don't include the dollar sign when using this command. E.g.:

```
$ printenv USER
Nigel
```

If you run the `printenv` command without specifying any variable name, it shows you a list of all environment variables along with their current settings.

Working with environment variables

Any time you include these variables as part of a command, Unix will work with the *contents* of the variable rather than the actual variable name. This means that it is possible to use environment variables with regular Unix commands – e.g., if Nigel wanted to list the contents of his home directory, he could type any of the following four commands:

```
$ ls /home/nigel
$ ls ~nigel
$ ls ~
$ ls $HOME
```

The last option shows how it can sometimes be useful to work with environment variables when using other Unix commands. For the most part, you will not need to know about many of the different environment variables that might be present as part of your Unix account. Occasionally you will want to check what the current setting of a certain variable is, and you will sometimes need to change that setting.

| We command you to check out these options

So far we have only introduced you to a small handful of Unix commands and we have shown you how to run these commands to achieve their default behavior. Usually, the default behavior of a command is all we want, but sometimes we would like to modify the behavior and/or output of the command. For many Unix commands we can produce alternative output by specifying what are known as *command-line options* when we run the command.

Revisiting the ls command

Let's assume that our user Nigel has just bought a new computer. The first thing he wants to do is see what the root level of his hard drive looks like.[30] He opens up his terminal application and runs the following command:

```
$ ls /
Applications   System     cores   mach_kernel   tmp
Developer      Users      dev     net           usr
Library        Volumes    etc     private       var
Network        bin        home    sbin
```

The ls command does a fine job of showing us the names of everything in Nigel's root directory,[31] but that's about all it does. The default output doesn't show us any information about the sizes or modification dates of the files or directories, or who created them. It also doesn't make it clear as to which of the listed items are files and which are directories. Another limitation is that the default output is sorted alphabetically, which might not be what we want.

This is where command-line options can help us produce all of this extra information that we might need. Command-line options in Unix are specified by using a hyphen character (-) after the command name, followed by various letters, numbers, or words. If you add the letter 'l' to the ls command it will give you a *l*onger output compared to the default:

```
$ ls -l /
total 36494
drwxrwxr-x+  85 root   admin    2890 Jun 28   11:35
Applications
drwxrwxr-x@  15 root   admin     510 Oct 19    2009   Developer
drwxrwxr-t+  59 root   admin    2006 Jun 23   13:04   Library
drwxr-xr-x@   2 root   wheel      68 Jun 22    2009   Network
drwxr-xr-x    4 root   wheel     136 Jun 20   13:09   System
drwxr-xr-x    6 root   admin     204 Feb 22   09:45   Users
```

---

[30] We think this seems like a perfectly reasonable thing to do. If you disagree, then you probably haven't spent enough time working with Unix yet.

[31] This listing displays the standard contents of a Mac computer's root directory.

```
drwxrwxrwt@    5 root   admin         170 Jun 29   11:43   Volumes
drwxr-xr-x@   39 root   wheel        1326 Jun 20   13:04   bin
drwxrwxr-t@    2 root   admin          68 Apr 15   00:26   cores
dr-xr-xr-x     3 root   wheel        4901 Jun 29   11:10   dev
lrwxr-xr-x@    1 root   wheel          11 Oct 19    2009   etc -> private/etc
dr-xr-xr-x     2 root   wheel           1 Jun 29   11:11   home
-rw-r--r--@    1 root   wheel    18661932 Jun 10   16:19   mach_kernel
dr-xr-xr-x     2 root   wheel           1 Jun 29   11:11   net
drwxr-xr-x@    6 root   wheel         204 Oct 19    2009   private
drwxr-xr-x@   64 root   wheel        2176 Jun 20   13:04   sbin
lrwxr-xr-x@    1 root   wheel          11 Oct 19    2009   tmp -> private/tmp
drwxr-xr-x@   14 root   wheel         476 Apr 21   14:38   usr
lrwxr-xr-x@    1 root   wheel          11 Oct 19    2009   var -> private/var
```

As you can see, the simple addition of -l makes a lot of difference to the output. For each file or directory we now see much more information and all of the output is now arranged into columns with the file or directory name in the last column. We don't need to understand all of this information at the moment, but you can hopefully see that the penultimate column includes the last modification date and time of the file or directory.

One other thing about this long-form output that we'll mention now is that the very first character describes something called the *entry type*. If it starts with a dash (-), the item is a regular file – as we can see above, there is only one file at the root level of Nigel's computer. If it starts with a 'd', then the item is a directory; if it starts with an 'l', then it is something called a *symbolic link*.[32] There are some other possibilities as well, but these are the most common ones.

More command-line options for the ls command

The ls command has many different command-line options. Here are a few examples; we encourage you to run each of these commands in one or more directories on your computer. The text following each command (after the #) is just an explanatory comment, don't type it:

```
ls -t    # sort output based on file modification date
ls -S    # sort output by size
ls -r    # reverse-sort the output
ls -R    # recursively list output of all directories below current level
ls -1    # force output to be one entry per line
```

These examples illustrate that command-line options can use lower- or upper-case letters and also use numbers (such as in the last option).

---

[32] Symbolic links are similar to 'shortcuts' in the Windows OS and 'aliases' in the Mac OS. Basically, a symbolic link is a file that points to another file or directory (or even another symbolic link). Any action you perform against a symbolic link will produce the same result as if you performed the same action against the original item. The exception to this is that if you delete a symbolic link, you don't delete the original item. The long listing of the ls command will also show you what item the symbolic link is pointing to.

You can combine multiple options together (where appropriate) – e.g., maybe you want to list items in a directory in a long format, and then reverse-sort items based on their modification date. This would mean that the most recently modified items appear at the bottom of the list. We could do this as shown here:

```
$ ls -l -t -r /
total 36494
drwxr-xr-x@    2   root   wheel         68   Jun 22   2009     Network
lrwxr-xr-x@    1   root   wheel         11   Oct 19   2009     var -> private/var
lrwxr-xr-x@    1   root   wheel         11   Oct 19   2009     tmp -> private/tmp
lrwxr-xr-x@    1   root   wheel         11   Oct 19   2009     etc -> private/etc
drwxr-xr-x@    6   root   wheel        204   Oct 19   2009     private
drwxrwxr-x@   15   root   admin        510   Oct 19   2009     Developer
drwxr-xr-x     6   root   admin        204   Feb 22   09:45    Users
drwxrwxr-t@    2   root   admin         68   Apr 15   00:26    cores
drwxr-xr-x@   14   root   wheel        476   Apr 21   14:38    usr
-rw-r--r--@    1   root   wheel   18661932   Jun 10   16:19    mach_kernel
drwxr-xr-x@   39   root   wheel       1326   Jun 20   13:04    bin
drwxr-xr-x@   64   root   wheel       2176   Jun 20   13:04    sbin
drwxr-xr-x     4   root   wheel        136   Jun 20   13:09    System
drwxrwxr-t+   59   root   admin       2006   Jun 23   13:04    Library
dr-xr-xr-x     3   root   wheel       4901   Jun 29   11:10    dev
dr-xr-xr-x     2   root   wheel          1   Jun 29   11:11    net
dr-xr-xr-x     2   root   wheel          1   Jun 29   11:11    home
drwxrwxr-x+   85   root   admin       2890   Jun 30   10:49    Applications
drwxrwxrwt@    5   root   admin        170   Jul 2    15:58    Volumes
```

In this example we specify three different command-line options for the ls command. Note that we could put these options in any order. In these situations, you can combine all options by using the following syntax:

```
$ ls -ltr /
```

This achieves exactly the same as the previous command, but saves you a little bit of typing. Some Unix commands have options that are mutually exclusive; you will be warned if you attempt to use the options in an incompatible manner.

One other useful command-line option for the ls command is the -p option. This option simply adjusts the default output to additionally include a forward-slash character for any items which are directories. Compare the following output to what we saw before when we first used the ls command in this chapter:[33]

```
$ ls -p /
Applications/   System/   cores/   mach_kernel   tmp
Developer/      Users/    dev/     net/          usr/
```

---

[33] Note that the 'etc,' 'tmp,' and 'var' entries do not have a trailing forward-slash because they are symbolic links. Although these can sometimes act like directories, they are considered by Unix to be a special type of file.

```
Library/      Volumes/   etc      private/    var
Network/      bin/       home/    sbin/
```

Now it becomes a lot easier to distinguish the single item which isn't a directory ('mach_ kernel'). The -p option also works if you use the long listing (-l option).

Command-line options may require additional information

Let's take a very quick look at another basic Unix command. Do you want to know what the date and time is? Try the date command:

```
$ date
Fri Jul 2 17:17:45 PDT 2010
```

The date command has a few basic options available – e.g., we could display the current date in UTC format[34] with the -u option:

```
$ date -u
Sat Jul 3 00:19:39 UTC 2010
```

One of the other options for the date command is -r. Unlike the options we have seen so far, this option requires us to also specify some additional information. If we don't specify anything else, we will see something like this:

```
$ date -r
date: option requires an argument -- r
usage: date [-jnu] [-d dst] [-r seconds] [-t west]
       [-v[+|-]val[ymwdHMS]]
       [-f fmt date | [[[[[cc]yy]mm]dd]HH]MM[.ss]] [+format]
```

Buried in that cryptic-looking usage statement is a clue that the -r option requires some value which corresponds to 'seconds.' The -r option will give you a date that corresponds to the number of specified seconds that have elapsed since January 1, 1970.[35] E.g.:

```
$ date -r 123456789
Thu Nov 29 13:33:09 PST 1973
```

The point of this example is purely to illustrate that some command-line options work on their own, and others require data. You might be wondering how you can find out about what command-line options are available for any given command. Luckily, we are going to address that in the next chapter.

Problem 3.12.1 Try listing the contents of your home directory and the root directory using various command-line options for the ls command. Make sure you try combining different options together and try to see how the output changes.

---

[34] UTC refers to Coordinated Universal Time which is approximately equivalent, but not identical to Greenwich Mean Time (GMT).

[35] No, we're not quite sure why anyone would want to be able to do this, but we guess that some people must need to do it, otherwise the command-line option wouldn't exist. If you didn't already know, January 1, 1970 is the date that all Unix systems use (and many other computers too) as the starting point for what is called 'Unix time.'

Man pages 3.13

Time to man the battle stations

If Unix commands have so many options, you might be wondering how you find out what they are and what they do. Thankfully, every Unix command should have an associated 'manual' which is just a formatted page of text that describes everything about the command. These manuals are more commonly known as 'man pages.'[36]

Example 3.13.1 We can view man pages by using the Unix man command – e.g., if we want to see what the whoami command does, we just type:

```
$ man whoami
```

Explanation

What happens next is that Unix sends the contents of the manual to a Unix text-viewing program, which gives you basic controls for scrolling through the document and searching for specific text. The text-viewing program will almost certainly be a program called less.[37] This is what you should see if you type the above command:

```
WHOAMI(1)         BSD General Commands Manual              WHOAMI(1)
NAME
whoami -- display effective user id
SYNOPSIS
whoami
DESCRIPTION
The whoami utility has been obsoleted by the id(1) utility, and is equivalent to "id -un".
The command "id -p" is suggested for normal interactive use.

The whoami utility displays your effective user ID as a name.

EXIT STATUS
The whoami utility exits 0 on success, and >0 if an error occurs.
SEE ALSO
id(1)
BSD                       June 6, 1993                        BSD
(END)
```

Unix man pages are formatted in a standard layout which includes sections for 'name,' 'synopsis,' and 'description.' The man page for the whoami command is very short because this command does one thing, and one thing only ... it tells you who you are.[38] If this command had any command-line options then they would all be described in the 'description' section.

---

[36] For readers who were hoping that man pages would help them understand men, there is no such help from Unix. Additionally, there are no woman pages of any kind.

[37] Though bear in mind that it is possible to change the program that is used as the man page viewer.

[38] This is not as stupid as it may seem. Certain Unix users (e.g., system administrators) may be in charge of multiple accounts on multiple machines. It is entirely possible to forget which user account you have logged in as.

Controlling the man page viewer

As already mentioned, the default man page viewer uses a Unix program called `less`. This is something we will return to in a later section (Chapter 3.20), but for now you just need to know the following essential keyboard shortcuts for controlling things:

\<space>	Scroll down one page
b	Scroll up one page
j	Scroll down one line
k	Scroll up one line
q	quit the `less` program.

Some Unix commands have very long manual pages, and some are full of computer jargon, which can make them hard to follow. It is typical to always list the command-line options early on in the documentation, so you shouldn't have to read too much in order to find out what any particular option is doing.

Example 3.13.2 Try using the `man` command to look at the man page for the `ls` command:

```
$ man ls
```

Explanation

You should see that the 'synopsis' part shows a *lot* of command-line options:

```
LS(1)        BSD General Commands Manual        LS(1)
        NAME
    ls -- list directory contents
SYNOPSIS
    ls [-ABCFGHLOPRSTUW@abcdefghiklmnopqrstuwx1] [file ...]
```

The 'synopsis' part of the man page uses square brackets to denote things that are optional – i.e., the set of 38 command-line options are by definition optional, as is the need to specify the name of a file or directory after the `ls` command.

Whenever you are unsure about how to use a command, your first port of call should always be the man page for that command.

Searching man pages

One other useful thing to know about the `man` command is that you can use it to search through the documentation of all Unix commands. This can be very helpful when you can't remember the name of the command that you are looking for. To achieve this behavior, you just need to use the `-k` command-line option, followed by a keyword that you want to search for.[39]

---

[39] You can also use a completely different command (`apropos`) to search man pages. The `apropos` command does not require a -k command-line option. Unix sometimes allows you to achieve the same goal in two different ways.

Example 3.13.3 If you wanted to see which commands might be related to compressing files, you could try the following command:

```
$ man -k compress
compress(1), uncompress(1)    - compress and expand data
gzexe(1)                      - compress executable files in place
gzip(1), gunzip(1), zcat(1)   - compress or expand files
zcmp(1), zdiff(1)             - compare compressed files
zip(1)                        - package and compress (archive) files
zlib(3)                       - compression/decompression library
znew(1)                       - recompress .Z files to .gz files
zopen(3)                      - compressed stream open function
```

Explanation

The output consists of a list of other Unix commands that feature the keyword, along with a brief description of that command. Don't worry too much about the numbers that appear in parentheses after the command names.[40] Note that we are only showing some of the output from this command; one of the problems of searching man pages is that a simple keyword can appear in the text of dozens of different man pages. If you have more than a page of output, you'll have to use the man page viewer commands to exit back to the command-line prompt (i.e., press q to quit).

Problem 3.13.1 Go back and look at all the commands we have seen so far (cd, pwd, etc.) and try viewing the man page for each one. You don't have to read every part of each page, but try to get used to the general layout of all of the man pages.

---

[40] Unix man pages are divided into different numbered sections, and it is the number of the section that appears after the command name. Regular Unix commands are in section 1 (you will probably never need to know any more about what the other sections cover).

Make directories, not war

Where should you create files for this book?

Over the last few chapters we've had to make some digressions in order to teach you about some important Unix concepts. Now it is time to actually start doing 'stuff,' which you will hopefully find a little more enjoyable. But before we can do anything we should first ask you to choose a place to store all of the files and directories that you will create as you work through the rest of this book. This place should ideally be a *single* directory[41] that will subsequently have more subdirectories added. For now, we suggest creating that directory in your home directory (details of how to do this will follow shortly), but you may want to save it somewhere else (external hard drive, flash drive, etc.) if you frequently work on different computers.

Naming directories

When working with a graphical file manager, we are accustomed to including spaces as part of file or folder names (e.g., 'My important text file.txt'). You can do this on a Unix system too, but it does present an additional complication. Up till now we have used the space character to separate out different parts of a Unix command – e.g., if you had to type a Unix command that requires two arguments, we would separate those arguments with one or more space characters.

So how can we have a file or directory name that also includes a space? How does Unix know that the space belongs to the directory name and isn't just separating arguments? Well, rather than tell you the solution we will instead suggest that any time you create a file or directory in Unix, you use underscore characters rather than spaces. This is a very common practice in Unix, and if we take our prior example of a text file, this would now become 'My_important_text_file.txt'.[42]

Making directories

To make a new directory we can use the appropriately named mkdir command.

Example 3.14.1 Let's create a 'Unix_and_Perl' directory, which we will use throughout the rest of this book. We'll create this for our fictional user 'Nigel,' and so the output you're going to see is for Nigel's computer. Your output will differ (especially if you choose a different directory name):

---

[41] Of course, if you prefer a little chaos in your life, you might want to instead create 36 differently named directories in 36 different locations on your hard drive.

[42] If you really need to know how to do this, you need to include a backslash character before every space character you use. This will make much more sense when you get into the Perl parts of the book later.

```
$ cd
$ pwd
/Users/nigel
$ mkdir Unix_and_Perl
$ ls -lp
drwx------+   5  nigel  staff  170   Feb  5   12:48  Desktop/
drwx------+   4  nigel  staff  136   Oct 19    2009  Documents/
drwx------+   4  nigel  staff  136   Oct 19    2009  Downloads/
drwx------+  27  nigel  staff  918   Jan 14   13:43  Library/
drwx------+   3  nigel  staff  102   Oct 19    2009  Movies/
drwx------+   3  nigel  staff  102   Oct 19    2009  Music/
drwx------+   4  nigel  staff  136   Oct 19    2009  Pictures/
drwxr-xr-x+   6  nigel  staff  204   Jan 14   13:44  Public/
drwxr-xr-x+   5  nigel  staff  170   Oct 19    2009  Sites/
drwxr-xr-x    2  nigel  staff   68   Jul  8   14:10  Unix_and_Perl/
```

Explanation

Notice that we first run the cd command to navigate to Nigel's home directory[43] and then run the pwd command to confirm its location. Then we run the mkdir command and simply specify the name of the directory that we want to create. Finally, we use the -lp options of the ls command to get a long listing of the directory, which also adds forward-slashes to the ends of any directory names.

Example 3.14.2 Now that we have a container directory for all of the Unix and Perl material that we might create, we can also make some more specific subdirectories:

```
$ mkdir Unix_and_Perl/Code
$ mkdir -p Unix_and_Perl/Temp/Inside_temp
$ cd Unix_and_Perl
```

Explanation

There are two things to note from this. First, the mkdir command can be used to create directories inside existing directories. Of course we could have also just changed directory into the 'Unix_and_Perl' directory before creating the 'Code' subdirectory. Second, you can create nested directories in one go if you use the -p command-line option.[44] This allows us to create a directory 'Inside_temp' plus its parent directory ('Temp') in one operation. If we had not specified the -p option, we would have seen an error like so:

```
mkdir: Temp: No such file or directory
```

---

[43] This is output from an actual Mac computer. Don't confuse this with our 'fictional' filesystem hierarchy that we showed in earlier chapters in which Nigel's home directory was located in /home.

[44] Of course, you can always find out more about this (and any other options for the mkdir command) by looking at its man page.

We will use the 'Code' directory for storing any Unix or Perl scripts that we write in subsequent chapters. The 'Temp' directory will be a place for trying out various Unix commands.

Removing directories

We only wanted to create the 'Inside_temp' directory to illustrate the usefulness of the `mkdir` command's -p option. Now we should remove it by using the `rmdir` command.

Example 3.14.3 Let's navigate into the 'Temp' directory and then remove the 'Inside_temp' directory:

```
$ cd Temp
$ ls
Inside_temp
$ rmdir Inside_temp
```

Explanation

In this case we removed the 'Inside_temp' directory while being located just one level above it. However, we could have also removed the directory without having to use the `cd` command by doing either of the following:

```
$ rmdir Temp/Inside_temp
$ rmdir /Users/nigel/Unix_and_Perl/Temp/Inside_temp
```

Hopefully you will realize that the first of these examples uses the *relative* path to the 'Inside_temp' directory, whereas the second example uses the absolute path (revisit Chapter 3.8 if you need a refresher on absolute and relative paths).

Note that the `rmdir` command will only remove *empty* directories (we'll cover how to remove directories that contain files later on). Also, you can remove a directory from anywhere *except* if you are inside the directory you want to remove.[45]

---

[45] Unix won't let you remove the ground from beneath your feet.

Or the art of accomplishing more by typing less

When we interact with the world of Unix, most of that interaction occurs via the keyboard. If you can reduce the amount of typing you have to do in order to accomplish a task then this is good for you in two ways. First, it makes you more productive because you are spending more time running commands and getting results, and less time typing their names. Second, and more important, it will help you minimize the amount of typing you do, which in turn will help lessen the risks of developing a repetitive strain injury (RSI).[46] In addition to having very short command names, Unix offers a few other ways to ease the load on your digits.

Command-line completion

Perhaps the most important time-saver to learn is something called *command-line completion*. This allows you to automatically complete the names of files, directories, and programs as you type them. If you type enough letters that uniquely identify the name of something and then press tab[47] … Unix will do the rest.

Example 3.15.1 Type the letters 'tou' and then press the tab key on your keyboard. In the following examples, we'll use `<tab>` as a shortcut way of saying 'press tab.'

```
$ tou<tab>
$ touch
```

Explanation

If this works, you should see the letters 'ch' become magically added to the three letters you have already typed. This forms the name of the Unix `touch` command (which we will learn more about in a later chapter). In this case, command-line completion will occur because there are no other standard Unix commands that start with the letters 'tou'. If this didn't work it might be because you have a non-standard Unix command on your system that also starts with 'tou', or that there is a file or directory in your current location that starts with 'tou'. When there are no possible completions for the letters you have already typed, you may hear a beep.[48]

Command-line completion can be used to save time when typing program names, but it is equally useful when working with files and directories. If you have yet to type the name of a Unix command, then tab-completion will attempt to complete

---

[46] Please do not underestimate the health risks that can result from overusing (or incorrectly using) a keyboard and mouse. Unix and most programming languages make heavy use of various punctuation symbols on your keyboard, and this usually means your fingers end up being stretched a little more than if you were just typing 'regular' text. If you routinely experience pain or discomfort while using a keyboard, you should *stop* typing and see your doctor.

[47] It is possible that in some Unix shells, other keys will be used to trigger the completion, but the tab key is the most common trigger key, which is why command-line completion is also known as *tab-completion*.

[48] Some, but not all, Unix systems will use a beep sound to indicate errors or to serve as a warning/reminder.

from a list of all known *commands and programs* on the system. However, if you have already typed the name of a valid command, then you can use command-line completion to finish the name of whatever *file or directory* is required by the command:

Example 3.15.2 We will navigate into our 'Unix_and_Perl' directory using the cd command, but we want to save as many keystrokes as possible when typing the directory name:

```
$ cd U<tab>
$ cd Unix_and_Perl/
```

Explanation
In this example we only have to type five characters (c + d + space + U + tab) rather than the full 16 characters that we would otherwise have had to type. If this doesn't work for you, it is most likely because you have another file or directory in your current directory that starts with the letter 'U'. Try typing one more letter at a time, and pressing tab after each successive key press.

Another way you can use command-line completion is to press tab *twice* to show a list of all possible completions. The contents of such a list will depend on how many characters of a file, directory, or program name have already been typed. It will also depend on whether you have yet typed a full Unix command. You can even press tab twice before typing anything at all. If you do this you will probably see a warning message like this:

```
Display all 1712 possibilities? (y or n)
```

This suggests that, on our computer, there are 1712 different Unix commands and programs that can be run.

Example 3.15.3 Now we will use tab-completion to browse through some directories that are at the root level of the computer:

```
$ ls<tab><tab>
ls  lsbom   lsdistcc      lsm     lsof    lsvfs
$ ls /u<tab>
$ ls /usr/b<tab>
$ ls /usr/bin/auto<tab><tab>
autoconf       autom4te     automake-1.10   autoreconf
autoupdate
autoheader     automake     automator       autoscan
```

Explanation
The first part of this example shows that even when you have typed the name of an existing command (in this case ls), there may be other Unix commands that match the

combination of letters that you have typed; in this case by pressing tab twice we see the six Unix commands that all start with 'ls'.

Next we start listing the contents of anything that begins with the letter 'u' at the root level of the filesystem. This could be a mixture of files or directories, but in this case there is only one directory ('usr') and a single press of the tab key is enough to complete the name. We then type 'b' and press tab to autocomplete the single matching directory ('bin'). We then type four more letters and press tab twice to see all of the items in this directory which start with 'auto'. If there was only one item that started 'auto' then we would only need to press tab once.

Using tab-completion can save you a lot of time when working with Unix. Imagine if you had to type the following:

```
$ ls -l /home/nigel/Work/Project_x/important_file.txt
```

Without tab-completion this requires typing 52 characters. *With* tab-completion, this might only involve typing 17 characters, less than one-third. Of course, this assumes that each directory and file name can be completed after typing only a single character, which may not be the case. Another major reason to use tab-completion is that it reduces the possibility of making typos – imagine if you had typed the above instruction in full, but had made a mistake and typed the following:

```
$ ls -l /home/ngiel/Work/Project_x/important_file.txt
```

The directory 'ngiel' does not exist and Unix would complain:

```
/home/ngiel/Work/Project_x/important_file.txt: No such file or directory
```

Unix tells you that there is a problem, but doesn't specifically tell you *where* the problem might be. If you use tab-completion to type out every directory name, you will never make a typo as the only directory names that you can type are those that you have managed to tab-complete. We can't recommend tab-completion enough.[49]

> *Always use tab-completion when typing any Unix command!*

Command history

Another great time-saver derives from the fact that Unix stores a list of all the commands you have typed during each session. You can see the history of all of your commands by using the appropriately named `history` command:

---

[49] From many years of teaching Unix to students, we have found ourselves repeating this tip more than any other. This is particularly because we have noticed that a huge proportion of mistakes that are made by newcomers to Unix are because of typos – typos which would have been avoided if tab-completion had been used.

```
$ history
    1 ls
    2 cd /home/nigel/Work/Project_x
    3 ls -l
    4 ls -l /usr
    5 ls /bin
    6 pwd
```

In this example we see the six previous commands we have typed. You can easily reproduce any of these commands by typing the appropriate number in the history list after an exclamation mark – e.g., to reproduce the second command, just type:

```
$ !2
cd /home/nigel/Work/Project_x
```

At this point Unix will print out a reminder of what the full command is and will then re-run that command.

Keyboard navigation of command history

It is common to want to repeat commands that were run recently. Rather than typing `history` and then selecting the appropriate command, it is often easier to access your command history by using the up and down arrows. Press your up arrow once and you should be able to access the last command you typed. Press it twice to access the second from last command, and so on. You can then press the down arrow to move forwards through your history. This can save you a lot of time! There are many other useful ways of accessing items from your command history, and you should read the documentation for whatever shell you are using to find out more.

Quick keyboard navigation of the command line

If you use the previous tip and access the last command that you typed by pressing the up arrow once, you will note that the full command appears at your command prompt. However, the command is not executed and the cursor will be placed at the *end* of the command. Sometimes you want to change one small aspect of what you typed – e.g., imagine that you retrieved the following from your command history:

```
$ ls /home/nigel/Work/
```

Maybe you ran this command but then realized that you actually wanted a *long* listing. As the cursor will be placed at the end of the command, you would have to press the back arrow several times to move the cursor to just after `ls` in order to add `-l`. A quicker way of jumping to the start of a line is to press Ctrl + a (hold down the control key and then press the letter 'a' once). This should make the cursor jump to the start of the line. This shortcut is one of a number provided by a software library called *readline*. There are many other shortcuts provided by readline; here are just a few which might be most useful to you:

Ctrl + a   move to start of line
Ctrl + e   move to end of line
Ctrl + w   delete previous word on command line
Ctrl + l   clear screen[50]

Problem 3.15.1 Practice navigating through some directories using the cd command, but make sure you use tab-completion to type every directory name.

Problem 3.15.2 Practice accessing your command history (using either the history command, or up-arrow navigation). Make sure you can repeat an older command.

---

[50] Note that there is also a simple Unix command, clear, which does the same thing.

Or how you can learn to move heaven and Earth

The next few chapters, including this one, will deal with Unix commands that help to work with files, i.e., commands that will allow us to move, copy, rename, delete, and view files. In order to learn how to use these commands, we will need to have some files to play with. The Unix command touch lets us easily create some empty files that we can work with.[51]

Example 3.16.1 Let's create two new files in your newly created 'Unix_and_Perl' directory. Remember to always use tab-completion when typing file and directory names:

```
$ cd ~/Unix_and_Perl
$ ls
Code    Temp
$ touch heaven.txt
$ touch earth.txt
$ ls -l
total 0
drwxr-xr-x 2  nigel   staff  68  Jul  8  14:55  Code
drwxr-xr-x 2  nigel   staff  68  Jul  8  16:19  Temp
-rw-r--r-- 1  nigel   staff   0  Jul 19  15:10  earth.txt
-rw-r--r-- 1  nigel   staff   0  Jul 19  15:10  heaven.txt
```

Explanation

We first ensure that we are in the 'Unix_and_Perl' directory, which in this case is one level below the home directory. Knowing this allows us to navigate there by using the ~/Unix_and_Perl syntax (see Chapter 3.9 for more information on using the tilde character [~] to refer to your home directory).

We then use the touch command to create two files which we give a .txt extension.[52] The ls -l command shows a long listing which confirms that the two new files exist and that they are zero bytes in size.[53] If we had wanted to, we could have also created both files in one step:

```
$ touch heaven.txt earth.txt
```

The mv command

If you want to move a file in Unix, you need to use the mv command. Whenever we use commands like mv, we must always bear in mind the two concepts of *source* and *target*. The source is the location of the file (or directory) that we want to move; the target is the

---

[51] This command does some other things as well as creating blank files. Look at its man page if you want to know more.

[52] This is not strictly necessary at this stage, because the files do not contain any text. Indeed, they do not contain anything at all.

[53] The size of files (in bytes) is listed in the fifth column of the long output. Every character in a file increases its size by one byte.

location of the place where we want to move the file. Commands such as mv will always expect you to specify a source and target location (in that order).

Example 3.16.2 Let's move these two new files into the 'Temp' directory that we created previously (Chapter 3.14):

```
$ ls
Code            Temp        earth.txt      heaven.txt
$ mv earth.txt    Temp/
$ mv heaven.txt   Temp/

$ ls
Code        Temp

$ ls Temp/
earth.txt          heaven.txt
```

Explanation

We use the mv command twice, once for each file that we want to move. The *target* location for the move is the 'Temp' directory. After moving the files, we confirm that the current directory no longer has the files and that they are instead in the 'Temp' directory. If you use tab-completion to type the name 'Temp' Unix will automatically add the trailing forward-slash character. This helps to remind you that 'Temp' is a directory and not a file.[54]

Renaming files

In the last example, the destination for the mv command was a directory name ('Temp'). However, the target could have also been a (different) file name, rather than a directory. This is how you can use mv to rename files.

Example 3.16.3 Let's make a new file and move it while renaming it at the same time:

```
$ touch rags
$ ls
Code        Temp      rags
$ mv rags   Temp/riches
$ ls
Code      Temp
$ ls Temp/
earth.txt          heaven.txt      riches
```

---

[54] When working with directories, you do not need to add a trailing forward-slash character to the name of the last directory in any path, but it doesn't hurt. This is another reason why using tab-completion is such a good habit to get into, because it will always add the slash character for you. See the next chapter for more details.

Explanation

In this example we create a new file ('rags') and move it to a new location. In the process we change the name (to 'riches'). Hopefully, you are still using tab-completion rather than typing out 'rags' in full, but notice that you can't use tab-completion for the word 'riches'. This is because until you press enter, the newly named file does not exist.

The mv command can rename a file at the same time as moving it. The logical extension of this is using mv to rename a file without moving it; you have to use mv to do this as Unix does not have a separate 'rename' command.

Example 3.16.4 Let's navigate into the 'Temp' directory and rename the 'riches' file.

```
$ cd Temp/
$ ls
earth.txt        heaven.txt       riches
$ mv riches rags
$ ls
earth.txt        heaven.txt       rags
```

Explanation

In this example, the mv command is being used to rename a file ('riches' to 'rags'). You should be aware that if there was already an existing file called 'rags' then this file would be overwritten by the move operation. The mv command doesn't ask you for confirmation if the result of moving a file would overwrite an existing file. However, this is an option that can be turned on by using an appropriate command-line option (see the mv man page).

In the last example we first change directory in order to rename the file, but note that we could have stayed in the 'Unix_and_Perl' directory and used the following command:

```
$ mv Temp/riches Temp/rags
```

The point of this is that as long as you know where the source and target file is located, you don't have to be in the same directory to move or rename those files. This means that you can always move or rename files from *anywhere* in your filesystem, as long as you know the full paths to the source and target file:

```
$ mv /Users/nigel/Unix_and_Perl/Temp/rags /tmp/
```

In this case both the source and target locations are specified using absolute paths (i.e., starting from the root level)[55], and so this command would work from *any* directory on the filesystem. Hopefully you will now start to see the advantage of typing Unix commands rather than using a mouse and interacting with a graphical file manager.

---

[55] See Chapter 3.8 for a refresher about absolute and relative paths.

Using wildcard characters

In Example 3.16.2, above, we moved two files in two steps. But it is usually possible, and often desirable, to move many files in a single step. We can do this by using something called *wildcard characters*. Using wildcard characters is yet another way Unix can help you save yourself from a lot of typing. The most common wildcard character in Unix is the asterisk (*), and it is used to mean 'match anything.' We could have used an asterisk in Example 3.16.2 to move both files in one go:

```
$ mv *.txt Temp/
```

You can read this as 'move any files whose names start with anything but end with '.txt'. This would move *any* file from the current directory into the Temp directory as long as its name ended with '.txt'. Given that we only had two files in the current directory, we could have also used the following commands:

```
$ mv *t Temp/
$ mv *ea* Temp/
```

The first example would work because there are no other files or directories in the directory that end with the letter 't' (if there was, then they would be moved too). Likewise, the second example works because the two files are the only items in the directory that contain the letters 'ea' in their names. This example also reveals that the asterisk wildcard can match 'nothing at all' as well as 'anything.' This means that *ea* will also match any files that were named only 'ea'.

Problem 3.16.1 The second most common wildcard character is the question mark (?). Try to understand what the question mark does by first using the touch command to create four files called 'fat', 'fit', 'feet', and 'feat' inside the 'Temp' directory. Then see what happens when you run the following commands:

```
$ ls *
$ ls f*
$ ls f*t
$ ls *at
$ ls f?t
$ ls f??t
```

Moving and renaming directories 3.17

Don't be embarrassed if this chapter moves you to tears

Moving and renaming directories is done in exactly the same way that we move and rename files (see previous chapter). This means we can use much of our knowledge of working with files when working with directories. However, we will also show you that there is one feature of directory names that distinguishes them from regular files.

Example 3.17.1 We will create a new temporary directory, then rename it, and then move it inside the existing temporary directory:

```
$ cd ~/Unix_and_Perl
$ mkdir Temp2
$ ls -p
Code/      Temp/  Temp2/
$ mv Temp2 Temp3
$ ls -p
Code/      Temp/  Temp3/
$ mv Temp3 Temp
$ ls -p Temp
Temp3/          fat     feet      heaven.txt
earth.txt       feat    fit       rags
```

Explanation
After creating a new directory ('Temp2'), we first confirm that it is indeed a directory by using the -p option of the ls command. We then use the mv command to rename the directory to 'Temp3'. Finally, we use the mv command one more time to move the new directory *inside* the existing 'Temp' directory. Note that if the 'Temp' directory didn't already exist, this action would simply rename the directory. It is important that you understand this distinction.[56]

Should directory names always end in a slash character?

You may have noticed that directories sometimes include a forward-slash character after their name, and sometimes they don't, such as in the following two commands:

```
$ ls Unix_and_Perl
$ ls Unix_and_Perl/
```

The two examples are not quite identical, but they produce identical output. So does the trailing slash character in the second example matter? Well, not really. In both cases we have a directory named 'Unix_and_Perl' and it is optional as to whether you include the trailing slash.

---

[56] It can sometimes seem confusing because it may not always be obvious whether you are moving or renaming something. Once again, tab-completion is your friend. If the target file or directory name does not exist, then you can't tab-complete its name, and so you must be renaming an item.

Returning to the last step of Example 3.17.1, you should now appreciate that all of the following commands are equivalent:

```
$ mv Temp3    Temp
$ mv Temp3/   Temp
$ mv Temp3    Temp/
$ mv Temp3/   Temp/
```

The advantage of adding a trailing slash character is that it then becomes obvious that the item in question must be a directory. This can help you more easily distinguish the names of files and directories in long commands. When you tab-complete any Unix directory name, you will find that a trailing slash character is automatically added for you. This becomes useful when that directory contains subdirectories which you also want to navigate to. Imagine if you were trying to access a buried directory that was seven levels below your current directory:

```
$ cd aaa/bbb/ccc/ddd/eee/fff/ggg/
```

If tab-completion *didn't* add the trailing slash characters, you'd have to type the seven slashes yourself. This might not seem like much, but every character counts; if tab-completion is kind enough to offer to do the work for you, it would be rude to refuse such assistance!

Problem 3.17.1 Change directory to make sure that you are 'in' a directory that is somewhere outside your home directory (e.g., '/usr' or '/bin'). *Without leaving this directory*, perform the following operations:

(1) Create a directory called 'Tears' inside your home directory
(2) Create a directory called '123' inside '/tmp'
(3) Create another directory ('XYZ') inside '123'
(4) Move the nested directory '123/XYZ' inside the 'Tears' directory
(5) Confirm this move happened successfully
(6) Now move just the 'XYZ' directory to '/tmp'
(7) Finally, remove all three temporary directories

How to remove files 3.18

Welcome to the most dangerous Unix command you will ever learn!

In Chapter 3.14 we showed you how you can remove a directory with the rmdir command, but rmdir won't remove directories if they contain any files. So how can we remove the files that we have created in the 'Unix_and_Perl' directory? To do this, we will need to use the rm command, which is used for removing files.[57]

Please read the next section VERY carefully. Misuse of the rm command can lead to needless death and destruction!

The rm command

Potentially, rm is a very dangerous command; if you delete something with rm, you will not get it back. It does not go into the trash or recycle can, it is permanently removed.[58] With rm, it is possible to delete everything in your home directory (all files, directories, and subdirectories);[59] that is why it is such a dangerous command.

Let us repeat that last part again. If you misuse the rm command, it is possible to delete *every* file you have ever created! Are you scared yet? You should be. Luckily there is a way of making rm a little bit safer. We can use it with the -i command-line option, which will ask for confirmation before deleting anything:[60]

Example 3.18.1 Let's see the difference that the -i option makes to the rm command. To do this, we'll create two more temporary files inside the 'Unix_and_Perl' directory:

```
$ touch test1   test2
$ ls
Code   Temp    test1      test2
$ rm test1
$ ls
Code   Temp    test2
$ rm -i test2
remove test2? y
```

Explanation

We use the touch command to make two test files. The first is deleted by rm *without* using the -i option. Notice how there is no warning – the file is deleted as soon as we

---

[57] It can also be used for removing directories, which can be very useful but also very dangerous.

[58] Of course, you might have a backup system in place, but this will rarely be successful at recovering a deleted file that you had only just created.

[59] If you have administrator privileges on your computer then it is possible to delete *everything*!

[60] You can also use the -i option with the mv command; it behaves in the same way and will warn you before you potentially overwrite a file.

press enter, never to return. When we try deleting the second file, we instead use the `-i` option and this time we are prompted as to whether we really want to remove the file.

Making the `rm` command even more dangerous

Just as we can use wildcard characters with the `ls` command, we can also use them with the `rm` command. For example, we could have removed both temporary files in Example 3.18.1 by using the following:

```
$ rm test?
```

In this case, the question mark would represent any single character, so this would also remove any files called 'test5', 'testX', or even 'test?'[61] The more dangerous example of the `rm` command would be (and *do not type this!*):

```
$ rm *
```

Any time you ever consider using the asterisk character in conjunction with the rm command, you should first step away from the keyboard and count to ten. After doing so, check and double-check that you really have typed the command correctly *before* pressing enter. The above command will delete all files in your current directory (assuming that you have the permissions to do so). If you must type this command, please consider typing the safer version:

```
$ rm -i *
```

Problem 3.18.1 Use the `rm` and `rmdir` commands to delete all of the additional temporary files and directories you have created over the last few chapters. Remember that you can only use `rmdir` to remove a directory once you have first removed all files in that directory. Also remember that you can't delete a directory if you are inside it (first navigate to one level above it). Make sure that you *don't* remove the 'Code' and 'Temp' directories.

---

[61] It is generally considered to be a *bad idea* to include wildcard characters as part of a file or directory name. Consider this a warning!

How to copy files and directories 3.19

Two files are better than one

The way you can copy files and directories in Unix is almost identical to how you move files. The cp (copy) command behaves in a very similar way to mv, though there are some important differences.

Example 3.19.1 Let's make a file and then copy it to a different location, and then make a copy of the copy (renaming it in the process):

```
$ pwd
/Users/nigel/Unix_and_Perl
$ touch test1
$ cp test1 Temp/test1
$ ls
Code Temp test1
$ ls Temp/
test1
$ cp Temp/test1 Temp/test2
$ ls Temp/
test1 test2
```

Explanation
We make a test file in the 'Unix_and_Perl' directory and copy it to the 'Temp' directory. We then remain in the same directory but make a copy of our test file and rename it in the process (from 'test1' to 'test2').

Because the act of copying could potentially overwrite an existing file, there is a -i command-line option which works in exactly the same way as the -i option for the mv and rm commands (remember, you can check the man page if you want to know more).

Copying a file without specifying a target name
In the first part of Example 3.19.1, we copied a file to another directory and kept the name unchanged. If you are not planning to change the name of a file when copying it, you can omit the target file name. The following two commands are equivalent:

```
$ cp test1 Temp/test1
$ cp test1 Temp/
```

The second form may or may not be more intuitive to you (think of it as saying 'copy the 'test1' file *into* the 'Temp' directory.' Remember, the trailing forward-slash after a directory name is optional. This means that the following command is also equivalent to the previous two:

```
$ cp test1 Temp
```

This is a situation where omitting the trailing slash can make things more confusing. Are you copying the file *and* renaming it to a *file* called 'Temp' or are you copying it into a *directory* called 'Temp'?[62] Including the forward-slash character makes it clearer that you are copying to a directory. Of course, if you are using tab-completion, then the forward-slash will be added automatically.

Copying a file to the current directory

Let's imagine you want to copy a file ('test.txt') from the 'tmp' directory of your filesystem to your current directory (e.g., '/Users/nigel/Unix_and_Perl/Temp'). One way you might think of achieving this would be something like:

```
$ cp /tmp/test.txt ~/Unix_and_Perl/Temp/
```

However, it becomes tedious to have to type out the full path of the target directory when you are already inside that directory (i.e., when the target directory is the current directory). Fortunately, Unix provides a shortcut. You can always refer to your current directory by using a single dot.

Example 3.19.2 Let's see the new dot character in action. Remember, any time you see the dot as part of a file operation, think of it as being identical to whatever the current directory name is.

```
$ ls -p
Code/   Temp/   test1
$ cp Temp/test2 .
$ ls
Code   Temp   test1   test2
$ cd Temp
$ ls -l
total 0
-rw-r--r-- 1 nigel staff 0 Jul 28 16:33 test1
-rw-r--r-- 1 nigel staff 0 Jul 28 15:43 test2

# WAIT 1 MINUTE

$ cp ../test1 .
$ ls -l
total 0
```

---

[62] Appreciate that sometimes you will be working in directories that contain hundreds of other files and directories, so you might not know the names of all the other items that are there. You might mean to copy a file and give it a new name of 'Temp,' but might be unaware that there is already a *directory* called 'Temp' inside your current directory. One consequence of this issue is that you can't have a file *and* directory with the exact same name in the same place on a Unix OS.

```
-rw-r--r-- 1 nigel staff 0 Jul 28 16:34 test1
-rw-r--r-- 1 nigel staff 0 Jul 28 15:43 test2
```

Explanation

The first copy operation reads as 'copy the file 'test2' from the 'Temp' directory to the current directory.' We then navigate into the 'Temp' directory and perform a long listing before waiting for one minute. Then we copy the 'test1' file from the parent directory (`../`) to the current directory.

Hopefully, you noticed that we already had a file called 'test1' in our current directory ('Temp'), so what happened? In this case the 'test1' file from 'Unix_and_Perl' was copied to the 'Temp' directory and overwrote the existing 'test1' file that was already there. We prove this by using the long listing format to see the file modification times. As long as you wait for at least one minute, you will see that the modification time for 'test1' will be changed. Remember, if you want to avoid accidentally overwriting files, you can use the `cp -i` syntax.

How you can use the dot character to represent the current directory

The dot character will always be taken to represent the current directory. This means that you can issue pointless commands such as 'change directory to the current directory':

```
$ pwd
/Users/nigel/Unix_and_Perl/Temp
$ cd .
$ pwd
/Users/nigel/Unix_and_Perl/Temp
```

You can also use it in conjunction with the `ls` command. Because the dot character is referring to a directory, we also know that we can always append a forward-slash to it. This means the following three commands all achieve the same thing:

```
$ ls
$ ls .
$ ls ./
```

In Example 3.19.1 we issued the `ls` command twice: Once to see the contents of the current directory, and once more to see the contents of the 'Temp' directory. The `ls` command can list the contents of multiple directories at once, so we can replace the use of two commands with just one:

```
$ ls . Temp/
.:
Code      Temp      test1      test2

Temp/:
test1      test2
```

This reads as 'list the contents of the current directory and then the 'Temp' directory.' When listing the contents of multiple directories, the output from ls adds the name of each directory before showing its contents. We will see other useful ways to use the dot character later on.

Copying multiple items

You will sometimes want to copy *everything* in your current directory to another location (frequently your home directory). We can use an asterisk character on its own to represent 'all files or directories.' As we also know that the tilde character (~) is used to represent your home directory, we can copy everything to our home directory with the following command:

```
$ cp * ~
```

This can be a dangerous command to use because your current directory could potentially contain thousands of files and/or directories. Copying large numbers of files in this way could take a long time, and could also contribute to you running out of available disk space. It is common to use the asterisk with some other characters to restrict the copying to files of a certain type:

```
$ cp *.txt ~     # only copy text files
$ cp file* ~     # only copy files that start with 'file'
```

Copying directories

If you want to copy a directory, you need to make a slight change to how you run the cp command.

Example 3.19.3 Let's try the most obvious way of copying a directory, and then see how to fix the problem that will occur:

```
$ cp Temp/ Temp2
cp: Temp is a directory (not copied).

$ cp -R Temp Temp2

$ cp -R -v Temp Temp3
Temp -> Temp3
Temp/test1 -> Temp3/test1
Temp/test2 -> Temp3/test2

$ ls -p
Code/     Temp/     Temp2/     Temp3/     test1     test2
```

Explanation

The first attempt at copying fails. This is because the default behavior of Unix is to not let you simply copy directories in the same way as files.[63] The solution is to use the -R command-line option, which allows you to *recursively* copy a directory.[64] This means if the source directory contains many subdirectories, these would also get copied (along with all the files in those subdirectories).

Notice that we make another copy of 'Temp', but this time we also include the -v command-line option. This turns on verbose mode, which just means Unix will confirm all of the items that are copied.[65] It is often desirable to check on the progress of a command as it is running (especially if it is copying thousands of files). Note that many Unix commands support a verbose mode.

Problem 3.19.1 It is important that you are comfortable copying files and directories. Experiment by making more copies of the temporary files and directories that we have already created. Make sure you practice copying items to the current directory, and try to use wildcard characters to only copy certain items. When you have finished, delete all of the temporary files and directories (except the 'Code' and 'Temp' directories). In future, we'll assume that you have tidied up after yourself – we promise not to nag you any more about this!

---

[63] I wish we could definitively tell you why this is the default behavior, but we can't. Most likely it is to prevent the accidental copying of very large directories.

[64] The ls command has a similar option. Try it and see.

[65] Of course, you probably already know about this mode because you've read the man page. You *have* read the man page, right?

A good example of where less is more

One of the most common activities you will want to do on a computer is view the contents of a file. On any Unix system it is very easy to view regular text files, also known as *plain-text* files.[66] The plain-text file format is the *lingua franca* of the computing world, and as such can be viewed on *any* type of computer. Programs that are written in any scripting language (Perl, Python, etc.) always use plain text, and it is common to see plain text used for documentation, particularly installation instructions.

How to tell which files are text files

The name of a file will help, particularly if it ends with .txt or .text. However, just because a file is named something like 'info.txt' doesn't absolutely mean that it will be a text file. It is perfectly possible, though pointless, to take a PNG image file and instead give it a .txt file suffix. Luckily, there is a Unix command called file, which can help you determine whether something is a text file or not. Here is some example output to show you how you can use the file command:

```
$ file Unix_and_Perl
Unix_and_Perl: directory

$ file manual.pdf
manual.pdf: PDF document, version 1.4

$ file stuff.gz
stuff.gz: gzip compressed data, was "stuff," last modified: Thu Jul 29

$ file test.pl
test.pl: a /usr/bin/perl script text executable

$ file README
README: ASCII text
```

Notice that the file command can often give you some very specific information about different file types. It even informs you when you have a text file which is also an executable script (as in the penultimate example with a Perl script). Finally, notice the name of the last file in the list. 'README' (in upper-case) is commonly used in Unix as the name for files that contain some useful information. This might be installation instructions for a program, or it might be a description/summary of all of the files in the current directory.

---

[66] Note that plain-text files are different from 'rich-text format' files, which additionally let you store basic formatting information (alignment, bold and italic fonts, etc.). Plain-text files do not contain any additional formatting. Note that certain programs like Microsoft Word may create *.doc files which are actually stored in plain text, but which have a lot of metadata associated with them. This means that while you can use regular Unix text viewers to look at them, they might not make much sense as any meaningful content in the document will be buried in a sea of XML tags.

Why you should avoid using text-editor applications to view text files

From time to time you will come across a text file, and your first thought will be to switch to the graphical file browser and double click on the file in order to open it. This is a bad idea for a few different reasons. First, we are trying to teach you to use Unix, so you should try to avoid using the mouse and graphical file managers.

The second reason is that you might not know what your computer will use as the default application for opening a text file. It might be Notepad (on Windows) or TextEdit (on a Mac), but it's possible that your computer may be set up to use a program like Microsoft Word to open text files. Using Word to view the contents of a file which may contain only one or two lines of plain text is like using a sledgehammer to crack a nut.

Another issue is that you won't always know the size of the text file you are trying to open. In science, and bioinformatics in particular, it is common that a lot of raw data is stored in very large text files – e.g., a file containing the human genome sequence might be about 3 GB in size. The aforementioned text editors will try to open the entire file at once, which can cause problems if your computer does not have sufficient memory. This may even crash the program – even if it doesn't, it might make that program run very, very slowly.

A final reason for not using text editors to view text files is that text editors can also *edit* the file. If you accidentally slip on a key while viewing a file you might change the contents, and as many editors have an autosave function, you might be *permanently* changing the contents. For data files used in science, this is generally known as a *very bad thing*! If you only want to view a file, use a text viewer, not a text editor.

The less program

Once upon a time the most popular text-viewing program on Unix was a program called more.[67] But then someone came along with a similar program that offered more functionality, and they decided to name it less. Any time you need to view a text file in Unix, we suggest you *always* try using the less program. One nice feature of less, is that it handles very large text files very easily, and if the file is very large, it doesn't try to read the entire contents of the file into memory. To view a text file with less, simply type less followed by the file (or files) that you want to view:

```
$ less somefile.txt
```

We have already seen the less program in use; any time you view a man page you are probably seeing less in action (see Chapter 3.13). We documented the most basic keyboard controls in that earlier chapter. Some additional keyboard controls that are useful to know are:

h   access 'help' page (press 'q' to quit help)
g   jump to start of file
G   jump to end of file

---

[67] The 'more' program is still installed on most Unix systems.

/	start searching forwards in the file
?	start searching backwards in the file

If you try searching you must also specify a pattern to search for – e.g., if you are viewing a file with `less` and you type:

```
/cheese<enter>
```

it will search forwards in the file for the next occurrence of the string called 'cheese'; it will also highlight all matches to this string. Once you have performed any forwards or backwards search, you can jump to the next match simply by typing / or ? (followed by enter).

Problem 3.20.1 Find some text files on your computer. If you already have some text files then that's great, otherwise download them or create them using a text editor. Check that these files really are text files by using the `file` command. Use this command to inspect other types of files that you find in your Unix filesystem.

Try using `less` to view your text files and ensure you can properly navigate around those files. You must be able to move backwards and forwards (by a line at a time or a page at a time). Can you jump to the end of the file? Can you find a certain word and search for all occurrences of that word? Finally, look at the help page within `less` and try to use one or two keyboard controls that we haven't mentioned here.

Or, learn how Samuel Clemens became Mark Twain

Hopefully you are starting to appreciate that there are a *lot* of commands in Unix. You are probably already having trouble remembering the names of many of them, let alone all of the options that many of those commands possess. Fortunately, Unix provides a way for you to create alternative names for any command you want. One use of these *aliases* is to create easier-to-remember versions of existing commands (e.g., 'move' instead of mv or 'copy' instead of cp).[68] If you do this, you should choose a name for the alias which doesn't already exist as a command. Another common use of aliases is to make a shortcut for those commands where you typically use lots of command-line options. For instance, if you find yourself frequently typing ls -ltr (or even ls -l -t -r), then maybe you can save yourself some typing by creating an alias called 'ltr' which does the same thing.

Creating aliases

Aliases are created using the alias command, but the syntax can differ slightly depending on which shell you are using.[69] If you are using any of the bash family of shells (sh, bash, ksh, or zsh), then the syntax for the alias command is:

```
alias new_command_name='old_command_name'
```

If you are using csh or tcsh, the syntax is slightly different and you can omit the equals sign and any quotation characters:

```
alias new_command_name old_command_name
```

Example 3.21.1 Let's create the alias that we mentioned earlier. We want to be able to type 'ltr' and get a long listing from the ls command, reverse-sorted by modification date. We'll stick with the bash syntax in this chapter:

```
$ pwd
/Users/nigel/Unix_and_Perl
$ ltr
bash: ltr: command not found

$ alias ltr='ls -ltr'

$ ltr
total 0
drwxr-xr-x  2 nigel  staff  68 Jul  8  14:55  Code
drwxr-xr-x  2 nigel  staff  68 Jul  28 17:17  Temp
```

---

[68] It is true that these aliases are longer than the commands they are linked to, but you may consider that a worthy tradeoff if it means you don't have to keep on looking up the name of a command in a book.

[69] This is one of the few instances where we need to show you how to do things in more than one way, depending on which shell you are using. See Chapter 3.10 for a reminder about Unix shells.

Explanation

We start by checking that 'ltr' isn't already an existing command. We then proceed to define our new alias using the `alias` command. As soon as we have typed the alias command we have effectively made a new command `ltr`, which will do exactly the same thing as typing `ls -ltr`. We confirm this by running our new alias.

How to see whether an alias already exists as a command

In the previous example we tried seeing if a command exists by typing its name. This is potentially very dangerous! What if the `ltr` command already existed and did something really nasty? Typing a command to see what it does is a bit like placing your hand in a pan of water to see how hot it is! It is better to use the `which` command to see if a potential alias already exists as a Unix command. The `which` command is normally used to tell you *which* version of a program you will get when you type its name (it is possible to have multiple versions of Unix programs installed). However, if your proposed alias name doesn't exist, `which` will return nothing at all, which tells you it is safe to use that name as an alias:[70]

```
$ which ls
/bin/ls
$ which which
/usr/bin/which
which commandthatdoesnotexist
$
```

Example 3.21.2 If you do not specify any information to the `alias` command, then you can use it to see a list of all of the aliases that have been created. Let's create one more alias and then check that we can see details of both of the aliases:

```
$ which p
$
$ alias p=pwd
$ p
/Users/nigel/Unix_and_Perl

$ alias
alias ltr='ls -ltr'
alias p='pwd'
```

---

[70] More specifically, the `which` command is searching your $PATH to find commands. So it is possible, but unlikely, that there is a command on your system but which isn't in your $PATH. Furthermore, the `which` command doesn't tell you if the command exists but only as an alias.

Explanation

This time we start by using the `which` command to check that 'p' isn't already an existing Unix command. Because `which` doesn't find any match, we assume it is safe to create an alias called 'p'. We then make 'p' an alias for `pwd` and then test that it works. Note that if the name of the alias does not include spaces, then you don't need to surround it with quotation marks. Finally, we run the `alias` command without any arguments and it then shows us details of both aliases.

You might not think it very useful to effectively shorten a three-letter command down to one letter, but if you typed `pwd` 100 times in a day (which is not such an unlikely scenario) then you would be saving yourself 200 keystrokes. Other aliases can save you from even more typing.

Creating aliases that replace existing aliases

Sometimes you might want to change how an alias behaves. Alternatively, it's possible that you might try creating an alias which uses the same name as an existing alias that you didn't know about. If you try this, then the last alias you create replaces the functionality of any earlier aliases that have the same name.

Example 3.21.3 We'll make an alias, use it once and then change it to do something slightly different.

```
$ which saycheese
$ alias saycheese='echo cheese'
$ saycheese
cheese

$ alias
alias saycheese='echo cheese'
alias ltr='ls -ltr'
alias p='pwd'

$ alias saycheese='echo CHEESE'
$ saycheese
CHEESE

$ alias
alias saycheese='echo CHEESE'
alias ltr='ls -1'
alias p='pwd'
```

Explanation

We first create a new alias ('saycheese') which simply prints the word 'cheese'. After testing this, we modify the alias to instead print 'CHEESE' in upper-case characters. This new functionality replaces the old behavior and we confirm that there is only one alias called 'saycheese' by running the `alias` command.

Aliases that change existing commands

If you have been thinking 'Hey, can I make an alias which has the same name as an existing command?' then the answer is 'Yes you can, but are you sure you really want to?' This is something that has both advantages and disadvantages. One simple advantage is that you can effectively change the default behavior of a command to include one or more command-line options – back in Chapter 3.12 we introduced you to ls -p which makes it easier to tell whether an item is a file or directory. Some people might always like to know this, and might think that the extra info you get from this command-line option should be the default. No problem! Simply type:

```
$ alias ls='ls -p'
```

If you do this then you will have changed the default behavior of the ls command. The ability to do this can be really useful as it allows you to make certain commands safer than normal. For instance, if you make rm alias to rm -i, then you will no longer be able to delete a file without first confirming that you *really* want to delete it. Sounds good, right? The problem, though, is what if you forget you have made an alias? Or worse, what if someone else is using your computer who doesn't know you have changed the default behavior of their favorite command? There is also the potential for causing deliberate mischief by making aliases perform undesirable actions,[71] so please bear in mind that:

> *Changing the behavior of common Unix commands with aliases can be dangerous!*

This isn't to say that you should never use aliases to modify the default behavior of commands (the authors do it all the time), but please tread carefully (especially if other people might use your computer). Of course, the other danger of becoming overly dependent on using aliases is that when you have to use someone else's computer you discover, to your horror, that their rm command deletes things without asking!

Temporarily turning off aliases

If you create an alias that changes the default behavior of an existing command, it is easy to temporarily restore the normal behavior. Simply prefix the command with a backslash or place it between quotation marks. For instance, if we assume that we created the ls alias from the previous section that will automatically include the -p option:

```
$ ls
Code/          Temp/

$ \ls
Code           Temp

$ 'ls'
Code           Temp
```

---

[71] Just imagine what would happen if someone made an alias called mv that actually ran the copy command (cp) instead?

The last two examples (temporarily) restore the default behavior of the ls command, meaning the forward-slash character that appears after directory names is now omitted.

How long do aliases last?

If you create aliases using the alias command, you will find that they are short-lived. They are not saved. As soon as you close your terminal window, the alias is gone. This also means they only exist in the terminal window in which they were created. If you have two terminal windows open and create an alias in one of those windows, it will not be available in the other window. In a couple of chapters' time, we will learn how to overcome this limitation and make aliases work in all your terminal windows, all of the time.

Problem 3.21.1 Create three aliases that make *safer* versions of the rm, mv, and cp commands. Make all of these commands include a -i option by default. Use touch to create some test files to test that these aliases are all working.

It's time to fire the editor

So far we have only learned how to create empty files using the touch command. Now we will learn how to edit any plain-text file using a Unix text editor. We will show you how to use a basic, but functional, text editor but will also mention some of the other commonly used Unix text editors that are out there.

How are Unix text editors different from word processors?

The Unix text editors that we are talking about are those that are run from *within* the terminal window.[72] These editors do not have any mouse control and all editing must therefore be done using keyboard shortcuts. You will not be able to do things such as italicize text, add tables, or add images. Indeed, you will not be able to do many of the functions that you might be used to doing with a typical word-processor program. Instead, you will be able to create and edit *text*, though such editing may include a wide variety of tools and manipulations.

You might be thinking that Unix text editors sound rather primitive and limiting. That is not the case at all. They excel at doing what needs to be done when working with plain-text files, and some programs contain many complex features. The most common Unix editors are available on just about any Unix machine you will ever use. This means that if you take some time to learn the subtleties of one (or more) of these programs, then you will be set up for editing files for the foreseeable future.

Which Unix editor to use?

We should preface this section by warning you that the subject of *which* editor to use is a topic that inspires a vibrant debate within the Unix community. In fact, two of the most venerable Unix editors inspire an almost religious fervor among their followers.[73] One of the oldest editors is called vi and dates back to 1976. This is the editor that you can guarantee will be installed on all Unix systems. When you use vi, you are either in something called *insert mode* or *normal mode*. This distinction can be very confusing to the newcomer and vi has a steeper learning curve than other editors. Despite this, it remains one of the most popular and widely used editors.

In 1991, an extended form of vi was created called vim (Vi IMproved). Compared to vi, there are many enhancements in vim, and it offers a far higher degree of customization. By and large, the operation of vim is very compatible with vi, and it remains a very popular editor for Unix users.

---

[72] Though there are also many Unix text editors that can be run within a windowed environment and which support mouse control.

[73] We are not kidding about this. When meeting other Unix users, you should be very careful when disclosing what editor you use. If they discover that you use what they consider to be the *wrong* editor, you might find that they stop returning your phone calls.

After vi, the next most popular editor is called 'Emacs' (though the Unix command is just emacs in lower case) and like vi, it was also developed in 1976. As is the Unix way, there are many different versions of Emacs, though the main functionality is common to all variants. Emacs has less of a learning curve than vi, and is very customizable. It is the rivalry between vi and Emacs that has become a famed part of Unix culture.[74]

For the purposes of this book, we just want to get you up and running, using Unix as quickly as possible. Therefore we are not going to tell you any more about vi, vim, or Emacs, though we do suggest that you try learning the basics of at least one of these editors. Instead, we are going to introduce you to a basic editor called nano.

The nano editor

Compared to the editors mentioned in the last section, nano is a very young program (developed in 1999) and it lacks a lot of the advanced tools that the other editors possess. However, it is easy to learn and is also installed on many different Unix systems. It was initially created as a clone of another editor called 'pico' (which is still available on many Unix systems). The pico editor was originally part of a Unix email client called 'Pine,' but became popular as a text editor in its own right. In the event that your system doesn't have nano, check to see whether pico is installed.[75] Even though nano is more basic than some other editors, it still has a reasonably complete set of features.

How to use nano

When you run nano, you can use it in three slightly different ways:

(1) to edit an existing file;
(2) to create a new *named* file and start editing it;
(3) to create a new *unnamed* file and start editing it.

The last option is invoked by simply typing the command nano with no further options. If you do this, your terminal window becomes the nano editing window, and you will no longer be able to type terminal commands until you exit the editor. The editor should look like this:

---

[74] See Wikipedia's page on 'Editor War' for a good overview.
[75] Remember, you can use the which command to see whether a specified program is installed.

The gray cursor in the top-left of the window indicates that you can start typing some text. The arrow keys on your keyboard can be used to move the cursor (assuming you have first typed some text). At the bottom of the window is a series of keyboard shortcuts that you can type. The ^ character means *control*; to exit the program you would use control + x; to jump forward a page of text you would use control + v; and to jump backwards a page, control + y. The set of available keyboard shortcuts will often change depending on what you are doing – e.g., if you try searching for some text with the control + w option, then you will see a new range of options related to searching for text. The top line of the window (in black) shows you the name of the file you are working with. In this example we didn't specify a file name when we launched nano, which is why it says 'New Buffer'.

If you have typed any text in a blank nano window and you then try exiting the program, you will be asked if you want to 'Save modified buffer' and will be given yes/ no choices (type 'y' or 'n'). If you type 'y' you will next be asked to type the name of the file that you wish to create. By default, this will be saved in the directory in which you launched nano.

Some of the time you will want to use nano to create new files and you can do this more easily by specifying the name of the new file at the same time you run nano:

```
$ nano newfile.txt
```

This will tell nano to get ready to create a new file in the current directory called 'new-file.txt'. Note that it won't actually create this file until you specifically tell nano to save it (control + o) or you try exiting the program and then tell nano to save.

Most of the time you will probably use nano to edit existing text files. Remember, as long as you know where a file is *relative* to your current location, you don't need to first change directory to another location:

```
$ nano file.txt        # edit file called "file.txt" in current
directory
$ nano ../2nd.txt      # edit file called "2nd.txt" in parent
directory
$ nano /tmp/tmp.txt    # edit file called "tmp.txt" in /tmp
```

For now, this is about all you need to know about nano, but please feel free to read more about it (access the nano help page by pressing control + g).

Problem 3.22.1 Use nano to create a new text file called README in your 'Unix_and_ Perl/Temp' directory. Put some random text in this file and then save it and exit. Check that you can view the contents of this file using the less program. Try to find an existing text file that you can open with nano (but be careful to not make any changes to it), and practice navigating through the file one page at a time. Finally, try using the control + w keyboard shortcut to search for a specific word.

The source of how to automate commands is the `source` command

Chapter 3.21 introduced the concept of aliases. We ended that chapter by saying that ideally we would prefer to not have to retype our aliases every time we open a new terminal window. Over the next few chapters you will learn: How to store Unix commands in files; how to treat those files as programs; and how to get things to happen automatically as soon as you open a new terminal. The combination of these topics will allow you to put your aliases into a file which will be read automatically every time we open a terminal.

Reading the contents of text files with the `source` command

There is an easy way of telling Unix to read the contents of a text file and treat each line of the file as a Unix command. You can do this by simply using the `source` command followed by the name of a text file that contains your instructions.

Example 3.23.1 Use nano to create a text file in your 'Unix_and_Perl' directory that simply contains one line with the following text:

```
ls -R -l
```

You don't need to press enter at the end of the line, though it doesn't hurt if you do this. Save this file as 'recursive_long_listing', exit nano, and then check that you can view the contents with the `less` program. Now type the following command and observe the output it produces:

```
$ source recursive_long_listing
total 8
drwxr-xr-x   2 nigel  staff   68 Jul  8   14:55    Code
drwxr-xr-x   2 nigel  staff   68 Jul 28   17:17    Temp
-rw-r--r--   1 nigel  staff    9 Aug  4   10:01
recursive_long_listing

./Code:

./Temp:
```

Explanation
We have taken a Unix command with a couple of command-line options and saved that to a text file. The `source` command then reads that file and acts on its contents, meaning that it runs the `ls -R -l` command. As a result, we get a long, recursive directory listing of the current directory.

If you use the `source` command, then you should only try reading files that contain valid Unix commands.[76] If you replace the contents of the file from Example 3.23.1 with

---

[76] You can also include the names of programs that you (or others) have written in Perl (or other languages). But that's another story for another day.

the text 'cheesecake', and then try re-running the source command, you will see the following:

```
$ source recursive_long_listing
bash: cheesecake: command not found
```

There is no Unix command called 'cheesecake' and so Unix, quite understandably, complains that it can't find it. If you so desire, you can place multiple commands in a file and source will work through each command in turn. You could create a file that contained:

```
ls /
date
echo $HOME
```

If you ran the source command against the contents of such a file, it would: list the contents of the root directory; print the date; and then show the location of your home directory. In a few chapters' time we will show you a more advanced way of doing the same thing.

Adding aliases to an alias file

Now we know about the source command, we can do something useful such as put a list of useful aliases into one file called 'aliases'. Then, each time we open a new terminal we can make them all available by typing:

```
$ source aliases
```

This is still not a perfect solution as you have to remember to type the command, but at least it saves you from typing them all individually. There may be some aliases that you only want to use when working on specific projects, so you can also create multiple alias files if desired.

Example 3.23.2 Open a new terminal window (to ensure that no aliases will be available) and use nano to create an 'aliases' file in your 'Unix_and_Perl' directory. For now we suggest you include the following in this file, but feel free to add others. Also note that we are using the bash-family syntax for the alias command:[77]

```
# Add slashes to directory names
alias ls='ls -p'

# make copy, move, and remove commands safer by adding -i option
alias cp='cp -i'
alias mv='mv -i'
alias rm='rm -i'
```

---

[77] This syntax will work if you are using bash, sh, ksh, or zsh shells. To make it compatible with csh and tcsh shells, you need to remove the equals sign and any quotation characters.

After saving this file check that it works by running the following commands:

```
$ source aliases
$ alias
alias cp='cp -i'
alias ls='ls -p'
alias mv='mv -i'
alias rm='rm -i'
```

Explanation

Note that we added a mixture of blank lines, and *comments* to the 'aliases' file. Comments are any line of text that start with a hash character (#). The source command will simply ignore these lines.[78] Commenting files is a very useful habit to get into.[79] If you create aliases for complex commands you may not always remember what that command is doing. Adding some simple comments to your aliases file will make it much easier to look up what an alias is doing.

After running the source command, we confirm that the aliases have all been loaded by running the alias command.

---

[78] This is because anything you type after a hash character on the command-line will also be ignored by Unix. Try it and see!

[79] We shall return to the subject of commenting in much greater detail in the Perl parts of this book.

Peekaboo!

Most of the files that we work with in Unix are visible to you, i.e., they appear in the output of the ls command. Occasionally, though, you might want to make some files invisible. They will still exist, but won't be shown in the default output of ls. As we will see, Unix uses some of these *hidden* files for special reasons.

How to hide a file

Hiding a file is very easy. Just change the file's name to include a dot character as the very first character.

Example 3.24.1 Let's try hiding the 'aliases' file that we created in the last chapter.

```
$ ls
Code/    Temp/    aliases

$ mv aliases .aliases

$ ls
Code/    Temp/

$ cat .aliases
# turn on display of directory listing
alias ls='ls -p'

# make copy, move, and remove commands safer by adding -i option
alias cp='cp -i'
alias mv='mv -i'
alias rm='rm -i'
```

Explanation

We first confirm that the 'aliases' file exists. Note that the output from the ls command is adding slashes to directory names. This is because we have used source to read the contents of the aliases file (so that ls acts like ls -p).[80]

After using the mv command to rename 'aliases' to '.aliases' it no longer appears when we run ls. We then use a command called cat to display the contents of the hidden file (to prove that it is still there). The cat command does nothing more than print the entire contents of a file (or files). It is useful for quickly looking at small files. It also returns you straight back to the command prompt (you don't have to exit from cat like you would do with less).

---

[80] We didn't include this step in the example because we figure that you're wise enough now to not have to be told every single thing that we do.

How to see hidden files

Now that you have learned to hide files it is only fair that we show you how you can find hidden files (if any are present). If you have previously spent any time looking at the man page for the ls command, you may have noticed the -a option. This is all you need to know in order to see any hidden files.

Example 3.24.2 Let's see that hidden file in all of its invisible glory!

```
$ ls -a
./   ../    .aliases    Code/    Temp/
```

Explanation

The most important thing to notice from this output is that the addition of the -a option now allows us to see the .aliases file. This option simply shows you any files that start with a dot in addition to the normal ls output.

You will also notice that we now see two additional items in the directory (. and ..). What are these doing? Well, the other thing to notice is that they both include a / character after their name, which tells us that these are directories. We have already seen both of these items before: they refer to the current and parent directory.[81] This might seem a little confusing to you because we don't normally think of these things as actual directories. Don't worry about them too much for now, just remember that as they both start with a dot, we will always see them when we run ls -a. By extension, this means that *any* directory will also be hidden if its name starts with a dot.

One final thing to note about this output is the fact that our ls alias still works as ls -p even though we actually ran ls -a. From Unix's point of view, our aliases file has specified that ls is always run as ls -p. This substitution will happen *before* you add any further command-line options. If you want to override this behavior, remember you can always get the original version of a command by starting the command with a backslash, like so:

```
$ \ls -a
.        ..       .aliases       Code    Temp
```

Why hide files?

The main use of hidden files and directories is to store configuration settings for various programs. Configuration information is usually generated and modified by the program in question, and so it makes sense to hide it so as to prevent accidental modification (or deletion). Such files are typically stored in your home directory as this is the *only* directory that a program can assume will always exist. Some configuration files are also used to contain preferences and settings for the user. In the next chapter we will learn about an important user configuration file; by adding information to this file we can automate the reading of our aliases.

---

[81] We first saw . in Chapter 3.19 and .. back in Chapter 3.7.

Problem 3.24.1 Make a new hidden directory inside your 'Unix_and_Perl' directory and then move it into your 'Temp' directory. The point of this is to realize that all of the file/directory manipulations you have learned so far will work equally well with hidden items (this includes tab-completion). When you have finished, remove the hidden directory.

Customize yourself ... without tattoos or piercings

Now that we have shown you how to create both aliases and hidden files, we can finally achieve the goal of putting all of the aliases into a special configuration file that will be loaded automatically when you open a new terminal. As we shall see in later chapters, this configuration file can be used to do more than just store aliases, and we will also use it to store other important settings.[82] Just remember that instructions that are added to this file are always executed *every time* you open a new terminal window.

Where to store the configuration file

As we mentioned in the previous chapter, hidden configuration files (sometimes called 'dot files') are always stored in your home directory. This is because it is the one directory that all users have; your Unix system doesn't need to know the exact name of your home directory because it can always access it through the $HOME environment variable (see Chapter 3.11). When you create the new master configuration file, you will have to do this in your home directory and *not* in the 'Unix_and_Perl' directory (or any other directory that you are using for this book).

What is the configuration file called?

The answer to this question very much depends on what type of Unix shell you are using (see Chapter 3.10). Different shells tend to use differently named files to store your configuration settings, and there can also be multiple files involved:

Shell	Name of configuration file	
	Login	Other
Bourne shell (sh)	.profile	
Bourne-Again shell (bash)	.bash_profile *or* .bash_login *or* .profile	.bashrc
C shell (csh)	.login	.cshrc
TENEX C shell (tcsh)	.login	.tcshrc
Korn shell (ksh)	.login	
Z shell (zsh)	.zlogin *or* .zprofile	.zshrc

You may feel justified in finding this all a bit confusing.[83] The reason for some of this confusion is that there is a distinction between something called your *login shell* and other shells known as *interactive shells*. You really don't need to worry too much about this though. Just identify your current shell (by running echo $SHELL) and then be prepared to create the corresponding file from the 'login' column of the above table.

---

[82] Such as setting or modifying some of the environment variables that we learnt about in Chapter 3.11.

[83] Though you should also thank us as we have taken the liberty of simplifying things a little. There are also global configuration files that affect all users, as well as configuration files that are read at login time but take effect at logout time.

For two shells (bash and zsh) there are potentially multiple files that you can use and these actually have an order of precedence. However, as a newcomer to Unix we expect that you will only need to create one configuration file at this point. Depending on your shell, we suggest choosing .login, .profile, or .zprofile.[84]

Creating your configuration file

We are finally ready to set up your configuration file. This is a happy day and may very well change your life! We should point out that you may already have a configuration file present. This partially depends on who set up your account for you. In many business environments it is common to add some default configuration settings to all user accounts. Let's check:

Example 3.25.1 Make sure you are in your home directory:

```
$ pwd
/Users/nigel
$ ls -a
.                   Desktop         Pictures
..                  Documents       Public
.CFUserTextEncoding    Downloads        Sites
.DS_Store           Library         Unix_and_Perl
.bash_history       Movies
.lesshst            Music
```

Explanation

We start by confirming that we are in the home directory of our user, Nigel. We then list all hidden files in the home directory. Notice that there are a few hidden files already present. Your home directory may look very different, but don't worry too much about any hidden files that you see there (apart from configuration files). In our case, we have a relatively new account on a Mac computer (with a bash shell) and there are no existing configuration files. Check your directory for any of the possible configuration files that might be there. If you find one, have a look at it using less.

Assuming you don't already have an existing configuration file, we can build a new one using the existing .aliases file that we made in the last chapter. Remember, our aliases file was written using the correct syntax for the bash shell. If you are using csh or tcsh as your shell, you'll need to use the alternative syntax for the alias command (see Chapter 3.21).

---

[84] Once you become a Unix guru, you will want to revisit the setup of your configuration files. For this book, we are only ever working with the login shell. Many Unix programs allow you to access a shell without having to leave the program. These interactive shells, depending on your choice of shell, might read a different configuration file to your login shell. Interactive shells are also the default when you do things like use the su command to change to another user.

Example 3.25.2 *Don't follow this example if you already have an existing configuration file.* Instead, you can just manually add information from the .aliases file to your pre-existing configuration file using *nano*.

```
$ cp Unix_and_Perl/.aliases .profile
$ ls -a

.                  .profile      Music
..                 Desktop       Pictures
.CFUserTextEncoding    Documents     Public
.DS_Store          Downloads     Sites
.bash_history      Library       Unix_and_Perl
.lesshst           Movies
```

Explanation

We simply copy our existing .aliases file to the home directory and rename it to .profile. Because '.profile' is the name of one of the *bash* configuration files, it will now be read automatically for us *every time* we open a new terminal window. However, it is not read automatically after creating or modifying it, so any time we make changes to it we'll need to re-read it using the command:

```
$ source .profile
```

Or, if we want to ensure that we re-read it without first changing directory to our home directory, we can more explicitly refer to it using the tilde (~) syntax to refer to our home directory:

```
$ source ~/.profile
```

Editing your configuration file

Over time your configuration file will grow in size. You will probably add several more aliases to it, and there will be various environment variables that you will want to modify. Some configuration files often contain instructions to print a welcome message or specify that another program or Unix command should always be run each time a shell is opened. Remember, everything in this file will be read and acted upon by Unix in every new terminal window. This means that if the first thing you do after logging in is to *always* change directory to somewhere like 'Unix_and_Perl' you can simply add this instruction to your configuration file.

Example 3.25.3 Let's make just a few more changes to our configuration file. We'll add a welcome message that reminds us of the date, and then add some comments that will become more useful later on. Use *nano* to edit your .profile file to look like the following:

```
#########################
#
# .profile file for Nigel
#
#########################

# FIRST STEPS
MYDATE=`date "+%H:%M:%S %m/%d/%y"`
echo "Welcome $USER, the current time is $MYDATE"

# ENVIRONMENT VARIABLES
# (blank for now)

# ALIASES

# turn on display of directory listing
alias ls='ls -p'

# make copy, move, and remove commands safer by adding -i option
alias cp='cp -i'
alias mv='mv -i'
alias rm='rm -i'
```

Explanation

We first use comments to add a header that makes it clear who this configuration file belongs to. Next we do something using the date command which might look strange, but it's not too complex. Effectively, we run the date command specifying a certain format for the date and the result of this command is assigned to a variable which we call MYDATE. This is another instance where the choice of shell determines the syntax. The example above shows the correct syntax for bash-family shells. For csh and tcsch shells you need to use the set command:

```
set MYDATE=`date "+%H:%M:%S %m/%d/%y"`
```

We next use the echo command to print out a short message that relies on using the environment variable $USER and our new variable $MYDATE.[85] You should note that the whole date command is enclosed between a pair of *backtick* characters.[86] Finally, we add suitable comments to delimit the various sections of this file ('First steps,' 'Environment variables,' and 'Aliases').

To see the effects of these modifications, save the file and then run:

```
$ source ~/.profile
```

You should see something like:

```
Welcome nigel, the current time is 14:28:20 08/05/10
```

---

[85] If this part sort of makes sense to you then you are on your way to understanding some basic aspects of programming.

[86] These are also known as *backquote* characters and are yet another character which occupies various locations on different types of computer keyboard.

This is what should happen every time you open a new terminal window from now on. One final complication to be aware of is that some terminal applications let you choose whether the shell used by the terminal is a *login shell* or not. Sometimes the default may be that it is *not* a login shell. As the .profile file is only read by login shells, you might find that your terminal application is not reading the contents of .profile when you open a new terminal window. If this is the case, open the settings/properties/preferences for the terminal application and see if there is an option to enable the default shell to be a login shell.

Customizing your profile

You should feel free to customize your .profile file to have a different welcome message or no welcome message at all if you prefer. Some Unix users have .profile files that contain pages and pages of instructions. It's not uncommon to add instructions to a .profile file to make it run scripts that do things such as produce a fortune for the day,[87] or tell you the weather forecast for where you live. One of the most frequent 'customizations' that you will see in .profile files are commands that change the format of your command prompt. As mentioned in Chapter 3.3, some users like the command prompt to include useful information such as their user/machine name and/or the current directory. We won't go into the full details of how to do this in this book, other than to tell you that in the bash shell, you can customize your command prompt by modifying the $PS1 environment variable.

---

[87] Some versions of Unix still include the fortune command, which will produce a fortune for you.

Stick to the script

The goal of this book is to teach you enough Unix to make you comfortable writing Perl scripts in a Unix environment. However, it is worth spending just a little bit of time to show you that you can also write programs using Unix. As we have already demonstrated, it is possible to put a set of Unix commands into a file and then execute those instructions by using the source command. The next step is to treat the file containing the instructions as if it was any other Unix command. We want to be able to type the name of the file in order to run it and execute the instructions contained within it.

Running your first Unix script

The first thing you should learn to do when learning *any* programming language is to write a very simple program, just to check that the programming environment is working. Traditionally, this program will do nothing more than print the words 'hello world'.

Example 3.26.1 Use nano to put the following two lines into a new file. Save the file as 'hello.sh' and make sure it is saved in your ~/Unix_and_Perl/Code directory.[88]

```
# my first Unix shell script
echo "Hello World"
```

Now try running the following commands and observe the output:

```
$ cd Unix_and_Perl/Code/
$ hello.sh
bash: hello.sh: command not found
```

Explanation

We have made a two-line 'program.' The first line is a simple comment to describe the program, and the second line will just use the Unix echo command to print out some text. The program name ends with '.sh' as this is a common suffix for what are known as *shell scripts* (programs that run in the Unix shell).

You will have noticed that our newborn program doesn't actually work. When you type hello.sh as if it was a regular Unix command, you get a complaint back from the shell that it can't find this command. In a moment we will explain what you need to do in order to make this work, but it is important to first understand that it is not enough just to place Unix commands into a text file.[89]

---

[88] We will use the 'Code' directory to store all of our programs.
[89] Unless you're happy to always use the source command.

How to turn files into programs

There are two reasons why the 'hello.sh' program did not work. The first reason is that, in Unix, you have to explicitly give permissions to files that are going to be run as programs. Specifically, you need to add something known as the *executable* permission. The second reason concerns the *location* of the program. Unix will only look in certain locations for files that can be run as programs. You either have to make Unix know about the location of the 'Code' directory or perform a little shortcut in order to force Unix to look for programs in the current directory. We will address both of these issues in the following chapters.

We permit you to read this chapter

Files and directories in Unix have three main types of permission:[90] read, write, and execute. A file or directory can have any combination of these permissions and they are either 'on' or 'off.' If you create a new file it will automatically gain read and write permission,[91] but it won't have execute permission; this is something we have to add manually. Further complicating the issue is the fact that there are three different levels of permissions: user, group, and other.[92] In Unix, any user can belong to one or more groups of similar users (e.g., staff, grad students, administrators, etc.). It is sometimes useful to have permission to read, execute, or write to files owned by other people in the same group as you. Conversely, you might not want people who are *not* in your group to have permissions to do anything with your files. This is what the 'other' level of permissions is for. You can think of this level of permissions as those that control what files and directories are publicly accessible.

Unix has a single command called chmod that controls the read, write, and executable permissions for any file or directory at all three levels (user, group, and other). We can see the current permissions for a file or directory by simply running ls -l:

Example 3.27.1 Let's look at the permissions of some files and directories:

```
$ cd Unix_and_Perl/
$ ls -l
total 0
drwxr-xr-x   3  nigel  staff   102   Aug 6   10:01   Code/
drwxr-xr-x   2  nigel  staff   68    Aug 4   14:26   Temp/

$ ls -l Code/
total 8
-rw-r--r--   1  nigel  staff   49    Aug 6   10:01   hello.sh
```

Explanation
The first character in the long output from the ls command tells you whether something is a file or directory ('-' or 'd'). The next nine characters tell you the read, write, and execute permissions at the level of user, group, and other. Read, write, and execute permissions are denoted by 'r', 'w', and 'x', respectively. Lack of permission is denoted by '-'. Following the permissions is information related to the user ('nigel') and the primary group that they belong to ('staff').

We can now see that the 'Code' and 'Temp' directories are readable and executable to anyone (user, group, or other) but are writeable only by the user. Directories must have executable permission for you to be able to navigate into them with the cd command.

---

[90] These permissions are more technically called 'modes,' but we figured that 'permissions' would make more sense to you.
[91] It would be odd if this didn't happen as otherwise you wouldn't be able to view, modify, or remove your own files.
[92] You can think of 'user' permissions as those that concern the owner of the files or directories in question.

When we look at the long listing for our 'hello.sh' file we see that it lacks any executable permission, is readable to anyone,[93] but is writeable only by the user.

The `chmod` command can be run in a number of different ways, but the simplest usage can be understood with the following examples (using a fictional file called 'file.txt'):

```
$ chmod   u+x  file.txt   # add executable permission at the "user" level
$ chmod   g-r  file.txt   # remove read permission at the "group" level
$ chmod   o+w  file.txt   # add write permission at the "other" level
```

Just use '+' or '−' to add or remove permissions and 'u', 'g', or 'o' to denote the level. You can also change permissions for all levels in one go, and also change permissions for multiple files or directories:

```
$ chmod   a-w file.txt   #   remove write permission at all levels
$ chmod   a+w *          #   add write permission at all levels to all files
```

As you work through this book you will create many Perl scripts (and a few Unix scripts too), and you will only ever need to add executable permission to a file at the user level. Please be aware that `chmod` is potentially a dangerous command if misused. Not only can you do stupid things[94] like this:

```
$ chmod u-r hello.sh
$ less hello.sh
hello.sh: Permission denied
```

but you also have the potential to let all users on your computer or network have the ability to read and/or delete your files.

Always check the results of using the chmod command!

Running your first Unix script revisited

Let's fix the permission on the 'hello.sh' script and then see what happens when we try to run it.

Example 3.27.2 Note that after adding executable permission, we will try running the script from *two* different locations:

```
$ cd
$ chmod u+x Unix_and_Perl/Code/hello.sh
```

---

[93] The default set of permissions for new files varies between different Unix OSs and can also be changed. On some systems (such as Macs), any files created *outside* of the default set of directories (Desktop, Documents, Music, etc.) are readable by any other user accounts on that computer.

[94] Actually, this might be useful for paranoid types who want to ensure that people who have access to their computer can't easily see their files.

```
$ ls -l Unix_and_Perl/Code/hello.sh
-rwxr--r-- 1 nigel staff 49 Aug 6 10:01 Unix_and_Perl/Code/hello.sh

$ Unix_and_Perl/Code/hello.sh
Hello World

$ cd Unix_and_Perl/Code/
$ hello.sh
bash: hello.sh: command not found
```

Explanation

We first change directory to the home directory and run chmod to add executable permission to the file. This is then confirmed by running ls -l.

Next, staying in the home directory, we try to 'run' the program. We just type the full path to the 'hello.sh' file. The result is that the contents of the, now executable, file are processed by the Unix shell and we see the 'Hello World' output of the echo command.

We then change directory to the same directory that contains the 'hello.sh' script ('Unix_and_Perl/Code') and attempt to run the program again by typing its name. However, this time the program fails to run and we see the same error message as in the example from the previous chapter.

You might be scratching your head over this last example. In particular, you may be wondering what the difference is between trying to run a program while being in a *different* directory from that program, and trying to run it from the *same* directory. The solution awaits in the next chapter!

File permissions on files from other computers

If you transfer files between different computers, e.g., from a Unix/Linux machine to a Windows PC, you may experience some surprises regarding file permissions. Put simply, a Windows PC won't respect any file permissions that were set under Unix. The reasons for this are not important; it just means that you must be careful when transferring files to a Unix/Linux machine from a PC, because all permissions might be turned on. This situation can also apply when you use a Unix machine to access files on an attached USB flash drive – e.g., if you plug a USB drive into a Mac and navigate to the flash drive in the terminal,[95] then all files and directories will be readable, writeable, and executable. A fictional USB flash drive called 'USB' which contained a directory and a file might look like this:

```
$ ls -l /Volumes/USB
total 32
drwxrwxrwx 1  nigel  staff  16384  Dec 14  15:33  tmp_directory
-rwxrwxrwx 1  nigel  staff      0  Dec 14  15:33  tmp_file
```

---

[95] All attached drives will be accessible from the /Volumes directory.

How to specify which directories contain programs

Or why it's good to stray from the $PATH

The last chapter revealed that you can run a program while being in another directory, but *not* if you are in the same directory as the program. To solve this apparent dilemma, we have to learn a little bit more about Unix paths and about how Unix keeps a list of specific directories that it expects to contain programs.

Can a script have the same name as an existing Unix command?

Let's imagine what would happen if we created a new Unix shell script that just happened to share the same name as an existing Unix command – e.g., imagine that instead of being called 'hello.sh', our script from the last chapter was called 'pwd'. If you type 'pwd' into a Unix terminal you should rightfully expect to be able to run the pwd command. But if you were in the same directory that contains our newly named Unix script 'pwd', what should happen? Should Unix run its own pwd command or our 'pwd' script? If you try this you will find that Unix will run the pwd command. This is a good thing. If you create a file which coincidentally shares the same name as a Unix command, you should expect the pre-existing command to always take priority.

However, if we were in a *different* directory from our newly created 'pwd' script, we could get the script to work as long as we specified an absolute or relative path to the location of the program.[96] If we made a script called 'pwd' and it was in our 'Code' directory then we could run the script from *anywhere* in the filesystem by typing:

```
$ ~/Unix_and_Perl/Code/pwd
```

This would work because we are describing an absolute path to the location of a specific file in our filesystem and Unix would recognize that this file ('pwd') is different to the pre-existing pwd command. All you really need to understand is that typing the name of a program, even when you are in the same directory as that program, will not run the program. There is an easy fix to this problem. You might remember from previous chapters that a single dot character is used by Unix to refer to the *current* directory. This provides a way to run a script even when we are in the *same* directory as that script:

Example 3.28.1 Compare the output from the two following variants:

```
$ pwd
/Users/nigel/Unix_and_Perl/Code

$ hello.sh
bash: hello.sh: command not found

$ ./hello.sh
Hello World
```

---

[96] See Chapter 3.8 for a refresher on absolute and relative paths.

Explanation

The first example fails because Unix assumes we are trying to run a command called 'hello.sh' so it goes off to look in all of the places where it knows commands live (more on this in a moment). The second example, with the addition of ./ before the file name, works and Unix runs the script. You can read this second example as 'run the file called hello.sh that's in *this* directory and don't bother looking for any programs called 'hello.sh' anywhere else.'

Don't worry too much if you find this a little bit confusing. The important lesson from this is that once you have made a script executable you can always run it by prefixing its name with a path (even if the path is just ./). This is a better solution than having to use the `source` command, but it is still one stop short of being able to run our scripts as if they were like any other Unix command. However, we are not finished yet!

Modifying your path

The final piece of this puzzle is to understand that any time you try running a program, Unix will check through a list of predefined directories to see if that program exists in any of those locations. If it finds a match, then it will try running that program and will stop looking in any other directory. If, after looking through all of the directories, it cannot find a match, it will report the dreaded 'command not found' error. We can see the list of places that Unix looks[97] for programs by printing the $PATH environment variable.[98] Let's see what this list looks like for our user, Nigel:

```
$ echo $PATH
/usr/bin:/bin:/usr/sbin:/sbin:/usr/local/bin:/usr/X11/bin
```

This output contains a list (separated by colons) of six directories on Nigel's computer which contain programs. When we say 'programs' we just mean any file that has executable permission. Unix doesn't care what these programs do, or even if they work. All of the Unix commands we have seen so far are just files with executable permissions that live in one of these directories. The order of directories is important as it determines the order in which Unix will search for programs.

It is easy to modify the $PATH variable to include other directories, which is exactly what we want to do. If we can add the 'Code' directory on to the end of this list, then Unix will know to also look in this directory. Let's try modifying $PATH:

Example 3.28.2 Note that the following code includes two mutually exclusive options depending on whether you are using the bash family of shells (sh, bash, ksh, zsh)

---

[97] Of course, Unix isn't really looking and we don't want to give you a false impression that Unix is a sentient being. It just makes it easier to describe. Otherwise we would have to say really dull things like 'code in the computer's memory determines whether the name of the specified file is an exact match to the names of files that are stored in designated arrays that correspond to locations in the filesystem.'

[98] See Chapter 3.11 for a refresher on environment variables.

or the `csh` family of shells (`csh` and `tcsh`). Make sure you only type one of the two possible commands that are shown (`export` or `setenv`).

```
$ echo $PATH
/usr/bin:/bin:/usr/sbin:/sbin:/usr/local/bin:/usr/X11/bin

# For shells in the bash family
$ export PATH=$PATH:~/Unix_and_Perl/Code

# For shells in the csh family
$ setenv PATH $PATH\:~/Unix_and_Perl/Code

$ echo $PATH
/usr/bin:/bin:/usr/sbin:/sbin:/usr/local/bin:/usr/X11/bin:/Users/
nigel/Unix_and_Perl/Code

$ cd
$ pwd
/Users/nigel
$ hello.sh
Hello World
```

Explanation

We first use the `echo` command to confirm the contents of the $PATH environment variable. Then we use one of two commands, depending on which shell we are using. However, both the `export` and `setenv` commands achieve the same thing, albeit with slightly differently syntaxes.[99] They both add the location of the 'Unix_and_Perl/Code' directory to the end of our current $PATH. You should read this assignment as 'Take the current value of $PATH, and add this new directory on to the end of the list.' It is important to give an absolute path here.

After adding a new directory to the $PATH, we confirm that this has happened by printing the value of $PATH again. The full path to the 'Code' directory gets appended to the end of the list and '~/' gets expanded to '/Users/nigel/'.

We then navigate to Nigel's home directory and try running the 'hello.sh' program. This time it works!

Advantages of modifying $PATH

Once you have added a new directory to your $PATH variable, you can use that directory to store all of your scripts. Any program in that directory can then be run from *anywhere* in the filesystem (as long as the program file is executable). You also gain the very useful

---

[99] Note the difference in syntax between these two commands. The `export` command requires that there are no spaces either side of the equals sign. The `setenv` command doesn't need an equals sign, but does require that any colon characters are 'escaped' with backslashes. Alternatively, you could use either of the following syntaxes to avoid having to use backslashes (where 'whatever' is the name of some directory to be added to $PATH):

```
setenv PATH ${PATH}:whatever
setenv PATH $PATH":"whatever
```

advantage that you can use tab-completion when typing the program name. Your scripts will be treated like any Unix command.

If you wanted to, you could add multiple directories to $PATH (e.g., if you wanted to maintain Unix and Perl scripts separately). You can always see where Unix finds your programs by using the which command:

```
$ which hello.sh
/Users/nigel/Unix_and_Perl/Code/hello.sh
```

This becomes important because you might want to develop two different versions of the same program. If these different versions exist in different directories, then the first directory that occurs in the $PATH variable will be the one that Unix uses.

Bear in mind that, at the moment, the changed $PATH variable will only exist in the current terminal. Therefore, we need to move the command that changes your $PATH into the .profile file:

Example 3.28.3 This change will ensure that any terminal window you open will 'know' about the location of your 'Code' directory. This also means that you should not store scripts anywhere else (unless you also add other directories to your $PATH). Modify your .profile file to include the following and, depending on your shell, make sure to only include the export *or* setenv command, but not both.

```
# ENVIRONMENT VARIABLES
# for bash-type shells...
export PATH=$PATH:$HOME/Unix_and_Perl/Code
# for csh-type shells...
setenv PATH $PATH\:$HOME/Unix_and_Perl/Code
```

Explanation
Note that we use another environment variable ($HOME) to specify a full path to the 'Code' directory. We could also have used the '~/Unix_and_Perl/Code' format or even '/Users/nigel/Unix_and_Perl/Code'. However, the last example is not always the best choice. The physical location of home directories can sometimes change (e.g., if your home directory is moved to an external disk). Using $HOME avoids the hard-coding of a disk location into your $PATH variable, and will ensure that things will continue to work even if your home directory moves.

It's time to come out of your shell

So far we have created a single shell script that does nothing more than print the words 'Hello World'. This chapter will show you how to write some scripts that may actually be useful in some 'real world' scenarios.

How to make a 'proper' shell script

So far we have seen that we can put a series of Unix commands into a file, make that file executable, and then run it as if it was a program. This works, but lacks one feature that we should strive to include in all future shell scripts. You may remember that there are a number of different Unix shells (see Chapter 3.10) and that some Unix instructions differ between the different shells. When we write any shell script we can – and should – specifically indicate which type of shell we want to use to process the instructions. Basic Unix commands will be understood by all types of shell, but many shells use some specific syntax for commands that won't work in other shells.

To ensure that a shell script will be processed by a specific type of shell, we just need to add one line to the start of the script. This line is known as the *interpreter directive* and should always occur in the first line of the script. Additionally, it must always start with the characters #!.[100] These two characters are sometimes referred to as 'hash bang' or 'shebang.' They should be followed by a valid path to where the relevant shell command is installed. Let's revisit our 'hello.sh' script and make it into a proper shell script by adding an interpreter directive.

Example 3.29.1 Use the nano editor to make your 'hello.sh' file, which should still be inside your Code subdirectory, resemble the following:

```
#!/bin/bash
# my first Unix shell script
echo "Hello World"
```

Note, that the only change we are making is to add a new line to the start of the script. Save the changes, and exit the editor. We can now run this script by simply typing its name:

```
$ hello.sh
Hello World
```

---

[100] These characters must be the first two characters in the file. No spaces or blank lines should go before them. This is the one situation where a line starting with a hash character (#) is not treated as a comment.

Explanation

Some people will use '.sh' as a default file extension for *any* shell script, no matter what the actual shell is, so you shouldn't feel you have to end bash scripts with .bash, tcsh scripts with .tcsh, etc.[101]

In this example we edited the 'hello.sh' file in order to add the interpreter directive (the shebang line). This line includes the location of the bash shell (/bin/bash). When you try running any file that starts with a shebang line, Unix reads this as saying 'Aha, you want to use the program located at /bin/bash to interpret all of the instructions that follow.' If you don't specify an interpreter directive, the instructions in a script are handled by the default shell. In this case, the default shell is also bash, but it might not have been.[102]

Making your shell script do something useful

It is (finally) time for an example that might actually be of use to you.

Example 3.29.2 Use touch to create a new file called 'status.sh' in your Code directory. You will then need to use the chmod command to make the script executable. Finally, use nano to add the following contents to the file:

```
#!/bin/bash
HOST=`hostname`
KERNEL=`uname -rs`
UPTIME=`uptime`

echo "Username = $USER"
echo "Hostname = $HOST"
echo "Kernel details = $KERNEL"
echo "Uptime = $UPTIME"
```

The first line of this script should ideally be appropriate for the type of shell you are running.[103] This script will provide some simple system status information. Run the script by typing the file name; you should see four lines of output, which will be quite different to what is shown here:

```
$ status.sh
Username = nigel
```

---

[101] Naming the file extension after the shell will probably only be useful if you write (or use) lots of scripts written in different shells. Furthermore, some people might also use .script or .scr to name their shell scripts; it's also the case that some people won't bother to give any file extension to their shell scripts at all. These are bad people.

[102] Adding a shebang line means you can send your bash scripts to a friend who uses a different shell by default. Their shell might not understand all of the instructions in your bash script, but as long as your friend's computer has the bash shell installed, your friend will be able to run your script.

[103] If you are running the zsh shell, then this line could be changed to #!/bin/zsh. You can find out 'where' your shells are installed by using the which command – type which tcsh. Having said that, it may not matter as long as the other shell in question is also installed on your machine – i.e., you can use csh as your default shell but still run bash scripts if it is installed.

```
Hostname = stonehenge.ucdavis.edu
Kernel details = Darwin 10.6.0
Uptime = 11:16 up 4 days, 2:36, 3 users, load averages: 0.57 0.45 0.51
```

Explanation

Shell scripts can store data in something called variables. These are named containers that can store different items (words or numbers) and behave like the environment variables we have already seen (Chapter 3.11).

After the interpreter directive line (#!/bin/bash), the next three lines each assign the output from a different Unix command to a variable. The first example is where the output from the hostname command is stored in a variable called $HOST. If you want to capture the output of a Unix command, you need to place that command between a pair of backtick characters.[104] The next two lines capture the output from the uname and uptime command (we use a couple of command-line options for the former). Remember that these are just regular Unix commands, and you can try running them on their own.

After capturing the output from three commands and storing this output in three different variables, we use the echo command to output the information. When you want to access the content of a variable, it must be prefixed with a dollar sign. Notice that the first thing to be printed is actually the contents of an environment variable (again, see Chapter 3.11).

This is a very simple example which in some ways does nothing more than run a few different Unix commands. However, there are times when that is exactly what you want to do and having a script to run them all in one go can be a great time-saver.[105]

---

[104] Not to be confused with apostrophe characters!

[105] If you needed to access this information several times per day, then the convenience of running *one* shell script as opposed to several different Unix commands can be significant.

Unix summary

Is this the end of the $PATH?

We have covered many different Unix commands in this part of the book, albeit a small fraction of all of the commands available. We hope you are now comfortable in basic directory navigation and in dealing with the essential aspects of file manipulation (copying, moving, renaming, deleting, etc.). We recommend that at some point you take the time to become familiar with one of the common Unix text editors (vi, Emacs, nano, etc.); it also pays to learn some of the features of the less command.

We reiterate one final time that if you really want to learn Unix, then you should, where possible, try to stay in the world of the terminal. Try to refrain from using your graphical file manager too much. Believe it or not, there are Unix programs available that will do a lot of the regular activities you would otherwise use a graphical program for. Want to check your email, play some music, or even browse the web? All such activities are available, with some limitations, from the comfort of your Unix terminal.[106]

From the point of view of learning Perl, one of the most important Unix commands you need to know is chmod (to make your scripts executable[107]). It will also be useful to be able to modify your configuration file (.profile, .login, etc.) so as to properly configure your $PATH variable to let Unix know about the directory you use to store your scripts.

What next?

You are now equipped with enough Unix skills that you can proceed to Part 4 of this book and start learning 'Essential Perl.' However, if you are enjoying learning Unix, then feel free to jump ahead to Part 5 ('Advanced Unix') before returning to learn Perl. Part 5 will focus much more on working with some file types that are very commonly used in bioinformatics. It will also introduce many commands that have direct equivalents within the Perl programming language. Finally, it will show you how you can chain commands together in powerful combinations.

---

[106] It's possible that your Unix computer might not have all of these tools pre-installed, but they are available for most flavors of Unix.

[107] Though when you learn Perl, we will start by showing you another way of running Perl scripts that does not require the scripts to be executable.

Essential Perl 4

Your first steps into the world of Perl programming

The first program most people learn to write in any programming language is usually a 'Hello World' program that just prints the words 'Hello World' (surprise, surprise) to the screen. This isn't a legally binding requirement, and you could instead try to make your first program calculate the largest unknown prime number. However, all we ideally want to do at this stage is to check that the programming environment is working properly, so it makes sense to keep your first Perl script as simple as possible. Note that the words 'program' and 'script' are often used interchangeably. If you don't worry about the difference, then neither will we.

Example 4.1.1 Enter the text below into your text editor (see Part 2 for details about text editors), but *do not include the numbers*. The numbers are there only so we can reference specific lines. Your first Perl program is going to contain just two lines. This might not seem like much, but it means you will have written more Perl code than the vast majority of people in the world, who haven't written any!

```
1.   # helloworld.pl by _insert_your_name_here_
2.   print("Hello World!\n");
```

Understanding the script

Line 1 starts with a hash character (#). Most of the time Perl sees a #, everything that follows on that line will be considered a comment (we'll deal with exceptions to this in due course).[1] Programmers use comments to describe what a program does, who wrote the program, what needs to be fixed, etc. Think of comments as the notes you might write in the margin of a text book; they should aid your understanding. It's an extremely good idea to put comments in your code, especially as your programs grow larger. Perl will ignore everything after the hash sign so in theory you could embed a poem or even a recipe in the comments of your script, though you would be considered strange if you did this.

   Line 2 is the only line of this program that does anything. The print() function takes a list of *arguments* and sends them to your terminal so they can be read by some-one.[2] In this case, there is only one argument, which is the text "Hello World!\n". The pair of quotation marks are one way of letting Perl know that we are dealing with text. The funny \n at the end of the text is called a *newline* character, which is like a carriage return. We shall return to the newline character later. Most of the time, each line of Perl code will end with a semicolon. Think of this as like the period at the end of a sentence. There are some exceptions to this, the most obvious of which is in line

---

[1] All programming languages support commenting, and Perl uses the same hash character syntax as Unix (which we first saw in Chapter 3.23).

[2] An argument is just some piece of information that is provided to a particular piece of Perl code. We have already seen that most Unix commands support one or more command-line arguments and functions in Perl are analogous to this. If you find the word 'arguments' sounds strange, then you can think of them as 'bits of required information.' If you are still not happy about this then feel free to start an argument argument!

1: comments don't require semicolons because they are not interpreted as part of the script.

Use your text editor to save the program and make sure you name the file 'helloworld.pl'. It is common, though not strictly necessary, to add a '.pl' extension to the names of your Perl scripts. This will help you identify them more quickly, and it will also allow some editing programs to recognize that they are Perl scripts.

So, now we have written and saved our first Perl script, but how do we make it do something? To run the program, switch to your terminal application (see Chapter 3.2) and check that you are in the same directory as the script (ideally this will be a directory called 'Code', 'Scripts', or 'Perl', etc.). Being in the wrong directory will be a frequent cause of problems when working with Perl. For the rest of this book you should, if you want an easy life, save *all* of your scripts in the same location. If you would prefer a life of eternal frustration, then feel free to save every script in a different location … just don't come running to us when things go bad (and they *will* go bad). If you are in the correct directory, type the following command and press return.

```
$ perl helloworld.pl
```

What is happening here is that we are telling our terminal to first find the program called 'perl' (wherever it is installed on the computer), and then send the contents of our helloworld.pl file to the perl program. In this sense, `perl` is just like any other Unix command you've already seen (assuming you read Part 3 of this book). If this works, then great! You should now be basking in the glow of your success as a new Perl programmer. If it didn't work, then welcome to the world of Perl debugging. It is still good news when things don't work as planned, because learning how to fix broken scripts is an even more important skill than writing scripts that work first time. If it didn't work, then maybe you saw an error message like this one:

```
Can't open perl script "helloworld.pl": No such file or directory
```

This would suggest you may have forgotten to save the file, misspelled the file name, or saved the file to someplace unintended. You should *always* use Unix tab-completion (see Chapter 3.15) to prevent spelling mistakes when typing program names – don't type every letter of 'helloworld.pl'; instead type 'perl h' and then press tab to see if you can autocomplete the file name.[3] This is such an important point that will save you from hours, months, or even years of frustration (depending on your future career) that we will mention it again:

Always use Unix tab-completion when typing the names of Perl scripts!

---

[3] If tab-completion doesn't appear to be working, it's most likely because you are in the wrong directory.

Example 4.1.2

Let us explore the use of hash characters a little more. We have suggested that you can use hash characters to initiate a comment in your script. But what if you wanted to actually print a hash character?

```perl
1.   # hashcheck.pl
2.   print("This is a hash character: #\n");
3.   # this is a comment
4.   print("This is a print statement...\n"); # followed by a comment
5.   # print("This line will never be printed\n");
```

Understanding the script

As in the previous example, to run this script you need to be in the same directory as the file, and then just type `perl hashcheck.pl` (assuming you named the script 'hashcheck.pl').

The lessons to learn from this short script are three-fold. First, you can obviously print a # character, so Perl is smart enough to know that if a # character appears within a pair of quotation characters, you probably want to print the character, rather than start a comment. Second, you can have comments as entire lines by themselves (lines 1 and 3) or as add-on comments that immediately follow some code (line 4). Finally, you can add a hash character to the start of an existing line of code to effectively silence that line of code (line 5). This is known as *commenting-out* a line, and is frequently used when writing and debugging Perl scripts.

Problem 4.1.1 The world of programming would quickly become very tedious if your scripts only ever said 'Hello world' so modify the program to output three different lines of text. Note that there are two different ways to do this.

Problem 4.1.2 Try making a 'deleterious mutation' to your program. For example, leave off the semicolon or one of the parentheses. Observe the error message when you try running it. One of the most important aspects of programming is debugging. Probably more time is spent debugging than programming, so it's a good idea to start recognizing errors now.[4] Try making multiple mutations to your script and see whether you see multiple error messages. Finally, ensure that you can remove the errors and get your script working again.

---

[4] Part 7 of this book contains several chapters on debugging, though you should learn some more Perl before reading those sections.

Scalar variables

Variables make programming very able

Imagine that we want to write a program to calculate some property of a DNA sequence. The program might do something as simple as calculate the length of the sequence, or maybe it is going to do something more complex, such as count all of the dinucleotide frequencies in the sequence. Whatever it is we want to do, we clearly need to start with the DNA sequence itself. It is conceivable that we might only ever want to work with a single DNA sequence and therefore we could write a Perl script that is specific to that one sequence. This is unlikely, however. We should expect any DNA-processing program to work on *any* DNA sequence (within reason: very long sequences might take up too much memory). Therefore our program will need to be able to somehow store and manipulate the sequence, no matter what it is. That's where variables come in.

Variables hold data, and the data they hold are allowed to vary. In some cases you might want to change the data between separate runs of your program – e.g., the first run of your program processes sequence *A*, and the second run processes sequence *B*. In other cases, you might want to change the data during a single run of the program – e.g., you start with a DNA sequence but then you want to clip the ends of the sequence for some reason and work with the shorter, clipped version.

In Perl, the main variable type is called a *scalar variable*. A scalar holds one thing. This thing could be a number, some text, or an entire genome. We will see other data types later. You can always tell a scalar variable because it has a $ on the front (the dollar sign is a mnemonic for scalar). For example, a variable might be named $x. When speaking aloud, we do not say 'dollar x'; we just call it 'x'.

Example 4.2.1 Create a new plain-text document in your text editor. Enter the text below and save this program as scalar.pl in your Code directory (or wherever you are saving your Perl scripts).

```
1.   # scalar.pl by _insert_your_name_here_
2.
3.   $x = 3;
4.   print($x, "\n");
```

Understanding the script

Line 2 is deliberately blank. You should use spaces and blank lines to improve the readability of your code. In this case we are separating the header section of the script from the rest. In this script, the header is just a single line containing a comment; as we learn more Perl, this header section will get longer and longer.

Line 3 contains a *variable assignment* and it is actually doing two things. First, it's telling us that we're going to be working with a new variable called $x. , and second, it's assigning that variable a value of 3. Sometimes in Perl, we do these two steps (declaration and assignment) separately, and sometimes we do them together.

Line 4 prints the value of $x and then prints a new line. Notice the use of the comma inside the print() function. Although we frequently will use the print() function to print just one thing at a time, it can also print multiple things if they are separated by commas. A better way of putting this is that the print function *supports multiple arguments*.

Run the program by typing the line below in your terminal. Observe the output and go back through the code and the above line descriptions to make sure you understand everything.

```
$ perl scalar.pl
```

Example 4.2.2 Now try adding the following lines to your program:

```
5.   $s = "something";
6.   print($s, "\n");
7.   print("$s\n");
```

Understanding the script

Line 5 is another variable assignment, but unlike $x, our new variable $s gets a *character string*, which is just another term for some text.

Lines 6–7 print our new variable $s in two slightly different ways, though both ways include a newline character.

Save the script and run it again. You should see that although lines 6–7 are different they produce *exactly* the same output. The print() function can print a list of items (all separated by commas), but it often makes more sense to print just one thing instead. It would have been possible to rewrite our very first Perl script with the following:

```
print("H","e","l","l","o"," ","W","o","r","l","d","!","\n");
```

Hopefully you will agree that printing this phrase as one string and not 13 separate strings is a lot easier on the eye.

Example 4.2.3 Now try adding the following lines to your program:

```
8.   print "$s\n";
9.   print '$x $s\n';
10.  print "$x $s\n";
```

Understanding the script

Line 8 calls the print function but omits the parentheses. This introduces an important aspect of programming with Perl: Not all formatting is mandatory. You do not have to use parentheses for Perl functions, but they are often useful to keep a line organized and to aid understanding. However, in most cases you will see the print function without parentheses.

Lines 9–10 feature two print statements which appear to be printing the same content, but which are using either single or double quotation characters.

When you run this script you should notice that the output from line 9 may not be what you were expecting. Any text between single quotes will print exactly as shown. This is often desirable because you might want to print the string $x without it printing the value 3. Using single quotes also means that \n loses its special meaning as a newline character and it becomes a regular backslash character followed by the letter 'n'. This means the output from line 10 will follow immediately after the output from line 9. The script output should look like this:

```
$ perl scalar.pl
3
something
something
something
$x $s\n3 something
```

In contrast to using single quotes, any strings that are between double quotes undergo a process called *variable interpolation*. This means that variables are always expanded inside double quotes, and the print function will always show what values those variables contain.

Example 4.2.4 What happens if we try printing a variable that hasn't been assigned anything? Try adding the following two lines:

```
11.  $empty;
12.  print "This is a $empty value\n";
```

When you run the script you should find that line 12 prints nothing at all for the $empty variable (apart from the surrounding text). This sort of makes sense. If a variable doesn't contain anything then Perl can't print it (or do anything else with it). As we will learn later, we might want to be careful when trying to use these *unassigned variables*.

Naming variables

You can use (almost) anything for your variable names, though you should ideally try to use names which are descriptive and not too long. Variables must begin with letters or an underscore _ character. After the first letter, you can use numbers too. In general, you should use lower-case characters for your variable names. Upper-case variable names have a special meaning (see chapter 7.3). A quick example should illustrate what makes a good variable name. Which of the following is the best name for a variable that will store a DNA sequence?

```
$xyz = "ATGCAGTGA";                        # not very descriptive
$protein = "ATGCAGTGA";                    # misleading
$dna_sequence_variable = "ATGCAGTGA";      # needlessly long
$sequence = "ATGCAGTGA";                   # better
$dna = "ATGCAGTAGA";                       # even better
```

If you so desire, Perl allows you to assign a variable the same name as an existing function in Perl. However, this can sometimes lead to confusion – a variable named $print might look a bit too similar to the print() function. Sometimes, though, the choice of variable name is obvious: $length is often a good name for variables that contain the length of something, even though there is also a length() function in Perl (which we will learn about later).

As shown in the example above, variable names can contain underscore characters to separate 'words.' This is often useful and helps make things easier to understand:

```
$first_name  = "Harry";
$second_name = "Potter";
```

Hopefully you noticed that even though these two variable names are different lengths, the second-half of each line of code is aligned to each other. This was done by adding an extra space after $first_name. We hope that you agree that this makes things look tidier and easier to read. There are lots of situations where adding spaces will make your code appear clearer. Note that the following lines are all treated by Perl in exactly the same way:

```
$dna = "ATGCAGTGA";      # one space either side of the '=' sign
$dna="ATGCAGTGA";        # no spaces either side of the '=' sign
$dna   =   "ATGCAGTGA";  # lots of spaces!
```

Problem 4.2.1 Mutate your program. Try removing the $ from lines 3 and 5 and see what error message(s) you get.

Problem 4.2.2 Modify the program by changing the contents of the variables. Observe the output. Try experimenting by creating more variables.

Problem 4.2.3 What happens when you assign the contents of one variable to another variable? E.g.:

```
$x = 100;
$y = $x;
```

Problem 4.2.4 Perl allows you to assign a string to a variable in a few different ways. What do you think the difference is between the following (try printing the variable after each new assignment)?

```
$favorite_food =  cheeseburgers;
$favorite_food = 'cheeseburgers';
$favorite_food = "cheeseburgers";
```

Use warnings

You have been warned!

You've just written your first few Perl scripts and this may have caused you to get giddy and lightheaded as you experience the joys of seeing working Perl code in action. This section aims to bring you back down to earth, as we are now going to ask you to rewrite all of the scripts you've just written.

There are several habits that are widely considered to be 'good working practices' when writing Perl scripts. However, they are not always the easiest concepts to explain (especially when you have just started to learn Perl) and so we will introduce them in bite-sized chunks.

The first 'good practice' we want to introduce stems from the fact that Perl is a very tolerant programming language. It will often attempt to understand what you are trying to do even if you haven't written the code in the best possible way. This can be a good thing, particularly if you are a sloppy programmer. However, this can also be a bad thing as it will very possibly *make* you a sloppy programmer. Fortunately, Perl has a few ways of forcing you to write better code and by 'better,' we mean more accurate, understandable code that is less prone to causing unintended effects.

Example 4.3.1 The first, and perhaps easiest, way to get Perl to improve our code is by making it warn us when we are doing something stupid or undesirable. We can achieve this by adding a single line of code to the header section of our scripts:

```
1.   # helloworld.pl by _insert_your_name_here_
2.   use warnings;
3.   print("Hello World!\n");
```

Understanding the script
Line 2 is the only difference between this script and the original version. This line effectively tells Perl that we would like to be warned if we start writing certain types of 'bad' code. You can read this line of code as saying 'Before we go any further with this script, please turn on all of the extra checks that are specified in the set of rules called *warnings*.' Technically, these sets of rules are known as *pragmas* and Perl has more than 40 of them. However, in common practice you will probably only ever use a few of them.

When you run this script you might be disappointed to see that the new use warnings; line makes no difference whatsoever. This is because the script doesn't contain any code that would cause a warning. However, as you don't yet know what all of the bad things are that might cause warnings, you should get into the habit of including this line in *every* script you write. This is another point that bears repeating:

> Always, always, always include 'use warnings' in your Perl scripts!

Example 4.3.2 Let's write a script where the warnings pragma will actually be of use.

```
1.   # warn.pl by _insert_your_name_here_
2.   # use warnings;
3.
4.   $dna = actgagtag;
5.   print "DNA is $dna\n";
6.
7.   $peptide = 'MYAGWRREKP';
8.   print "Peptide is $peeptide\n";
```

Understanding the script

Line 2 contains the use warnings; statement, but note that it is deliberately commented out[5] so that Perl will ignore it (don't worry, we will uncomment it soon).

Line 4 assigns a string to a variable called $dna, and the following line attempts to print the contents of that variable. Notice that we don't use any quotation characters around the DNA string.

Line 7 assigns a string to a new variable ($peptide), but this time we use single quotes. Finally, line 8 attempts to print the contents of $peptide but there is a deliberate typo here and $peptide is misspelled as $peeptide.

Save this script as warn.pl. If you run this script you should find that Perl does not complain and just prints the following (expected) output:

```
DNA is actgagtag
Peptide is
```

However, if we now uncomment line 2 to turn on warnings and then re-run the script (making sure we save the changes first), we should see some new output:

```
Unquoted string "actgagtag" may clash with future reserved word at warn.pl line 4.
Name "main::peeptide" used only once: possible typo at warn.pl line 8.
Name "main::peptide" used only once: possible typo at warn.pl line 7.
DNA is actgagtag
Use of uninitialized value $peeptide in concatenation (.) or string at warn.pl line 8.
Peptide is
```

Among the four error messages that now appear, you should still notice that the script still includes the same two lines of output as before. So what is happening here? By turning on warnings, Perl is now telling us about various things that might be problematic, but it will still run the script.

First, Perl warns us that 'actgagtag' is an unquoted string (which it is) and it therefore might clash with a future reserved word. This just means that it might be possible that in some future version of Perl 'actgagtag' might actually become the name of a

---

[5] Remember that Perl ignores code that follows any hash character (see Chapter 4.1).

built-in function (like `print`). This is obviously unlikely with a string such as 'act-gagtag' but what if the string you were assigning to `$dna` was just 'tag'. It's easier to imagine 'tag' being used as the name of a function. So this warning is a reminder that we should always use quote characters (either single or double) when assigning a string to a variable. This makes it clearer to Perl (and hopefully to you too) that you are just working with a string.[6]

Next, Perl tells us that the variables 'main::peeptide' and 'main::peptide' are used only once. Please ignore the 'main::' part for now.[7] Generally when we introduce a variable in a script we should be doing something with it apart from just declaring it and assigning it a value. Perl spots when this might be the case and warns accordingly. If we correct the `$peeptide` variable name this error should disappear because we are now assigning a value to the variable `$peptide` *and* doing something with it. This warning message will help you catch lots of typos that you make (and you *will* make typos).

Finally, the warnings pragma gives us an error message that will become very familiar to you as you learn to program in Perl: 'Use of uninitialized value.' This will appear every time you try to do something with a variable that doesn't contain anything. We generally assume that variables should contain things; it makes no sense to try printing a variable if it is empty. This is one of the most common error messages you will see (assuming you have turned warnings on) and at times you might even regret asking Perl to warn you about such things. Just bear in mind that, overall, the usefulness of seeing all these warning messages far outweighs their annoyance. By including the `use warnings;` line in your scripts, Perl will help you catch many typos and will also alert you to errors in your code and to errors in the data you might be trying to process.

If your script produces error messages because of the warnings pragma, then you should always try to fix those errors straight away. *Do not ignore these errors.* Your code will be better code if you fix them, and in many cases your code might not work properly (or at all) if you don't fix them. This point is worth repeating:

A good script should not report any warnings!

Problem 4.3.1 It is important to put the `use warnings;` line near the top of your script (i.e., in the script 'header' section). Try moving line 2 from the above script to line 6 (making sure it is uncommented). What happens when you run the script?

---

[6] You code editor should highlight strings in a different color. This is very helpful because it helps you catch errors before you run the program. If you are not seeing text in different colors, examine the help documentation for your editor and look for syntax highlighting options.

[7] See Chapter 6.2 for more information.

Summing up, Perl is very functional

Perl, like most programming languages, supports a variety of mathematical operators and functions. This makes it easy to perform most common mathematical tasks that you might want to do. Let's experiment with some of these.

Example 4.4.1 Write the program below, save it as math.pl, and then run it.

```
1.  # math.pl
2.  use warnings;
3.
4.  $x = 3;
5.  $y = 2;
6.  $sum = $x + $y;
7.  print "Sum of $x and $y is $sum\n";
```

Understanding the script

Lines 4 and 5 declare two variables and assign a numerical value to each. Notice that we don't add quotation marks around the numbers. This is deliberate and we shall return to this issue shortly.

Line 6 declares a new variable and rather than assign it a fixed value or string, it is assigned whatever the result of adding $x to $y is.

It would be wrong if Perl only allowed you to add numbers. Like all programming languages, Perl lets you perform the full range of mathematical operations (addition, subtraction, division, multiplication, etc.). Furthermore, like most other programming languages, Perl uses the standard shorthand symbols for division (/) and multiplication (*). Let's see all of these operators in action.

Example 4.4.2 Add the following lines to your script, but don't add them all at once. Instead, this time we are going to take a slightly different strategy. The program is getting longer. If you type the whole thing and accidentally add lots of errors, it will become difficult to debug. So instead, write only a few lines, and then save the script. Run the program, and observe the output. Debug the script if there are any error messages. Try to get into the habit of checking your program as you write it, and not only at the end, when it is finished. As you get more experience, you will gain skill and confidence and will not need to check as frequently.

```
 8.  print "$x plus $y is ", $x + $y, "\n";
 9.  print "$x minus $y is ", $x - $y, "\n";
10.  print "$x times $y is ", $x * $y, "\n";
11.  print "$x divided by $y is ", $x / $y, "\n";
12.  print "$x modulo $y is ", $x % $y, "\n";
13.  print "$x to the power of $y is ", $x ** $y, "\n";
```

Understanding the script

Lines 8–13 perform a variety of mathematical operations with $x and $y. Notice that we are not assigning the results of each mathematical operation to any variable. Instead the results are printed directly to the screen. We can do this because the print() function allows you to print a list of arguments (separated by commas), and mathematical operations are processed before the arguments are sent to the print() function.

This means that when Perl sees line 8, it realizes that it has to print three separate things. First, there is the text string "$x plus $y is "; because this string is in double quotes, Perl will check to see if it contains any variables and then replaces the variable name with the contents of that variable. Next, Perl prints the result of $x + $y. Because this text *isn't* between quotes, Perl knows that you want to actually do the math rather than print it as a string. Finally, the third argument to the print function is the newline character.

Line 12 contains the modulo operator %, which is something you may not have seen before. Modulo gives the remainder after an integer divide. For example, 7%2 equals 1 because 7 divided by 2 is 3 with remainder 1.

Numeric functions

In addition to the mathematical operators we've just seen, there are a number of built-in numeric functions: e.g., abs(), int(), log(), rand(), sin(). So far we have only seen one Perl function, which is print(). Many functions in Perl are easy to understand because they perform one specific operation on one specific value. The next example illustrates this.

Example 4.4.3 Add the following lines to the program, run it, and observe the output.

```
14.  print "the absolute value of -$x is ", abs(-$x), "\n";
15.  print "the natural log of $x is ", log($x), "\n";
16.  print "the square root of $x is ", sqrt($x), "\n";
17.  print "the sin of $x is ", sin($x), "\n";
18.  print "a random number up to $y is ", rand($y), "\n";
```

Understanding the script

Lines 14–17 behave in an identical way to lines 8–13, but we are now using some mathematical functions to transform our value of $x. Note how we are just inserting a single value between the parentheses of the function. In this case the value in question is specified by a variable ($x), but we could also use actual numbers or pieces of math that would calculate a number, e.g.: sqrt(4), sqrt(4 + 12).

Line 18 uses the rand() function, which generates a fractional[8]random number between 0 and the number that you specify inside the parentheses. Being able to

---

[8] In a computer, numbers are generally stored either as integers or floating point numbers. Perl hides these details from you.

generate random numbers is very useful and provides a foundation for being able to perform simulations. You should test that each time you run the script, this line of the code is indeed producing a different random number.

Combining multiple functions at once

Functions are incredibly useful in Perl, particularly because we can sometimes place one function call inside another. We often want to generate random integers but as you have just seen, the rand() function generates fractional numbers. Luckily, Perl has an int() function to turn fractional numbers into integers.

Example 4.4.4 Add the following line to your script to make your code generate a random integer.

```
19.  print "a random integer up to $sum is ", int(rand($sum)), "\n";
```

Understanding the script

Notice how we place the rand() function call *inside* the int() function call. You might be wondering how Perl knows what to do with these nested function calls. It makes no sense to calculate the integer part of the code until you have first generated the random number. In situations like this, Perl will first attempt to resolve the inner-most part of the code. Once the rand() function returns a number, then this can be passed to the int() function.[9]

This part of the code could have also been written as int rand $sum. This is another example where you can omit parentheses if you like. But just because you can, doesn't always mean that you should.

Hopefully, you also noticed that this line reuses the $sum variable from earlier in the script. If we didn't reuse the variable we could instead write $x + $y instead of $sum. However, if we then changed our mind and wanted $sum to be the *product* of $x and $y we would have to change our code in three separate places. By using $sum as a container for whatever our desired mathematical operation is, we only have to change the one line which first defines $sum. The idea of reusing code is an important one and we elaborate on it in Chapter 7.4.

Operator precedence

Let's quickly discuss something called *operator precedence*. Some operators have higher precedence than others. We're used to seeing this in math where multiplication

---

[9] This whole line of code will be processed by Perl in the following order:

(1) Determine contents of $sum
(2) Calculate random number using $sum
(3) Pass random number to integer function
(4) Send integer value along with text string to the print() function

and division come *before* addition and subtraction: $3+2\times5=13$. If you want to force addition before multiplication, you can do this as $(3+2)*5=25$. Perl has a lot of operators in addition to the mathematical operators and there are a lot of precedence rules. Don't bother memorizing them. The universal precedence rule is this: Multiplication comes before addition, use parentheses for everything else.

Mathematical shortcuts

There are a few operations that are so common that Perl provides a shorthand way of using them. Consider the following:

```
$n = 2;
$n = $n + 5;
```

In this example we have a variable ($n) which we assign an initial value of 2, and then we add 5 to it and reassign that back to itself. The result being that $n now contains a value of 7. To save you some time when doing this, there is a simpler way:

```
$n += 5;
```

The += operator takes whatever variable is on the left of the operator and increments it by whatever value is on the right of the operator. There are similar operators for subtraction (-=), multiplication (*=), and division (/=). Here is a quick summary of these operators in action (the current value of $n is included as a comment):

```
$n = 8
$n += 7;      # $n is now 15
$n /= 3;      # $n is now 5
$n -= 1;      # $n is now 4
$n *= 2;      # $n is now 8
```

One other useful shortcut is for when you just want to increase or decrease the value of a variable by 1. Perl uses two operators to do this: the increment (++) and decrement (−−) operator:

```
$n = $n + 1;   # the longest way to write this
$n += 1;       # a shorter way of writing the same thing
$n++;          # even easier!
$n--;          # and this subtracts 1 from $n
```

Understanding Perl functions

Earlier we mentioned how the rand() function generates a random number between 0 and the number you specify inside the function. How did we learn that, and more importantly, how could you learn that? Whereas we sometimes like to think that programmers learn by osmosis, the more mundane truth is that every Perl function has an associated set of documentation. One place to find out about this documentation would be the book 'Programming Perl,' published by O'Reilly Media (this book also contains a whole lot of other stuff). But you can more easily find out about Perl functions by going to http://perldoc.perl.org/index-functions.html. This web page provides an A–Z list of all Perl functions, and you should really get into the habit of making sure you understand any function you use. The functions we have seen so far (print(), rand(), etc.) are relatively straightforward, but many other functions are slightly more complex.

   If you are learning Perl on a Unix-based system then you can also find out about Perl functions by viewing the perlfunc man page.[10] Type the following into your Unix terminal (you *do* remember that you have to press 'q' to exit the man page viewer, don't you?):

```
man perlfunc
```

Problem 4.4.1 Here is a short script that uses the rand(), int(), and sqrt() functions on three subsequent lines of code. Can you rewrite the script to use them all on just one line of code? Note that this code starts by creating a random number between 0 and 100.

```
1.   # 3_lines_to_1.pl
2.   use warnings;
3.
4.   $random = rand(100);
5.   $integer = int($random);
6.   $square_root = sqrt($integer);
7.   print "The answer is $square_root\n";
```

Problem 4.4.2 In the last chapter we mentioned how we should always make sure we enclose our strings inside single or double quotes. And if we have use warnings; in our scripts, then we will be warned if we don't do this. But in the scripts in this section we have assigned numbers to variables without using quotes, and Perl didn't complain. What would happen if you did the following?

```
1.   # strings_and_numbers.pl
2.   use warnings;
3.
4.   $x = 3;
5.   $y = '3';
6.   $z = "3";
7.
8.   $sum = $x + $y + $z;
9.
10.  print "x is $x, y is $y, and z is $z\n";
11.  print "The sum of x, y, and z is $sum\n";
```

---

[10] We return to other built-in methods of getting help with Perl at the very end of this book in Chapter 7.9.

Problem 4.4.3 What do you think the following script will print?

```
1.   # integers.pl
2.   use warnings;
3.
4.   $x  = 1.1;
5.   $int_x = int($x);
6.
7.   $y  = 1.9;
8.   $int_y = int($y);
9.
10.  print "int_x = $int_x, int_y = $int_y\n";
```

Stay on the $PATH

We've already asked you to go back and edit all of your Perl scripts once. Surely we wouldn't be so cruel as to ask you to go back and edit them again? Umm ... yes we would. Sorry. But trust us, it will be worth it.

Up to now, every time we have wanted to run a Perl script we have gone to a Unix terminal and typed:

```
perl helloworld.pl
```

Let's consider what is happening when we type this. Perl, the programming language, is executed on Unix systems by running perl, the command.[11] Like other Unix commands (ls, cd, grep, etc.) the perl command is just a file that lives somewhere in the Unix directory tree. On many Unix systems, the perl command (along with many other commands) lives inside the /usr/bin directory. If we wanted to, we could run a Perl script by typing:

```
/usr/bin/perl helloworld.pl
```

We can actually take the location of the perl command and include it within each Perl script. This means that, with a couple of other small changes, we can run Perl scripts just by typing their names.

Example 4.5.1 Edit your helloworld.pl script and add one new line to the top.

```
1.   #!/usr/bin/perl
2.   # helloworld.pl by _insert_your_name_here_
3.   use warnings;
4.
5.   print("Hello World!\n");
```

Understanding the script
This first line will be added at the top of every Perl script we write from now on. This line of code is very similar to the line that appeared at the top of our Unix shell scripts (see Chapter 3.29). It lets Unix know that the perl program in the /usr/bin directory[12] can read this file and understand the contents. This line isn't a comment, though it behaves like one because Perl effectively ignores the /usr/bin/perl part.[13] If you really wanted to get technical, this line is an example of an *interpreter directive* and we saw exactly the same sort of thing when we learned about Unix shell scripts (Chapter 3.29).

---

[11] This is a widely used convention for distinguishing between the programming language (Perl with an upper case 'P') and the program that interprets your code (perl with a lower case 'p').

[12] Note that perl is *usually* located in /usr/bin, but on some OSs it may be elsewhere. Some people therefore prefer using #!/usr/bin/env perl, which finds perl using the env command. However, there are potential security risks with this method. We prefer /usr/bin/perl. If your perl binary is located in another directory, you can simply add a symbolic link to /usr/bin (but, you'll need to read about symbolic links first ... see the ln man page).

[13] It is possible to add a number of optional arguments to this line which control various aspects of how Perl runs the program. So bear in mind that this line isn't always ignored.

Before we can run our Perl scripts by just typing their names, we also have to do two more things. First, we need to ensure the files containing the Perl code are made *executable* by using the chmod command (see Chapter 3.27):

```
chmod u+x helloworld.pl
```

Second, we need to make sure the directory containing our Perl scripts has been added to the Unix $PATH environment variable (see Chapter 3.28). If we have performed both of these steps then we can run the script just by typing:

```
helloworld.pl
```

You probably only need to type the letters 'hellow' and tab-completion should take care of the rest. If you try this and it doesn't work, you may see an error message that looks like this:

```
-bash: helloworld.pl: command not found
```

The 'command not found' error will appear if you have either mistyped the name of the command, *or* the specified file is not in any directory that is in your $PATH. This might also mean you have not changed your $PATH to include the directory containing your Perl scripts. If you don't see any error message then this means you should now be able run your helloworld.pl script (and any other scripts in your Code directory) from *any* directory on the filesystem. Test this by navigating to your root directory and checking that you can still run helloworld.pl just by typing its name.

If you understand this, then you are making progress

Conditional statements are one of the foundations of programming. In their simplest form, a conditional statement is simply: if *condition*, then do something. The condition is some kind of test that must evaluate to either being true or false (programming languages don't recognize the concept of 'maybe').

The if statement

The basic structure of an if statement in Perl is:

```
if (condition) {
    code to be executed if condition is true
}
```

Example 4.6.1 The easiest conditional statement to learn is one that just tests whether two numbers are equal to each other. Write the following code and save it as a new file (conditional.pl). Notice that the header section of our script has now grown to three lines and we have added a blank line after this section. Hopefully you will agree that the use of blank lines keeps the different parts of the code clearly separated. Also note that on line 9 there are two equals signs stuck together. This is not a mistake! Before you run this script, remember to make it executable by using the chmod command (see previous chapter).[14]

```
1.   #!/usr/bin/perl
2.   # conditional.pl
3.   use warnings;
4.
5.   $x = int(rand(2));
6.   $y = int(rand(2));
7.   print "x = $x, y = $y\n";
8.
9.   if ($x == $y) {
10.      print "x is equal to y\n";
11.  }
```

Understanding the script

Lines 5 and 6 randomly assign $x and $y either a 0 or 1. Line 9 contains the conditional statement and uses the == operator to test whether $x and $y are numerically equal. Half of the time you run this script, $x and $y will have equal values. When they do, line 10

---

[14] Remember, you will have to do this for *every* new script you create. Because file permissions are also copied when you copy a file, you might want to make a Perl script 'template' file. This will simply include the first few lines from the header section (but with the script name left blank). Make this template file executable and then simply copy it for each new script you work on.

prints their value. It is important to realize that line 10 is only evaluated *if* the condition on line 9 is true.

The condition is placed within parentheses and is then followed by an opening curly brace character '{'. This denotes the start of a *block* of code, and when we start a block we must always end it with a closing curly brace '}'; typically this is placed on its own line as is the case with line 11.

Note that this code between curly braces is *indented*. Indenting is used to show the logical hierarchy of your code. The spacing is achieved by using a tab character. Many code editors will be smart enough to put tabs in for you automatically.

Did you notice that not every line in this script ended with a semicolon? Lines with opening or closing braces do not get semicolons.

One of the most common errors of novice programmers is using a single equals sign to test for equality. For example, if ($x = $y). When this happens, $x is assigned the value of $y. Not only does this destroy the value of $x, but it also makes the conditional statement always true.[15]

Numerical comparison operators

In addition to ==, there are several other numerical comparison operators. These behave in the same way as their mathematical counterparts.

Operator	Meaning	Example
==	equal to	if ($x == $y)
!=	not equal to	if ($x != $y)
>	greater than	if ($x > $y)
<	less than	if ($x < $y)
>=	greater than or equal to	if ($x >= $y)
<=	less than or equal to	if ($x <= $y)

Try changing line 9 of conditional.pl to use other numerical operators. For example, print when $x and $y are not equal (you might want to change the print statement too).

9.   if ($x != $y) {

Alternation with if − else

Being able to do something *if* a certain condition is met is a very useful feature of programming languages. However, we might also want to do something else if the condition *isn't* met. For example, *if* the cinema *isn't* sold out *then* watch film, *else* go home. This is very straightforward in Perl.

---

[15] Operators and functions often return values that are interpreted as true or false. Variable assignments return true if they succeed, and they always succeed.

```
if (condition) {
    code to be executed if condition is true
} else {
    code to be executed if condition is false
}
```

We use `else` to start a new block of code that will only be executed if the condition is *not* met. It is important to understand that in any `if-else` statement, one of the two blocks of code *must* be executed – a condition is either true or false, it can never be true *and* false.[16] As we are creating a new block of code, we once again use opening and closing curly braces to denote the start and end of the block. Add these lines to your program from Example 4.6.1 (only lines 11–13 are new).

```
9.   if ($x == $y) {
10.     print "x is equal to y\n";
11.   } else {
12.     print "x is not equal to y\n";
13.   }
```

This new code now ensures that we always see some printed output no matter what the combination of $x and $y are.

Alternation with `if – elsif – else`

Sometimes we might want to specify more than one test condition. Perl allows us to do this very easily with the use of *elsif* statements.[17] The syntax for an `if-elsif-else` statement looks like this:

```
if (condition 1) {
    code to be executed if condition 1 is true
} elsif (condition 2) {
    code to be executed if condition 2 is true
} else {
    code to be executed if condition is false
}
```

We now have three blocks of code, but it is still the case that only one block of code will ever be executed. If condition 1 is true, then Perl doesn't even bother to look at what condition 2 is. As long as you have one `if` statement, you can follow it with as many `elsif` statements as you want. Although you don't have to end with an `else` statement, it is usually a very good idea to do so if you want to make sure at least one of the conditions is executed.

---

[16] Maybe if Schrödinger's cat wrote Perl, then something could be both true and false.
[17] Note the somewhat unusual spelling of `elsif`. You might find yourself writing 'elseif' by mistake.

If the results of the `if` and `elsif` conditional tests result in a single statement, it is sometimes better to break out of the typical block structure, and format your code slightly differently:

```
if     ($x == 1)    {$y = 1}
elsif ($x == 2)    {$y = 2}
elsif ($x == 3)    {$y = 3}
else               {$y = 0}
```

Notice that we have added spacing after `if` and `else` to ensure that all the conditions and assignments line up. This not only makes your code easier on the eyes, it also helps catch some kinds of bugs before they happen. Also note that there are no semicolons in this syntax.[18]

Nested conditionals

Conditional statements can contain other conditional statements. To show the hierarchy, we will increase the indentation of the code, again by using the tab character.

Example 4.6.2

```
1.   #!/usr/bin/perl
2.   # nested_conditional.pl
3.   use warnings;
4.
5.   $x = int rand 20;
6.   print "x is $x\n";
7.
8.   if ($x < 10) {
9.       print "$x has a single digit\n";
10.      if ($x % 2 == 0) {
11.          print "$x is an even number\n";
12.      }
13.  }
```

Understanding the script

Line 5 creates a random number which is then printed on line 6. Note how we have omitted the parentheses from the `int()` and `rand()` functions. Line 8 starts a nested conditional and then we add a second conditional statement on line 10. The print statement on line 11 will only be executed if $x is less than 10 *and* $x is an even number. In case you forgot what the % operator does, it performs an integer divide and returns the remainder.

---

[18] The last line of a block does not require a semicolon. Usually it is a good idea to add one anyway. Here, omitting the semicolons makes the code look better.

Note the extra indentation on line 11; it is common to see Perl scripts with many levels of indentation. With experience, it becomes easy to glance at a Perl script and see which parts of the code are at 'the same level.'

Evaluating truth in Perl

So far we have shown you that we can use conditional statements to test whether certain mathematical operations evaluate to true or false. But Perl, along with most programming languages, will let you test whether *anything* evaluates to true or false. This can initially seem strange, but it becomes a very powerful part of programming. Compare the following:

```
if ($x > 0) { code if true }
if ($x)     { code if true }
```

The second example might confuse you. What does it mean to ask whether a variable (or string) is true? Well, in Perl, most things evaluate as (logically) true, and it is therefore more important to first consider what is evaluated as false. Zero is logically false. An empty string is logically false. *All other values are true*.[19] So in the above example, $x could contain *any* number apart from 0, and it would evaluate as being true. It could also contain *any* text string apart from "" (an empty string).

Testing whether a variable evaluates as true is used a lot as part of checking data. If the contents of a variable were imported from a data file, then we might want to warn (or stop the script) if the variable is empty.

Logical (Boolean) operators

Perl provides the logical operators and, or, and not, which allow you to chain conditions together. These work a lot like they do in language. The conditional code from Example 4.6.2 could therefore be rewritten like this:

```
if ($x < 10 and $x % 2 == 0)
```

It can become confusing if you chain a lot of conditions together, so keep things simple for now and don't use more than one Boolean operator at a time. Later, we will see that Boolean operators are very convenient for chaining statements together.

In addition to and, or, and not, Perl also has some funny symbols that mean exactly the same thing:[20] && means 'and,' || means 'or,' ! means 'not.'

```
if ($x < 10 && $x % 2 == 0)
```

---

[19] Undefined values are also logically false, but you should never see undefined values because you should always include the use warnings pragma.

[20] They have lower operator precedence but otherwise work exactly in the same way. We prefer using the English-language versions and not the funny symbols.

The unless **statement**

There are times when you want to test if something is *not* true. In Perl, we can do this using two different syntaxes, using either not or !:

```
if (not $x)   { code if not true }
if (!$x)      { code if not true }
```

As an alternative, you can also use the unless statement. You cannot use elsif with unless, but you can use an else.

```
unless ($i_am_sick) {
    $go_to_work = 1;
} else {
    $stay_in_bed = 1;
}
```

Postfix notation

In English, we could say 'if I have money, go to the store' *or* 'go to the store if I have money,' and they both have the same meaning. Perl allows you to do the same thing. This is called *postfix notation*. You do not need the parentheses around the conditional statement in postfix notation. As you can see, this reads very well. You can't use elsif or else with postfix notation.

```
$go_to_work = 1 unless $i_am_sick;
$go_to_bed = 1 if $i_am_sick;
```

We like using postfix notation for one-line conditionals, but the typical if–elsif–else is the most common conditional construct.

Trinary operator

For very quick if–else statements, Perl also provides the trinary operator. Consider the following typical conditional construct:

```
if ($i_am_sick) {
    $go_to_work = 0;
} else {
    $go_to_work = 1;
}
```

The following code does the exact same thing:

```
$go_to_work = $i_am_sick ? 0 : 1;
```

We don't advocate using the trinary operator because it rarely makes code more readable. We only use it for very simple statements like the one above.

Numeric precision and conditionals

Although Perl hides the details, numbers in a computer are generally stored either as *integer* or *floating point* (decimal) numbers. Both *ints* and *floats* have minimum and maximum values, and floats have limited precision. You have probably run into these concepts with your calculator. If you keep squaring a number greater than 1.0 you will eventually run into an *overflow* error. In Perl, this will happen at approximately 1e+308. Similarly, if you repeatedly square a number less than 1.0, you will eventually reach an *underflow* error. In Perl, the closest you can get to zero is approximately 1e−308.

Floating point numbers do not have the exact value you may expect. For example, 0.1 is not exactly one-tenth. Perl sometimes hides these details. Consider the following:

```
$x = 0.1 + 0.1 + 0.1;
$y = 0.3;
$answer = $x - $y;
```

From this code you might expect $answer to store a value of zero. However, if you tried printing out the value you would see that it equals some very small number such as 5.55111512312578e−17. This is because adding the imprecise 0.1 three times does not result in the same value as the imprecise 0.3.

Since floating point numbers are approximations, you should not compare them in conditional statements. Never ask whether ($x == $y) if the values are both floats because, as we have just seen, 0.3 is not necessarily equal to 0.3. Instead, you should use the abs() function to ask if their *absolute* difference is smaller than some acceptable threshold value:

```
$threshold = 0.001;
if (abs($x - $y) < $threshold) {print "close enough\n"}
```

Use strict   4.7

A little bit of discipline can be good for you

We're afraid to tell you that, once again, we are going to introduce a topic that means you should go back and rewrite your existing scripts. But we promise you that this will be the last time we do this.[21] Compared to the last two set of changes that necessitated rewriting your scripts, this one is probably the most important, but it is also the most complex to fully explain. Since you just started programming, we will have to defer some of the discussion until later.

Imagine you are having a dinner party and you invite three friends: Alex, Jo, and Chris. You set the table for four, serve the meal, and start a conversation. Someone casually mentions their friend Dominique and suddenly Dominique is teleported into the house and you have another mouth to feed. In reality, this doesn't happen; but in Perl, it does! Whenever you mention a new variable name, it automagically[22] springs into life (and stays with your program even if you ask it to leave). Let's see this in action.

Example 4.7.1 Create the following program which flips a virtual coin. Half of the time, the program will print '<heads>' and half of the time '<tails>'. Note that use warnings is deliberately omitted for now (we will add it back in the next chapter).

```
1.   #!/usr/bin/perl
2.   # coin.pl
3.
4.   if (rand(1) < 0.5)   {$coin = 'heads'}
5.   else                 {$coin = 'tails'}
6.   print "<$coin>\n";
```

Understanding the script
There's something slightly magical happening here, though it might not seem so. The variable $coin is created either in line 4 or line 5 and its contents are printed on line 6.

Example 4.7.2 Now remove line 5 and run the program again.

```
1.   #!/usr/bin/perl
2.   # coin.pl
3.
4.   if (rand(1) < 0.5) {$coin = 'heads'}
5.
6.   print "<$coin>\n";
```

---

[21] Though we reserve the right to change our minds.
[22] We did not make up this word. It is used a lot in the Perl community.

Understanding the script

Half of the time, the program prints '<heads>' as before. The other half of the time it prints '<>'. In this case $coin does not exist until the print statement. This might not seem magical to you, but trust us, it is! Perl variables are created whenever you need them. And in the case of line 6, the need can be in the middle of a print statement.

Example 4.7.3 Just to make sure you really understand this, add line 5 again, but this time misspell $coin as $cool.

```
1.   #!/usr/bin/perl
2.   # coin.pl
3.
4.   if (rand(1) < 0.5)   {$coin = 'heads'}
5.   else                 {$cool = 'tails'}
6.   print "<$coin><$cool>\n";
```

Understanding the script

This program creates two different variables depending on the random number. Nobody would do this intentionally, yet there is apparently nothing wrong with it from Perl's perspective. You could *catch* this kind of error with use warnings, but it would be better to *prevent* such errors.

use strict

The strict pragma helps prevent a variety of common errors. There is a tiny amount of extra work involved, but it is completely worth it. If your program includes use strict, Perl no longer creates variables automagically. You must declare your variables with the my keyword before you use them. This is very simple. Let's revisit the coin-flipping program.

Example 4.7.4 This will be your first script with the use strict pragma in place, but it will definitely not be your last!

```
1.   #!/usr/bin/perl
2.   # coin.pl
3.   use strict;
4.
5.   my $coin;
6.   if (rand(1) < 0.5)   {$coin = 'heads'}
7.   else                 {$coin = 'tails'}
8.   print "<$coin>\n";
```

Understanding the script

We first turn on the strict pragma by including the use strict statement on line 3. This behaves just like the warnings pragma that we have already seen. As soon as Perl gets to this line, it will turn on a new set of rules that the rest of the script will have to comply with.

Line 5 declares that you want to create a variable called $coin. We do this by using the my keyword. Lines 6 and 7 no longer create $coin, they only set its value.

What exactly does my do? It asks Perl to create some memory. The memory is not assigned to any particular value. This happens later. You do this kind of thing in everyday life. For example, you grab a piece of paper for a grocery list. Later, you add items to the list. When you grab the paper, you are effectively saying 'Let this piece of paper be an empty grocery list.' In the same way, on line 5, you are telling Perl, 'let $coin represent an empty coin.' It gets filled with 'heads' or 'tails' on lines 6 and 7.

Let's see what happens when an uninvited guest tries to steal a meal at our imaginary dinner table when we use strict.

Example 4.7.5 Note that on line 7 we once again misspell $coin as $cool. Previously, $cool showed up (and didn't leave). Time to get strict!

```
1.   #!/usr/bin/perl
2.   # coin.pl
3.   use strict;
4.
5.   my $coin;
6.   if (rand(1) < 0.5)    {$coin = 'heads'}
7.   else                  {$cool = 'tails'}
8.   print "<$coin>\n";
```

Now when you run the program you will see the following error message:

```
Global symbol "$cool" requires explicit package name at coin.pl line 7.
```

Alright, an error message! But what are 'global symbols' and 'explicit package names'? Since we hated when our parents said 'we'll tell you when you're older,' we will give you a brief explanation at the end of this chapter. But if it seems a little complex, don't worry – you will get it when you're older. The main take-away message here is that misspelled variable names are now an error that will prevent your program from running.

Scope

Scope can be a confusing topic. If this section makes your head hurt, feel free to come back to it later. But please take this one piece of advice with you:

Always use strict!

Variables declared with my are called *lexical* variables. Lexical variables exist within a hierarchy defined by *block structure* (as denoted by use of curly braces). A lexical variable exists from the line where it is declared until the closing curly brace. Note that every file is also a block, so you can imagine that there are invisible curly braces at the start and end of every file. This means that all of your Perl scripts will always contain at least one block of code.

Example 4.7.6 Create the following program and run it. The output should be x=1 y=2.

```
1.   #!/usr/bin/perl
2.   # scope.pl
3.   use strict;
4.
5.   my $x = 1;
6.   if ($x == 1) {
7.        my $y = 2;
8.        print "x=$x y=$y\n";
9.   }
10.
```

Understanding the script

In this program, $x is created on line 5 and exists until the end of the program (which is also the end of a block). This is not true of $y, which is created on line 7 and exists only until line 9. At line 10, $y no longer exists. To use the technical jargon, we say that $y is *in the scope* of the block containing the conditional (if) statement.[23] One of the reasons we indent code inside blocks is to show that all the indented statements *belong* to the block. That is, $y belongs to the block from lines 6–9. It does not exist outside this scope.

To illustrate this concept further, let's intentionally break the program by violating the scope of $y. Let's try to make the program 'see' $y in a place where $y should never be seen.

Example 4.7.7 This program is the same as the previous one, except we have moved the print statement of line 8 to the end of the script. Observe what happens when you run the program.

```
1.   #!/usr/bin/perl
2.   # scope.pl
```

---

[23] If you find the use of the phrase 'in the scope of' confusing, you can replace it with 'seen by' or 'available to.'

```
3.   use strict;
4.
5.   my $x = 1;
6.   if ($x == 1) {
7.       my $y = 2;
8.   }
9.   print "x=$x y=$y\n";
```

You should see an error that reports:

```
Global symbol "$y" requires explicit package name at scope.pl line 9.
```

Understanding the script

We've seen this kind of error message before. But that was when a variable was mis-spelled. There isn't anything misspelled here. The reason for the error is that lexical variables live and die within block structure (memory is allocated to them when they are declared and freed at the closing brace in the same scope). The variable $y is 'born' on line 7 and 'dies' on line 8. You can't print $y at line 9 because $y doesn't exist anymore. It only existed within the scope of where it was declared. Note that $x is 'alive' from line 5 until the end of the file. $x does not die until the program ends.

A variable is said to have global scope if it is available to the entire program. A variable with local scope has a more restricted availability. In the examples above, $x has global scope and $y is local to the conditional statement. In general, you should declare your variables in the most local scope possible. That way they won't interfere with variables in some other part of the program. This helps prevent some of the worst kinds of bugs (those that are syntactically correct but semantically erro-neous). Now let's see what happens when two variables in different scopes have the same name.

Example 4.7.8 Run the following script.

```
1.   #!/usr/bin/perl
2.   # scope.pl
3.   use strict;
4.
5.   my $x = 1;
6.   my $y = 3;
7.   if ($x == 1) {
8.       my $y = 2;
9.       print "inside: x=$x y=$y\n";
10.  }
11.  print "outside: x=$x y=$y\n";
```

Understanding the script

In this program, $y is declared twice, once outside the conditional and once inside. The output shows that the two $y variables are completely separate. The declaration of $y on line 8 is local to the if block. Within this inner scope, the new $y masks the more global $y that was declared on line 6. It is common, and often desirable, to reuse variable names within a program. As long as those variables are in different blocks of code, i.e., they all have a different scope, this will not cause any problems.

Why strict is essential

With lexical variables, you can make sure whatever happens in a block stays in the block and doesn't unknowingly affect variables in a more global scope. This concept becomes increasingly important as your programs grow in length and complexity. If you use global variables, you have to keep track of all the variable names to make sure you don't unintentionally overwrite some variable used in another part of the program. With lexical variables, each block can have its own set of private variable names. Not only will this improve your ability to focus on the task at hand, it will also help you share code among your programs and with other people. The real power of programming comes when you join your code to other people's code, and that can only really happen if you get into the habit of using lexical variables.

Symbol tables

The last part of this chapter is potentially confusing. It might not make much sense right now. Consider this optional reading.

Every new *word* you create in your program is stored in a table. So far, all the new words you have created were scalar variable names. But later, you will create other things with names, such as arrays, hashes, and subroutines. These names you make up are called *symbols*, and they are stored in your program's *symbol table*, which is also called a *package*. The name of your program's package is *main*. This is sort of like a family name. A variable you define as $x is also known as $main::x. There are other families your program may interact with one day, and these all have their own family names. $Toolbox::x refers to a variable called $x in a symbol table called Toolbox. The Toolbox $x is a completely different variable from your $x. Variables in a symbol table are true global variables, available to the entire program.

Lexical variables (those created with my) are not stored in a symbol table. They live and die within blocks of code. Now let's revisit the error message from Example 4.7.5.

```
Global symbol "$cool" requires explicit package name at coin.pl line 7.
```

This error is saying that if you want to use $cool, you have to say what package (family) it belongs to. Otherwise, the assumption is that the variable is a lexical variable, and no such lexical variable exists.

Today is a good day to die()

We often write scripts where we end up with some data in a variable, but we don't always know exactly what that variable will contain – maybe we have used the rand() function to generate some data, or maybe the data are being read from a file that we never actually looked at. This raises the possibility that maybe the data is missing or not what you were expecting.

A poorly written program will carry on regardless, oblivious to the fact that the variable contains the wrong type of data. However, Perl provides the ability to stop a program at any point, and we will often want to do this to prevent a program from running if the data is incorrect. We can do this by using one of two functions: die() or exit().

Example 4.8.1 We use the die() function when we want to stop the script *and* print a warning message at the same time.

```perl
1.   #!/usr/bin/perl
2.   # death_and_taxes.pl
3.   use strict; use warnings;
4.
5.   my $tax_rate = int(rand(100));
6.   print "Tax rate is $tax_rate%\n";
7.
8.   if($tax_rate > 50){
9.       die "I can't afford to pay that, I quit my job!\n";
10.  }
11.
12.  my $salary = 100000;
13.  my $tax = $salary * ($tax_rate/100);
14.  print "I have to pay $tax\n";
```

Understanding the script

Note that line 3 includes both the strict and warnings pragmas on one line. We could also put these on two separate lines. From Perl's point of view, it treats these as two lines of code because of the two semicolons. All of our scripts from now onwards, will include both of these pragmas.

On line 5 we calculate a tax rate using the rand() function, and then ask (on line 8) whether that tax rate is above a certain level. If it is too high, then we use the die() function to stop the script at that point. You can think of the die() function as acting like print() but that it also stops the program. In this case, *if* the tax rate is too high we don't want to even consider looking at the rest of the code. The 'rest of the code' may only be three lines in this example, but it could be several hundreds of lines in other scripts!

Example 4.8.2 Substitute lines 8–10 in the above script for the following:

```
8.    if($tax_rate > 50){
9.        print "I can't afford to pay that, I quit my job!\n";
10.       exit;
11.   }
```

Understanding the script

The outcome of this script should be identical to the previous example. However, line 10 is now using the exit() function to stop the script. The exit() function can be used at any point when you want to make your script stop. Note that exit() doesn't print out anything, which is why we added a print statement on the preceding line.

There are some differences between how die() and exit() stop a program, but it's not worth discussing them here. The important point is that you should use die() when you want your script to stop *because of a problem*, and you should use exit() for all other occasions. You will most likely only use exit() when your scripts become more complex and end up being embedded in other scripts.

Example 4.8.3 The most common use of the die() function is for performing a series of validity checks on your data. When your scripts are working with biological data, you have to remember that Perl does not know anything about the rules of biology, so it's up to you to check that your data is correct. Let's write a short script to check whether a given DNA sequence is the correct length to be considered a gene, i.e., that the length is a multiple of three.[24] We can write a short script to check this.

```
1.    #!/usr/bin/perl
2.    # dna_checker.pl
3.    use strict; use warnings;
4.
5.    my $dna = "ATGCACGTAGGTAACACTGACTGAA";
6.    my $length = length($dna);
7.    die "$length is not a valid length\n" if (($length % 3) != 0);
```

Understanding the script

We calculate the length of the sequence and then use the modulo operator to divide the length by 3. If the remainder is anything but zero, then this might not be a valid DNA sequence, and so we have a die() statement to stop the script.

We can imagine that this program could be a lot longer and perform many subsequent operations on the $dna variable. But if $dna is not the right length, we might want to stop the script immediately and warn that there is a problem.

---

[24] If you remember your biology, you'll know that protein-coding genes consist of three-nucleotide *codons* that each specify one amino acid. If you don't remember your biology, consider this a free lesson.

What makes a good die() statement?

You don't actually need to include any statement with the die() function. The following code will stop a script, but not print anything (assuming the if condition evaluates as true).

```
die if ($pi != 3.14);
```

However, it generally makes more sense to make the die() function print something that will tell you what the problem was. You should make the error message as detailed as possible. All three of the following die() statements are better than not having any statement at all, but the last example is the most useful as it will actually tell you what the incorrect value of $pi was.

```
die "Wrong number\n"          if ($pi != 3.14);
die "Wrong value for pi\n"    if ($pi != 3.14);
die "Error, $pi is not 3.14\n"   if ($pi != 3.14);
```

Problem 4.8.1 Let's assume that your annual bonus is decided by the outcome of the rand() function. Write a script that declares a variable $bonus, and then assigns it a random integer between 0 and 3 (maybe this is your bonus in thousands of dollars, rupees, yuan, etc.). Then add a test that will stop the script with a suitably worded die() statement if the amount is zero.

How long is a piece of string? I don't know, but the `length()` function does

From algebra, we are used to the idea of using variables in math. But what about strings? Can you add, subtract, multiply, divide, and compare strings? Not exactly, but there are analogous operations.

Concatenation operator

We often have two strings that we want to join together to make one string. Imagine you have one variable which contains your first name, and another variable that contains your surname. Some operations would be easier to do if you could just join those variables together to form a new variable. In Perl, we do this with the concatenate operator, which is represented by a single dot (.) character.

Example 4.9.1 Create the following program and run it. This program introduces the concatenate operator and just uses it to join two variables.

```
1.   #!/usr/bin/perl
2.   # strings.pl
3.   use strict; use warnings;
4.
5.   my $s1 = "Hello ";
6.   my $s2 = "World\n";
7.   print $s1 . $s2;
8.
9.   my $s3 = $s1 . $s2;
10.  print $s3;
```

Understanding the script

In this example we join two strings ($s1 and $s2) and immediately print the concatenated result. It's important to realize that the concatenation happens *before* the `print()` function. This would be more obvious as `print($s1 . $s2)`. Always include spaces on either side of the dot character to improve the readability of your code.[25]

Line 9 takes the same two strings and performs another concatenation. However, this time the result is stored to a new variable before printing. If we wanted to, we could have also created an equivalent string to $s3:

```
11.  my $s4 = 'Hello' . ' ' . 'World' . "\n";
```

---

[25] In many languages, including the next major revision of Perl (version 6), a dot is used to access object attributes and methods. For example, given a dog object, we could access the name and breed attributes by using dog.name and dog.breed. Objects are a much more advanced topic, so for now just remember to put spaces between the things you want to concatenate.

In this case, we concatenate four strings: the word 'Hello,' a single space character, the word 'World,' and a newline character. Notice that we are using single quotes for the first three strings, but double quotes for the last string. The last string needs double quotes so it is recognized as a newline character and not just a backslash and 'n.' You will often see Perl scripts that use double quote characters when single quotes would be equally valid.

In the earlier section on mathematical operators we learned that there are shortcuts for when you want to add a value to a variable:

```perl
$n += 5; # this is the same as $n = $n + 5;
```

There is a similar shortcut for the concatenation operator (.=) and this can be convenient when you have a variable that behaves like a log file, collecting several different pieces of output:

```perl
my $log;
$log = "Captain's log\n";
$log .= "Perl - the final frontier\n"; # concatenates on end of $log
$log .= "These are the voyages of the Perl script Enterprise.pl\n";
```

Sometimes it is useful to 'grow' a string in this way and print it out at a later point in your script.

String comparison operators

Let's look at some other string operators that can be used to compare strings. It will hopefully make sense to you that you can compare strings to see if they are identical. However, Perl also allows you to compare strings to see if one is *greater* or *less* than another string.

Example 4.9.2 This script will use three of Perl's string comparison operators. Try to guess what is happening in lines 9 and 10.

```perl
1.   #!/usr/bin/perl
2.   # string_comparison.pl
3.   use strict; use warnings;
4.
5.   my $s1 = "Apples";
6.   my $s2 = "Oranges";
7.
8.   if    ($s1 eq $s2) {print "same string\n"}
9.   elsif ($s1 gt $s2) {print "$s1 is greater than $s2\n"}
10.  elsif ($s1 lt $s2) {print "$s1 is less than $s2\n"}
```

Understanding the script

Lines 5 and 6 assign two strings to $s1 and $s2, and then lines 8–10 print various lines of output depending on the outcome of comparing the strings using three different operators (eq, gt, and lt).

It might make sense to compare any two strings by their length, but that's not what is happening. All three of these operators compare strings based on something known as their 'ASCII value.' Put simply, this is a numerical value that is associated with every character on your computer keyboard and many more that are hidden.[26]

The eq operator on line 8 tests whether two strings are identical, but the lt and gt operators test whether one string is 'less than' or 'greater than' another. To do this these operators compare the ASCII value of the characters in each string. This means the letter 'A' of 'Apples' (ASCII value = 65) is considered 'less than' the letter 'O' of 'Oranges' (ASCII value = 79). The length of the strings being compared is not important – 'A' is less than 'BBB' and 'CCC' is less than 'D'. Due to the nature of ASCII values, all lower-case letters have ASCII values higher than their upper-case counterparts.

Note that for the sake of demonstrating all three string comparison operators, the above script does not end with an else statement. There is no need for the last elsif statement; if one string is not equal to or greater than another string, then it *must* be less than it.

Repetition operator

There are many operators and functions in Perl that you may not need to use very frequently, and the repetition operator is in this category. However, it fulfills a simple but useful role of just letting you make multiple repeats of any string. The repetition operator is represented by a single 'x' character.[27]

Example 4.9.3 As you write these scripts, try to predict what each line is going to do *before* you read the explanation below. Also, any time you write a script that uses the rand() function, you should make sure you run it several times to check you are getting different output each time.

```
1.    #!/usr/bin/perl
2.    # haha.pl
3.    use strict; use warnings;
4.
5.    my $number = int rand(20) +1;
6.    my $string = 'Ha ' x $number;
7.    print "$string\n";
```

[26] Wikipedia provides a good overview of ASCII along with a list of the ASCII values for many different characters: http://en.wikipedia.org/wiki/ASCII.

[27] Some people will see this and think it looks like a multiplication operator, but remember that Perl uses the asterisk character for multiplication.

Understanding the script

Lines 5 generates a random number from 1 to 20 and line 6 then creates a string of variable length. It does this by using the repetition operator (x). The left-hand operand of the repetition operator is a string, and the right-hand operand is a number. When you run the program, the program will report strings such as 'Ha Ha' and 'Ha Ha Ha Ha Ha' depending on the random number generated.

length() **function**

A string is a sequence of ASCII characters. It doesn't matter if the characters are letters, spaces, or special characters that don't even print.[28] You can find out how long a string is with the length() function.

Example 4.9.4 A simple script to calculate the length of two strings:

```
1.   #!/usr/bin/perl
2.   # length.pl
3.   use strict; use warnings;
4.
5.   my $s1 = "Hello World";
6.   my $s2 = "Hello World\n\n";
7.
8.   print length($s1), " ", length($s2), "\n";
```

Understanding the script

$s1 and $s2 are very similar, but $s2 contains two newline characters (remember that backslash-n represents a single character). This means that $s2 is two characters longer than $s1, and the output is therefore '11 13.'

---

substr() **function**

One of the most useful and powerful Perl functions is substr(), which is short for substring. This function lets you operate on parts of a string. For example, you could retrieve the first six characters from a string, or the last five, or something in the middle. The substr() function takes two or three arguments: a string, an *offset*, and an optional length. Let's see this in action.

---

[28] We will learn about several special characters in the next chapter.

Example 4.9.5 This script will demonstrate the different ways we can use substr()
to extract text from a string.

```perl
1.   #!/usr/bin/perl
2.   # substr.pl
3.   use strict; use warnings;
4.
5.   my $quote = "Strive not to be a success, but rather be of value";
6.   print substr($quote, 0, 6), "\n";   # Strive
7.   print substr($quote, -5), "\n";     # value
8.   print substr($quote, 19, 7), "\n";  # success
```

Understanding the script
Line 5 creates and assigns a string to a variable.[29] Line 6 uses substr() to extract the
first six characters of $quote, which are then printed. Notice that the offset, the position
in the string which we start extracting from, is set to zero. The first character in a string
is at position zero, not one.[30]

Line 7 prints the last five characters. If you specify an offset that is negative, you can
extract from the right-hand side of a string. We could also have written substr($quote,
45) here. Notice that the optional length parameter was not given, so the substring continues to the end of the string. If we had wanted to extract just the last five characters,
we could specify it with substr($quote, -5, 5). Line 8 extracts characters from
the middle part of the string.

Note that in the previous example, the original string in $quote was not changed by
lines 6–8. The substr() function copies part of the string and sends this to the print()
function. But substr() can also change strings if you want it to. For example, you
could delete the first five characters. You can also insert strings into strings and replace
strings with other strings. To perform these kinds of tasks, one simply puts substr()
on the left of an assignment operation as substr() = string.

Example 4.9.6 Add the following two lines to observe string replacement.

```perl
9.   substr($quote, 19, 7) = "clown";
10.  print $quote, "\n";
```

Understanding the script
Line 9 removes seven letters from $quote, beginning at position 19, and replaces them
with a new string. This results in the string 'success' being replaced by 'clown'. Note
that the strings are not the same length.

---

[29] This quote is from Albert Einstein.
[30] You might find this strange, but it is a common practice in computing to start counting at zero rather than one. We will
see this again later, when we learn about arrays.

Case conversion functions

Perl has four functions for changing the case of characters: uc(), lc(), ucfirst(), lcfirst(). The uc function converts an entire string to upper-case characters, whereas ucfirst converts just the first character. The lc equivalents convert strings to lower-case characters.

Example 4.9.7 This script uses three of these functions to manipulate a simple text string:

```
1.    #!/usr/bin/perl
2.    # shout.pl
3.    use strict; use warnings;
4.
5.    my $s = "HOMO sapiens\n";
6.    print uc $s;   # HOMO SAPIENS
7.    print lc $s;   # homo sapiens
8.    $s = ucfirst(lc($s));
9.    print $s;      # Homo sapiens
```

Understanding the script

The string on line 5 has mixed case and is therefore not the proper style for a taxonomic classification. Lines 6 and 7 print upper- and lower-case versions. Note that these lines do not change the underlying content of $s, only the printed output is converted.

Line 8 uses two functions to change the string to lower-case and then capitalize the first character. Note that Perl allows you to put $s on *both* sides of the assignment, meaning that the original value of $s becomes overwritten by the modified version.[31]

---

[31] As always, Perl will evaluate the right-hand side of an assignment first (i.e., everything after the equals sign), and will resolve functions from the 'inside-out' (the lc function must happen before the ucfirst function). With practice, it becomes common to see a line of code like this and read it in the same way that Perl will execute it, rather than simply reading it left to right.

Do you have an escape plan?

As we have seen, Perl gives a special meaning to a lot of characters on your keyboard. This is generally fine *until* we want to print some of those special characters. We've already learned that we can put things in single quotes so they lose their special meaning, but that is not always the best solution. Let's imagine we want to print out the contents of a variable called $salary and that we also want to include a dollar sign. We might think of doing this:

```
print "I earn $$salary\n";
```

However, this will not work and will produce an error. Perl expects that if there is a dollar sign within a pair of double quotes then this is going to signify the name of a variable. One solution would be to use single quotes for the dollar sign, but double quotes for the rest of the string:

```
print "I earn ", '$', "$salary\n";
```

This works, but requires that you print a list of three things, one of which has to be enclosed in single quotes. That seems to be a lot of extra work just to do something quite simple. Luckily for us, Perl provides an easy way to do what we want as you can ignore the special meaning of a character by *escaping* it. To escape any character in Perl, just prefix it with a single backslash character (\):

```
print "This is a dollar symbol: \$\n";
print "This is a quotation mark: \"\n";
```

The backslash is itself a special character. This then raises the question of how do you print a backslash character? The following looks like it might work, but it doesn't quite behave as expected.

```
print "This is a backslash: \ \n";
```

If you try printing this line, the output will just look like:

```
This is a backslash:
```

The backslash character affects whatever character comes *after* it, so in the last example, we are actually escaping the space character that precedes the \n, even though the space character doesn't need to be escaped. To escape a backslash character, we have to use another backslash character!

```
print "This is a backslash: \\\n";
print "This is a newline character: \\n\n";
```

This can end up looking a little confusing, but luckily we don't usually need to print too many backslash characters. Note that in the last example, we only need to escape the first backslash of \n. The 'n' character that follows is just a regular character with no special meaning and therefore needs no escaping.

The ability to escape characters in this way will become more important as there are many special characters in Perl, and sometimes you need to escape them in other situations apart from printing. We will therefore use escaping from time to time in the later sections of this book.

The tab character

There are not too many special characters that you will use frequently apart from \n. The other character you might need to use a lot is the tab character: \t. Anytime you print \t within a double-quoted string, it will behave just like you have pressed the tab character on your keyboard.[32] The following two lines show you an example line of code and what it will look like when printed to your terminal:

```
print "Three tabs: \t\t\t and a newline\n"; # this prints ...
Three tabs:                    and a newline
```

Even though we have to type two characters to represent a tab, it is still only treated as a single character by Perl (with a single ASCII value, just like \n). In this particular example, there are three tab characters surrounded by two space characters. This is therefore a good time to remind you:

> *Never assume that all whitespace only consists of space characters*

If you try counting the length of the string '12\t45,' it has a length of five characters even though it takes six characters to type it, and it will appear as even more characters when printed.

Tabs are often used to separate columns of text in many different data files. This makes them suitable for importing into spreadsheet programs, but it also makes them very easy to parse with a Perl script. Many data files use tabs to separate data in this way; such files are called *tab delimited* or TSV (tab-separated values).

---

[32] Historically, a tab character was equivalent to eight spaces on the keyboard of a mechanical typewriter. Many computer programs will still print eight-character tabs, but it is increasingly common that different programs use a different number of spaces. This is why pressing tab in different coding editors will sometimes indent code by different amounts. Just remember, it's all still one character internally.

You have now met your match

One of the most common tasks you may face as a programmer is to find a string within another string. Once you are able to do this, you can then ask questions such as 'Does this web page contain my email address anywhere within it?' or 'Can I extract all zip codes from this piece of text?' String matching is particularly common in various biological contexts – you might want to find a restriction site in a DNA sequence, or maybe you are trying to see if a protein sequence contains a particular peptide motif.

Sometimes we want to find an *exact* match to any given string, but other times we want to perform a so-called *fuzzy* search, using a search pattern that specifies one or more possibilities. For example, if we are trying to find whether 'Bob' is mentioned within a block of text that is stored in a scalar variable, we might want to find 'Bob', 'bob', or 'BOB'. When we have found a match to a pattern, we often then want to manipulate or modify the match. For example, maybe you want to correct a spelling mistake in a word, or remove the last five nucleotides from a DNA sequence. All of these kinds of operations are relatively easy to do in Perl.

The principle way in which Perl allows you to do all of this is with something called the binding operator (=~). Unlike other operators we have seen so far, the binding operator is used in slightly different ways, depending on what code follows it. The three main uses of it are for *matching*, *substitution*, and something called *transliteration*. Let's start off with some examples of how to match strings.

Example 4.11.1 Enter the program below and observe the output. The aim of this program is to see whether a DNA sequence contains a specific restriction enzyme digest site.[33]

```perl
1.    #!/usr/bin/perl
2.    # matching.pl
3.    use strict; use warnings;
4.
5.    my $sequence = "AACTAGCGGAATTCCGACCGT";
6.    if ($sequence =~ m/GAATTC/)   {print "EcoRI site found\n"}
7.    else                          {print "no EcoRI site found\n"}
```

Understanding the script
Line 5 assigns a short DNA sequence to a variable ($sequence). We then have an if–else statement on lines 6–7 which contains a slightly unusual-looking test condition compared to what we have seen previously.

The if–else statement contains the binding operator =~, and this signifies that we are going to do some string manipulation. We follow the operator with an m followed

---

[33] Restriction sites are just places in a DNA sequence that have a particular pattern of nucleotides and which can be targeted by something called a restriction enzyme, which cuts the DNA at that pattern.

by two forward-slashes that contain the pattern we are searching for. The m// pattern indicates that we are trying to match a string.[34]

You should read line 6 as 'If the string stored in the $sequence variable matches the pattern 'GAATTC,' then print a statement …'

Matching and not matching

The matching operator (m//) is very common in Perl, and it is used so frequently that the m can be omitted. The following two lines are treated as the same by Perl:

```
print "I've found love!\n" if ("ungloves" =~ m/love/);
print "I've found love!\n" if ("ungloves" =~ /love/);
```

We suggest you keep including the m for now as it will probably make it easier to differentiate it from the substitution operator (which we will introduce shortly). Note that when you use the matching operator, you are asking whether a string contains a pattern *anywhere* in that string. This also includes a match to the whole string itself:

```
$string = "ABC";
print "Match\n" if ($string =~ m/A/);     # true
print "Match\n" if ($string =~ m/B/);     # true
print "Match\n" if ($string =~ m/C/);     # true
print "Match\n" if ($string =~ m/AB/);    # true
print "Match\n" if ($string =~ m/BC/);    # true
print "Match\n" if ($string =~ m/ABC/);   # true
print "Match\n" if ($string =~ m/abc/);   # false
print "Match\n" if ($string =~ m/AB C/);  # false
```

The last two examples illustrate that matches are case-sensitive and that the pattern has to match *exactly*, i.e., if you add a space to the pattern, it will only match if a space occurs in the string.

We have previously shown how you can ask whether two numbers or two strings are *not* equal to each other. It also makes sense to ask whether one string does *not* match another. To do this we use the 'not match' operator (!~).

Example 4.11.2 This script takes a name of a species and tries to check whether that name might belong to a fruit fly species.[35] We will just assume that if the species name contains the string 'Drosophila' then it is a valid fruit fly species.

```
1.   #!/usr/bin/perl
2.   # fly_check.pl
3.   use strict; use warnings;
```

---

[34] As soon as you add m// after the binding operator, then we more commonly call the whole thing the *matching* operator, even if this really is a binding operator followed by a pattern match.

[35] Most fruit flies belong in the genus *Drosophila*; the most famous species of fruit fly is *Drosophila melanogaster*.

```
4.
5.    my $species = "Drosophila pseudoobscura";
6.
7.    if($species !~ m/Drosophila/){
8.        die "$species does not look like it is a fruit fly\n";
9.    } else {
10.       print "$species looks like it is a fruit fly\n";
11.   }
```

Understanding the script

Line 5 assigns a species name to a variable and then lines 7–9 ask if that name does *not* match the pattern 'Drosophila'. If the name doesn't match then we use the die() function to stop the script. If we didn't need to have the else statement we could have also achieved the same outcome by using an unless statement with the regular matching operator:

```
unless ($species =~ m/Drosophila/) {
    die "$species does not look like it is a fruit fly\n";
}
```

Identifying the location of the match

Consider the following code:

```
my $text = "I took a good look at the book";
print "Match" if ($text =~ m/oo/);
```

Clearly we would expect the print statement to happen because $text does match the pattern 'oo'. But the pattern 'oo' occurs four times in the input string, so which one is Perl matching, or is it matching all of them? Perl will always try to match at the earliest position possible, so in this case it is matching the 'oo' in 'took.'

In many cases you might only care that a match *does* occur and not *where* it occurs. However, sometimes you will want to potentially loop through all matches in a string, or identify the parts of the string that match (or the parts that don't match). We return to these issues in the later chapters that cover *regular expressions* (Chapters 4.26–4.28).

Matching operators in Perl

Below is a table that shows the full range of matching operators. We've just seen how to make a match and a 'not match.' Before we look at the substitution operator, try to guess how it will work based on the example in this table.

Operator	Meaning	Example
=~ m//	match	if ($s =~ m/GAATTC/)
=~ //	match	if ($s =~ /GAATTC/)
!~ //	not match	if ($s !~ m/GAATTC/)
=~ s///	substitution	$s =~ s/thing/other/;
=~ tr///	transliteration	$count = $s =~ tr/A/A/;

Substitution

In addition to matching things, we often want to substitute a pattern with a string; we can do this using the substitution operator. You can think of this as a match *and* substitute operator, as it can only substitute things if it first finds a match. The syntax is very similar to the matching operator, but we obviously need to specify a replacement string in addition to the pattern that we are trying to match. The basic syntax is as follows:[36]

```
$string =~ s/match/replacement/;
```

Note that we use an s instead of an m to indicate that this will be a substitution. The only other difference is that we include a second string that will replace the matching pattern. Let's see how this works in practice.

Example 4.11.3 This script turns one string into another, and then subsequently modifies just part of the new string.

```
1.   #!/usr/bin/perl
2.   # currency_conversion.pl
3.   use strict; use warnings;
4.
5.   my $currency = "pounds";
6.   print "I used to get paid in $currency\n";
7.
8.   $currency =~ s/pounds/dollars/;
9.   print "But now I get paid in $currency\n";
10.
11.  my $replacement = "nuts";
12.  $currency =~ s/llars/$replacement/;
13.  print "Ideally, I'd like to get paid in $currency\n";
```

This script should produce the output:

```
I used to get paid in pounds
But now I get paid in dollars
Ideally, I'd like to get paid in donuts
```

---

[36] The syntax of this operator very closely resembles the syntax of the Unix sed command, which performs a very similar function (among many other things). We will mention sed very briefly in Chapter 5.8.

Understanding the script

We start off with a string 'pounds' that is stored in the $currency variable. Line 8 then uses 'pounds' as a pattern and substitutes it with a string 'dollars'. More specifically, Perl is first asking 'Can I find the pattern 'pounds' anywhere within the variable $currency?' If there is a match, then Perl proceeds to the substitution step.

   Line 12 performs another substitution, but this time the replacement pattern is actually another variable rather than a fixed string. You can use variables in either the matching or replacement portion of the substitution operator (or both). Note that this second replacement only affects part of the input string.

Using options with the binding operator

Let's look at a very common problem that arises when you use the substitution operator in its default mode. Luckily, it is a problem that is very easy to fix.

Example 4.11.4 This script attempts to convert a DNA sequence into an RNA sequence.[37] Try to predict what the output of this script will be before you run it.

```
1.  #!/usr/bin/perl
2.  # dna_to_rna.pl
3.  use strict; use warnings;
4.
5.  my $dna = "ATGCAGGATGGGATCCTTTATTGA";
6.  my $rna = $dna;
7.  $rna =~ s/T/U/; # convert thymine to uracil
8.  print "$dna\n$rna\n";
```

Understanding the script

In this script we first make a copy of our $dna variable and assign that to a new variable ($rna). When you start writing Perl scripts that modify data, you will often want to keep the original, unchanged data as well.

   You should have noticed that the output did not change all of the 'T's to 'U's. In fact, the substitution operator only changes the *first* occurrence of the matching pattern.

   Clearly we would like to be able to change *all* occurrences of a pattern, not just the first. Something else that we frequently want to do is to be able to change (or match) text regardless of whether it is upper- or lower-case. Perl makes it very easy to do both of these things by allowing you to specify additional options after you specify the match or substitution. Here is how you can tell Perl to substitute *all* matches (a so-called *global* replacement) and to also ignore the case of the string being matched.

---

[37] RNA uses the nucleotide uracil (U) instead of thymine (T)

```
$string =~ s/T/U/g;     # global replacement of 'T' to 'U'
$string =~ s/T/U/i;     # substitutes the first 'T' or 't' in $string to U
$string =~ s/T/U/gi;    # global replacement of 'T' or 't' to 'U'
```

The g option tells Perl to replace globally, i.e., replace all occurrences of the match. The i option instructs Perl to ignore the case of the match. Notice you can use multiple options together (the order in which you specify the options does not matter). There are a few more options that are available, but these are the two that you will use most frequently. Time for another example!

Example 4.11.5 Hopefully, you are starting to get better at predicting what these scripts will do before you run them.

```
1.  #!/usr/bin/perl
2.  # revelation.pl
3.  use strict; use warnings;
4.
5.  my $text = "TREESWTREESOTREESOTREESDTREES";
6.  $text =~ s/TREES//g;
7.  print "$text\n";
```

Understanding the script

You should realize that line 6 of this script is performing a substitution. But notice that there is no replacement pattern. When you don't specify a replacement pattern you are effectively specifying 'nothing.' Therefore, this Perl code will replace all occurrence of the string 'TREES' with nothing – i.e., it will delete 'TREES'. Because we include the global option, we delete *all* occurrences of this string which then reveals a new word.[38]

Problem 4.11.1 What do you think this script will do? Will it print anything at all?

```
1.  #!/usr/bin/perl
2.  # will_I_print_anything.pl
3.  use strict; use warnings;
4.
5.  my $sequence          = "AACTAGCGGAATTCCGACCGT";
6.  my $empty_sequence    = "";
7.
8.  if ($sequence =~ m//){
```

---

[38] This example is playing around with the idiom 'Can't see the wood for the trees.'

```
9.       print "I am here, therefore m// must match '$sequence'\n";
10.  }
11.
12.  if ($empty_sequence =~ m//){
13.       print "I am here, therefore m// must match '$empty_sequence'\n";
14.  }
```

Transliteration is literally a bit like translation

The transliteration operator gets its own chapter as it is a little bit different to the two other matching operators that we introduced in the last chapter. If you have already worked through Part 5 of this book then you may have seen that there is a `tr` command in Unix (see Chapter 5.5). Perl's transliteration operator behaves in the same way as this command. It takes a list of characters and changes each item in the list to a character in a second list, though we often use it with just one thing in each list. It automatically performs this operation on all characters in a string (so no need for a 'global' option).

Example 4.12.1 This script demonstrates the full range of abilities of the transliteration operator. Notice how there are comments at the end of many of the lines. You don't have to type these comments, but adding comments to your scripts is a good habit to get into. Write the script and then run it.

```
1.  #!/usr/bin/perl
2.  # transliterate.pl
3.  use strict; use warnings;
4.
5.  my $text = "abc123abc123";
6.
7.  $text =~ tr/a/d/;      # changes any occurrence of 'a' to 'd'
8.  $text =~ tr/bc/xy/;    # 'b' becomes 'x', and 'c' becomes 'y'
9.  $text =~ tr/123/321/;  # 1 becomes 3, 2 stays as 2, 3 becomes 1
10. $text =~ tr/dxy/DXY/;  # capitalize the letters d, x, and y
```

Understanding the script

From experience, we would bet that many of you reading this have typed the script exactly as it appears, and have then run it, only to be puzzled when nothing happened. This requires a short diversion.

Computers vs. humans

Contrary to what many people think, computer programs are pretty faithful creatures.[39] More loyal than any dog, they will only ever do what you tell them to. This point bears repeating, and we will even go so far as to call it: *Don't blame the program: rule #1:*

Programs will only do what you tell them to!

---

[39] We are always surprised to see how many people think that programs deliberately do things to annoy them. Blame your family, blame your government, but don't blame your program.

If your program appears to do something stupid, then most of the time this will be because you (probably unknowingly) told it to do something stupid. From time to time you will also write scripts that don't do what you were expecting them to. You should then ask yourself: *Did I actually ask it to do the thing I was expecting it to do?* This leads us directly to *Don't blame the program: rule #2*:

> *Programs can work properly without appearing to do anything.*

You might one day write a program that will maybe solve the world's energy crisis, find a cure for the common cold, or work out why the 'World Series' in baseball is only contested by two countries. Your code may be immaculate and may even be the most powerful, well-written Perl code the world has ever seen. However, if your program fails to print out any information, then the secrets to these problems will remain forever unknown.

In the example above, we deliberately did not include any print statements. This was a test to see if you were paying attention. Without any print statements, the program will work and do everything you asked of it. However, it won't show you any information because you didn't tell it to. Okay, diversion over.

Understanding the script (part 2)

Add some print statements to your script so you can see what each instance of the tr operator is doing to $text.

```
1.   #!/usr/bin/perl
2.   # transliterate.pl
3.   use strict; use warnings;
4.
5.   my $text = "abc123abc123";
6.
7.   $text =~ tr/a/d/;       # changes any occurrence of 'a' to 'd'
8.   print "$text\n";
9.   $text =~ tr/bc/xy/;     # 'b' becomes 'x', and 'c' becomes 'y'
10.  print "$text\n";
11.  $text =~ tr/123/321/;   # 1 becomes 3, 2 stays as 2, 3 becomes 1
12.  print "$text\n";
13.  $text =~ tr/dxy/DXY/;   # capitalize the letters d, x, and y
14.  print "$text\n";
```

You should hopefully see that the tr operator is just converting one series of characters into another. Line 7 shows how you can just change all occurrences of a single character for another. In this sense, the tr operator behaves very much like the substitution operator (though no global option is needed). However, as lines 9–14 show, it becomes more powerful when you specify a series of different characters that you want to change.

It's important to understand that line 9 is saying 'Find any occurrence of b and change to x, and also find any occurrence of c and change to y.' It is *not* trying to find the string 'bc'.

Example 4.12.2 In addition to changing a string, the transliteration operator can also be used to count how many changes are made. This can be extremely useful when working with DNA sequences.

```perl
1.    #!/usr/bin/perl
2.    # nucleotide_count.pl
3.    use strict; use warnings;
4.
5.    my $sequence = "AACTAGCGGAATTCCGACCGT";
6.    my $g_count = ($sequence =~ tr/G/G/); # parentheses are optional
7.    print "The letter G occurs $g_count times in $sequence\n";
```

Line 6 may appear confusing. The first thing to note is that this line contains an assignment, so we know that everything on the right of the = sign has to happen first. It looks like the transliteration operator is changing the letter G to itself, and in fact this is exactly what is happening. This tells us that the $sequence variable will remain unchanged (which is what we want).

Taken as a whole, line 6 suggests that the tr operator is producing a value which is assigned to $g_count. The output of the script should indeed tell us that $g_count is equal to the number of G characters in $sequence. This reveals a useful feature of the tr operator, it keeps track of how many changes it makes. Normally it does nothing with this count, but if you ask Perl to assign the output of the transliteration to a variable, then it will store the count in that variable.

Alternative syntaxes of the binding operator

You will reach a point in your programming career when you want to use a binding operator to match and/or modify a pattern that includes forward-slashes. Imagine we have a web address stored in a scalar variable:

```perl
$text = "http://genomes2go.com";
```

Now let's imagine that we want to use the substitution operator to strip away the 'http://' part. Maybe you would first try this:

```perl
$text =~ s/http:////;
```

Hopefully you will already get the feeling that this might not work. How will Perl know that two of those forward-slashes are things that we want to delete and two of them are part of the substitution operator itself? To cut a long story short, Perl won't be able to process this. There are two solutions to this problem. The first involves the process of *escaping* a character. We first saw this in Chapter 4.10, when we were escaping certain printed characters. However, you can also escape characters in the matching operators:

```
$text =~ s/http:\/\////;
```

This tells Perl that the first two forward-slashes (which are escaped) are part of the pattern. This will work, but it can still be confusing to look at, particularly when the strings that you are matching include forward- *and* backward-slashes. Thankfully, Perl allows you to choose alternatives instead of the two forward-slashes.[40] The following examples all do exactly the same thing but each uses a different set of characters to act as the delimiters for the substitution operator:

```
$text =~ s!http://!!;
$text =~ s@http://@@;
$text =~ s#http://##;
```

The last example is another, relatively rare, occasion where a hash character doesn't denote the start of a comment.

If you use an alternative character that naturally comes in a pair (e.g., curly or square brackets, or parentheses), then the syntax is a little different and you have to make sure you use the opening *and* closing character to mark the matching part of the pattern, and then the same pair of characters to surround the replacement part of the pattern. This can actually make things a lot easier to understand, as this example with the tr operator illustrates:

```
$text =~ tr [abcdefgh]
            [hgfedcba];
```

In this case we use two pairs of square brackets to denote the two ranges of characters, rather than just three slashes. We have also wrapped one line of code across two lines of the editor so we can line up the matching and replacement patterns. The above example is functionally equivalent to the following two examples:

```
$text =~ tr [abcdefgh] [hgfedcba];   # one line, with spaces
$text =~ tr[abcdefgh][hgfedcba];     # one line, no spaces
```

Problem 4.12.1 Write a script that takes the name of a web address (stored in a variable) and then:

(1) Removes the http:// part of the name.
(2) Stops the script if the remaining name is longer than 25 characters, otherwise ....
(3) Converts the name to upper-case characters.
(4) Counts the number of As, Bs, and Cs that are in the name.
(5) Prints those counts.

---

[40] If you use alternative delimiters when using the matching operator, the 'm' character before the first delimiter is no longer optional. This is another reason why it is a good habit to always include the 'm'.

Because sometimes, scalars get lonely

Until now we have only worked with single variables or values, but hopefully you have seen that this still allows you to accomplish a lot of tasks. However, we often want to work with lists of things. In Perl, if you have multiple scalar values in parentheses separated by commas, this is known as *list context*. Actually, it's still list context even if you have one or even zero scalar values in parentheses, because a list can have one or even zero elements. A common use of lists is when we want to assign a set of values to a set of variables.

Example 4.13.1 Create the following short program and run it.

```
1.   #!/usr/bin/perl
2.   # list.pl
3.   use strict; use warnings;
4.
5.   my ($x, $y, $z) = (1, 2, 3);
6.   print "\$x=$x \$y=$y \$z=$z\n";
7.
8.   my ($a, $b, $c) = ("uno", $y, 'tres');
9.   print "\$a=$a \$b=$b \$c=$c\n";
```

Understanding the script
The code in line 5 takes a list of three values (1, 2, 3) and assigns them to a list of three variables ($x, $y, $z). Note how we use the backslash character in the print statement on line 6 in order to print the dollar sign. Without using lists, we would have to use three separate lines of code in order to declare and initialize each variable with a value.[41]

Line 8 is just meant to illustrate that lists can contain a mixture of strings and variables. After this list assignment, the new variable $b will contain whatever the current value of $y is.

Balancing list items
If you assign a list of values to a list of variables, it might make sense that you should have the same number of values as there are variables. Maybe we could call this a *balanced* list. However, Perl doesn't really care if one side of the list assignments contains fewer, or more items than the other side. Perl may not care, but you should!

---

[41] If you need to assign the same value to multiple variables then you could achieve this in two lines:

```
my ($x, $y, $z);
$x=$y=$z=0;
```

Example 4.13.2 Let's look at what happens if we have an unbalanced list assignment. Once again, try to predict what the output of the script will be *before* you run it.

```
1.    #!/usr/bin/perl
2.    # unbalanced_lists.pl
3.    use strict; use warnings;
4.
5.    my ($a) = (1, 2);
6.    my ($b, $c) = (3);
7.    print "A = $a, B = $b, C = $c\n";
```

Understanding the script
On line 5 we try assigning a list of two values to a (one-element) list.[42] As there is nowhere for the second value to go, it is effectively discarded by Perl. If you try assigning a set of values to a list, once the list is filled all subsequent values will be ignored.

On line 6, we try to do the reverse and assign just one value to a list containing two variables. In this case $c receives no value and therefore remains undefined. When we run the script you should see the dreaded Use of uninitialized variable error message.

What else can be assigned to a list?
When you make a list assignment, you will most commonly assign a list of values to a list of variables. Those values will typically be strings, but you can actually make a list from functions that return values from a set of strings. This will sometimes save you a few lines of code, but may not always be the best thing to do. As always, an example will help illustrate the point.

Example 4.13.3 Imagine we want to determine the lengths of a number of strings. Ultimately, we want to store the lengths of the strings in some new variables.

```
1.    #!/usr/bin/perl
2.    # crime_and_punishment.pl
3.    use strict; use warnings;
4.
5.    my ($a, $b) = ('crime', 'punishment');
6.    my ($length_a, $length_b) = (length($a), length($b));
7.    print "Lengths are: $length_a, $length_b\n";
```

[42] One-element lists can often confuse people as the lack of any commas can make you think that this is just a single scalar variable. Any time you see a variable surrounded by parentheses, you are looking at a list.

Understanding the script

On line 5 we have a standard list assignment with $a and $b being assigned two strings.

We then have a second list assignment on line 6 which uses the length() function to get the lengths from $a and $b and then assign these lengths to a new list. Note that the right-hand side of the assignment is still a list; it has a pair of parentheses containing comma-separated items. All that is different is that the items in the list are not values, they are functions that return a set of values.

By making line 6 use a list assignment we save ourselves one line of code – we would otherwise do this with two lines of code:

```
6.    my $length_a = length($a);
7.    my $length_b = length($b);
```

This is an example where you could argue that the effort to save one line of code is not really worthwhile. The end result is a single line of code that requires a little more inspection to work out what is going on. Remember that short and concise code is good, but clean and understandable code is better.

Swapping values

Overall we mostly use lists in Perl to assign values. A much more powerful way to work with lists of things will be addressed in the next section. One nice trick that you can do with lists is to swap two values. You can do this because assignments in lists occur simultaneously:

```
($x, $y) = ($y, $x);
```

Problem 4.13.1 Exchange the value of $x and $y without using list context. This is one of those problems that appears difficult at first, but once you see the solution, it will seem so obvious that you can't imagine how you didn't think of it immediately. Before you look at the solution, imagine how you would do it in real life. For example, given a glass of water and a glass of milk, how would you switch the liquids?

A ray of light at the end of the list

Lists are useful for declaring and assigning multiple variables at once, but they are transient. If we want to store the details of a list then we have to capture all the values into separate variables. Ideally, there should be a way of referring to all of a list in one go, and there should be a way to access individual items in a list. In Perl (and in most other programming languages) we do this by using something called *arrays*.

You can think of an array as a list that has been given a name. We refer to the contents of an array as a set of *elements*. Each array element is a scalar variable that we have seen so far, e.g., a number, letter, word, sentence, etc. In Perl, as in most programming languages, an array is indexed by integers *beginning with zero*. The first element of an array is therefore the zero-th element. This might confuse you but that's just the way it is. Arrays in Perl are named using the '@' character. Let's imagine we have an array called @cards that contains five playing cards; we can imagine that each card in the array would be stored as a text string such as '7D' for 'seven of diamonds.'

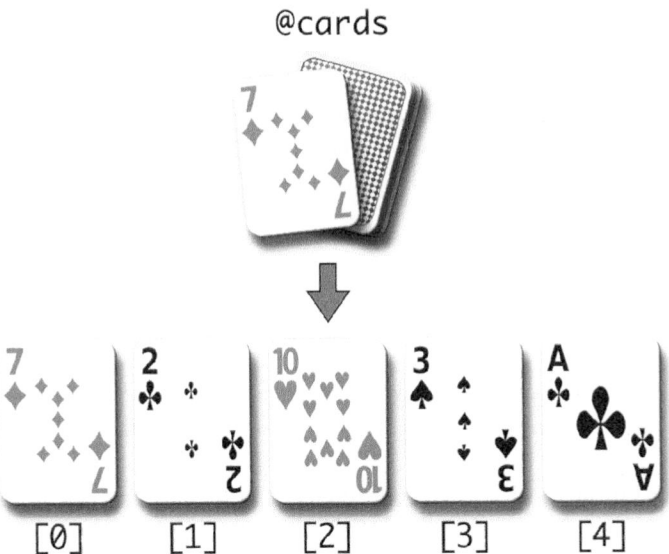

@cards

[0]   [1]   [2]   [3]   [4]

If we wanted to see what the individual elements of the @cards array were, we could access them at array positions 0 through 4. It's important to note that arrays always have a *start* (the so-called zero-th position), an *end* (in this case, position 4), and a *length* (in this case 5). Arrays may contain just one element, in which case the start and the end would be the same. Arrays may also contain no elements whatsoever. Arrays are containers that hold scalars. They can be empty, or they can contain millions of elements.

Example 4.14.1 Our first example with arrays will just show how to create an array and then print the contents.

```
1.    #!/usr/bin/perl
2.    # arrays.pl
3.    use strict; use warnings;
4.
5.    my @cards = ('7D', '2S', '10H', '3S', 'AC');
6.
7.    print "1st card in array is: $cards[0]\n";
8.    print "2nd card in array is: $cards[1]\n";
9.    print "5th card in array is: $cards[4]\n";
```

Understanding the script

Line 5 creates an array called @cards and assigns it a list of five strings. Because we are including use strict, we also have to declare arrays with my.

Lines 7–9 show how to access individual elements of an array. You specify a position in the array by putting an integer value between square brackets. This integer value is known as the 'array index.' Note the use of the dollar sign. More on this in a moment.

Lines 7–9 also shows that when you put a single array element within double quotes, Perl will interpolate the array element to show you the value. This is the same behavior as when we put a scalar variable between double quotes.

Scalars vs. arrays

We have previously been working with scalar variables, which are named using the $ sign. However, you will have noticed in the above example that we also use the $ sign to name individual elements of an array. This might appear confusing, but there is a consistent logic behind this.

An array can consist of many elements, but each element of the array is a scalar variable, so we write $cards[0] never @cards[0]. When you are working with a single variable in Perl, we will always use the $ sign (think $ = single). Therefore, there is no such thing as @cards[0] in Perl.[43] Writing @cards[0] is a common error of new programmers and if you ever try doing this Perl will produce a suitable warning:

Scalar value @cards[0] better written as $cards[0]

---

[43] Though this will change in Perl version 6, when such syntax will be valid.

Example 4.14.2 Let's look at some more ways we can assign values to an array.

```perl
1.   #!/usr/bin/perl
2.   # more_arrays.pl
3.   use strict; use warnings;
4.
5.   my @dna = ('A', 'T', 'C');
6.   print "DNA is $dna[0] $dna[1] $dna[2] $dna[3]\n";
7.
8.   my $nt = "G";
9.   $dna[3] = $nt;
10.  print "DNA is $dna[0] $dna[1] $dna[2] $dna[3]\n";
11.
12.  ($dna[0], $dna[3]) = ($dna[3], $dna[0]); # swap two values
13.  print "DNA is $dna[0] $dna[1] $dna[2] $dna[3]\n";
14.
15.  print "DNA is @dna\n";
```

Understanding the script

Line 5 creates an array (@dna) which initially consists of three nucleotides. Line 6 then tries printing the first four elements of this array. However, $dna[3] does not exist so Perl will print a Use of uninitialized value error.

Line 8 creates a new variable ($nt) and assigns it a value. On line 9 we then assign $nt as the fourth element of @dna. This illustrates that you can modify individual elements of an array, and that array elements can be scalar variables as well as strings.

Line 12 uses list context to swap the contents of two array elements (see the previous section to learn about list context).

Finally, line 15 shows that if you include an array name between double quotes, then the entire array interpolates. By default, Perl will add spaces between each element in the printed output. The output from lines 13 and 15 should be identical.

Example 4.14.3 A quick example to show you how you can save time when creating an array containing lots of strings.

```perl
1.   #!/usr/bin/perl
2.   # quoting.pl
3.   use strict; use warnings;
4.
5.   my @protein_A = ('M', 'A', 'R', 'W', 'P', 'C', 'S', 'E', 'R');
6.   my @protein_B = qw(M A R W P C S E R);
7.   print "@protein_A\n@protein_B\n";
```

Understanding the script

When you initialize an array and want to add lots of strings, it can be tedious to add all of those single quote characters around each element. Luckily, Perl has a 'quote words' tool to help you out. Just use the qw() function and write a list of strings, separated by spaces. Obviously, this is only useful if your array is going to include strings that are single words, letters, numbers, etc.

How long is ~~a piece of string~~ an array?

It is very useful to know how long an array is (i.e., how many elements it contains). You might want to use the length() function, but that returns the length of a string. Finding the length of an array is simple though – you just evaluate it in scalar context. This is most commonly done by simply assigning an array to a scalar variable.

Example 4.14.4 This is another script where you really should try to predict what is going to happen before running it.

```
1.   #!/usr/bin/perl
2.   # array_length.pl
3.   use strict; use warnings;
4.
5.   my $car = "Honda";
6.   print "Length of \$car is ", length($car), "\n";
7.
8.   my @cars = ("Toyota", "Ford", "Ferrari");
9.   print "Length of \@cars is ", length(@cars), "\n";
10.
11.  my $length = @cars;
12.  print "Length of \@cars is $length\n";
```

Understanding the script

Lines 5 and 6 are hopefully familiar to you. We take a string and count how many characters are in that string using the length() function.

On line 8 we introduce an array called @cars. Note that $car and @cars are two completely different things. Perl even allows you to give the exact same name to scalar variables and arrays (and other things too).[44]

On line 9 we try to calculate the length of the @cars array using the length() function, but we get a length of 1 rather than 3. The length() function is *not* meant to be used for arrays. We'll explain why it returns 1 in just a moment.

---

[44] Sometimes this is a good thing, sometimes this is a bad thing. It partly depends on how much it confuses you!

Lines 11 and 12 give us what we were looking for and $length contains the correct length of the array.[45] This introduces a very useful concept in Perl. If you assign a list or array to a scalar variable, then the scalar variable becomes the *length of the list*. This is very useful and you will use it a lot in your Perl code. You can think of this in another way: In any place where Perl is expecting a numerical value, you can usually specify an array instead.

Returning to line 9, what is happening here is that the length() function is expecting a single scalar value, and so in this context @cars is evaluated as the length of the array rather than the contents of the array. This means that the length() function is calculating the length of the string '3' (the number of elements in @cars), which is why it prints '1'.

Scalar context vs. list context

It is a common mistake to confuse the last two lines of code in the following example. Can you work out what the difference is?

```perl
my @cars = qw(Honda Ford BMW);
my $length;
$length    = @cars;
($length) = @cars;
```

The third line of code calculates the *length* of the cars array and assigns that to the $length variable, which gets the value 3. But when we add parentheses around $length, we are now making a list, and the last line of code is therefore a *list assignment*. It doesn't look much like a list because there is only one thing in it, but it is still a list. So the last line of code could be read as 'take the @cars array and assign all of its elements to a new list which will contain just one element, $length.' In this case, $length would contain 'Honda'.

In the language of Perl, we would say that in the third line above @cars is being evaluated in *scalar context*, and in the fourth line it's being evaluated in *list context*. This can become very important because certain functions and operators in Perl will actually behave differently depending on what context is being used. Perl even provides a scalar() function so you can force the evaluation of an array in scalar context. This provides another way you can determine the length of an array:

```perl
$number_of_elements = scalar(@array);
```

Array indexes

So far we have only been using integer values as array indexes. Let's look at other ways we can access specific positions within an array.

---

[45] Remember that $length is just a variable. Some people confuse $length with the length() function. We could also have called the variable $length_of_cars_array.

Example 4.14.5 This script will make a short array that contains the names of some different types of DNA and RNA molecule. It will then try printing out various elements from that array and indicate which array position it is trying to print. If you can guess what all of these `print()` statements will print, then you are well on your way toward becoming a Perl guru!

```perl
1.    #!/usr/bin/perl
2.    # array_indexes.pl
3.    use strict; use warnings;
4.
5.    my @molecules = qw(DNA cDNA mRNA tRNA rRNA);
6.    my $number = 1;
7.
8.    print "@molecules\n";
9.
10.   print "Position 0 = $molecules[0]\n";
11.   print "Position $number = $molecules[$number]\n";
12.   print "Position 2.01 = $molecules[2.01]\n";
13.   print "Position 2.99 = $molecules[2.99]\n";
14.   print "Position -1 = $molecules[-1]\n";
```

Understanding the script

Line 10 is using the standard array index notation that we have seen so far. Line 11 is just meant to illustrate that you can use a variable for the index position, as long as that variable contains a number.

Lines 12 and 13 show what happens when you use floating point numbers instead of integers. Both of these values are rounded *down*, meaning that both of these index positions will be equal to 2 (which is the *third* element of the array, remember).

Using a negative number for the array index position (line 14) has the effect of counting from the tail end of the array. Therefore, position −1 is always the last element of an array. This means that if an array only has one element, you can access it at position 0 *and* position −1.

Problem 4.14.1 This problem will just check that you have understood how to create lists and arrays, and how to get data from them. First take a list of three things, and assign them to three variables. Then copy those three variables into an array, and finally reverse their positions. The output from the script should consist of the three items before and after they are reversed.

Array manipulation

Don't try to push() me around

Perl arrays are *dynamic*. That is, they grow and shrink automatically as you add/remove data to/from them. It is very common to modify the contents of arrays, and it is also very common to start off with an array full of things, and then remove one thing at a time. Most of the time we add or remove things to either end of an array; Perl has four dedicated functions to do this:

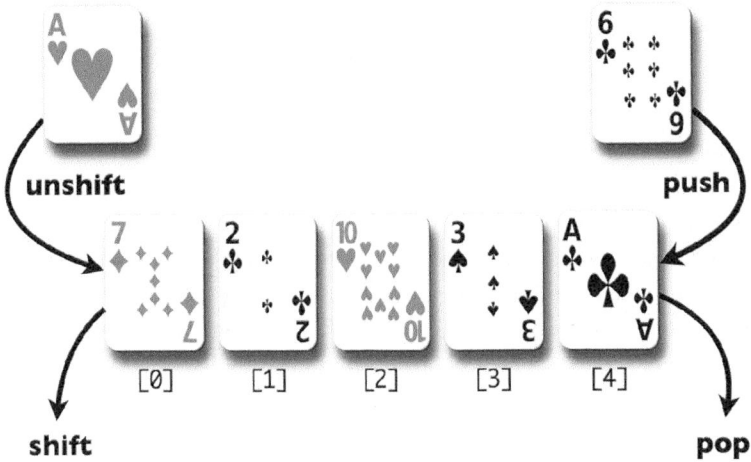

Example 4.15.1 To examine this dynamic behavior, we will first learn to use the push() function to add some new data on to the array. The push() function is used to add things to the *end* of an array. For now, we will consider the 'end' of an array to be the element with the highest array index position.[46]

```perl
1.  #!/usr/bin/perl
2.  # push.pl
3.  use strict; use warnings;
4.
5.  my @species = ('Homo sapiens', 'Felis catus', 'Bos taurus');
6.
7.  push @species, "Mus musculus"; # add one new element to end of array
8.  print "Fourth species is now $species[3]\n";
9.
10. my $animal = "Pan troglodytes";
11. my $plant = "Arabidopsis thaliana";
12. push @species, $animal, $plant, "Drosophila melanogaster";
13.
14. print "@species\n";
```

---

[46] For those of you who visualize arrays in a horizontal orientation, the 'ends' will be the left- and right-most edges. However, because there are probably some people who visualize arrays in a vertical fashion, we will try to avoid using left/right/up/down notations.

Understanding the script

After defining a three-element array (line 5), we then use the push() function on line 7 to add one extra item to the @species array. Note that the syntax of push() is to first specify an array and then specify the thing to be added. You can include parentheses if that makes it clearer for you:

```
7.   push(@species, "Mus musculus");
```

The thing that you push onto an array can be a string, variable, list of things, or even another array. On line 12, we push two variables and one string onto the end of the @species array.

Common array functions

We just saw push() as a way of adding an element to one end of an array. Naturally, you might also want to *remove* elements from that end of the list. You can do this with the pop() function. And of course, if you can add and remove elements to the end of a list, you might also expect to be able to do the same to the start of the list. Perl allows you to do this with two functions called unshift() and shift().[47] Together, these four functions allow you to add or remove items from the start or end of a list or array. However, if you want to do even more complex array manipulations, then you can use the splice() function. This function is the hardest one to understand, but also the most powerful because it allows you to add, remove, or substitute array elements at *any* position in an array, not just at the ends. The syntax of these operators are summarized in the following table.

Function	Meaning
push(@array, "some value")	Add a value to the end of the list
$popped_value = pop(@array)	Remove a value from the end of the list
$shifted_value = shift (@array)	Remove a value from the front of the list
unshift(@array, "some value")	Add a value to the front of the list
splice(...)	Everything above and more!

Example 4.15.2 Let's see if we can make a simple script which demonstrates the four ways in which you can manipulate the ends of an array. We can imagine that we

---

[47] Admittedly these names are not the most consistent choices of verbs that could have been used. If 'unshift' is the opposite of 'shift,' then you might expect 'unpush' to be the opposite of 'push' instead of 'pop.'

have an array which contains a shopping list, with items in the array listed in order of importance. This script shows how we can make changes to either end of that list.

```perl
1.   #!/usr/bin/perl
2.   # the_ends_of_the_array_are_nigh.pl
3.   use strict; use warnings;
4.
5.   my @list = qw(bread milk cheese apples);
6.   print "Starting list is: @list\n";
7.
8.   # don't need bread after all
9.   my $removed_item = shift(@list);
10.  print "Have just removed $removed_item from \@list\n";
11.
12.  # want oranges instead of apples
13.  pop(@list);
14.  push(@list, "oranges");
15.
16.  # but most importantly, get some chocolate
17.  unshift(@list, "chocolate");
18.
19.  print "Final list is now: @list\n";
```

Understanding the script

Line 5 uses the qw() function to easily add elements to the @list array without having to quote each character.

Line 9 then uses the shift() function to remove an item from the start of the list. Rather than just discarding this array element, we can optionally assign it to a new variable, which is what we do in this example.

Lines 13 and 14 then use the pop() function to remove the last element of @list and then add a new item in its place using the push() function. Note that we don't do anything with the popped value.

Line 17 uses the unshift() function to add an item to the front of the array.

Copying and erasing arrays

Two more aspects of arrays that we haven't yet covered concern making a copy of an array and deleting the contents of an array. Both operations are very simple.

Example 4.15.3 This is a simple example to show how you can combine, copy, and empty arrays.

```
1.    #!/usr/bin/perl
2.    # alphabet.pl
3.    use strict; use warnings;
4.
5.    my @a_to_m = qw(a b c d e f g h i j k l m);
6.    my @n_to_z = qw(n o p q r s t u v w x y z);
7.
8.    my @alphabet = (@a_to_m, @n_to_z);
9.    my @alphabet_copy = @alphabet;
10.
11.   @alphabet = ();
12.
13.   print "Original: @alphabet\n";
14.   print "Copy: @alphabet_copy\n";
```

Understanding the script

We first define two arrays which each hold half of the alphabet (lines 5 and 6). Next, we make a list assignment to copy both arrays to one new array (line 8). Understand that Perl will first expand all arrays on the right-hand side of the equals sign. Both arrays will become one long list, the contents of which then get assigned to @alphabet.

Line 9 makes a copy of the @alphabet array, simply by assigning it to a new array name.

On line 11 we empty the @alphabet array by assigning it an empty list (as represented by the open and close parentheses on their own). Note that this doesn't delete the array. Perl still knows that there is an array called @alphabet, it just no longer contains anything.

Finally, we print out both @alphabet and @alphabet_copy, but only the latter array now contains anything.

Making a string by joining together elements in an array

We have seen that if you try printing an array between double quotes, then Perl interpolates the array and prints each element separated by a space. What if we want something other than spaces? An easy way to do this in Perl is with the join() function; this allows you to create a string from an array and put whatever you want between the elements of the array. The syntax of the join() function is:

```
join('separator', @array)
```

The separator can be a single character or a longer string, or a variable that contains a string. The named array will be the thing you join together to form a string. You could also write a list of items rather than an array name, and you could even include a list of arrays:

```
join('separator', @array_A, @array_B, @array_C)
```

Example 4.15.4 Let's say we want to create a CSV (comma-separated values) formatted string from an array of gene names from the nematode *Caenorhabditis elegans*.

```
1.    #!/usr/bin/perl
2.    # array_to_string.pl
3.    use strict; use warnings;
4.
5.    my @gene_names = qw(unc-10 cyc-1 act-1 let-7 dyf-2);
6.
7.    my $joined_string = join(",", @gene_names);
8.    print "$joined_string\n";
```

Understanding the script
Line 7 uses the join() function to combine all of the elements in the @gene_names array. We specify that each element of the array should be joined with a comma character.

Note that the outcome of the join() function is assigned to a string. If we didn't need to store the string in a new variable, then we could have just used the following:

```
5.    print join(", ", @gene_names), "\n";
```

Also note that the join() function does not affect the contents of the array that is being joined.

Making an array by splitting a string at delimiters
The opposite behavior of join() is provided by the split() function. This divides a string into an array. But we have to tell it what character or pattern to use in order to split the array.

Example 4.15.5 A simple example of splitting a sentence into an array of separate words.

```
1.   #!/usr/bin/perl
2.   # string_to_array.pl
3.   use strict; use warnings;
4.
5.   my $sentence = "This is a sentence containing some words";
6.   my @words = split(" ", $sentence);
7.   print "First word = '$words[0]', last word = '$words[-1]'\n";
```

Understanding the script

The split() function on line 6 takes a string ($sentence) and splits it into an array using a space character as the delimiter.[48] Note that the space characters are not kept in the resulting array. Also note that the act of splitting a scalar variable does not destroy or modify the variable.

On line 7 we access the last array position of the @words array by using the −1 index position.

If we want to convert a string into an array and split the string at *every* possible position, we can use an empty string ("") in the split() function. More commonly, you will use split() to take lines from a comma- or tab-separated value file and split every item into an array; e.g., if we have read one line from a CSV file and stored it in a variable $line, then we can do the following:

```
my @fields = split(",", $line);
```

The splice function

The splice() function is very powerful and can be used to remove and/or add elements at *any* position within an array. It is quite difficult function to understand though, because like many Perl commands it can be used in different ways; the functionality depends on which of the optional arguments you provide. Each additional option adds another level of complexity to the function. Let's quickly look at the four ways we can use splice() on an array called @array. Imagine that in each of these examples, @array is reverted back to its starting condition after each use of the splice() function.

```
my @array = qw(a b c d e f g h);
splice(@array);           # gets rid of everything in @array
splice(@array, 3);        # gets rid of everything from position 3 onwards
splice(@array, 3, 2);     # gets rid of two elements from position 3 onwards
splice(@array, 3, 2, "D", "E");   # replaces 'd' and 'e' with 'D' and 'E'
```

---

[48] Later on we will learn much more powerful ways to represent space characters which will help us get round the problem of what would happen if a sentence accidentally contained two spaces in a row.

The first example uses the minimum number of arguments and is rarely ever used as it just removes *everything* from the named array. The second example provides more flexibility by specifying which position in the array you want to start removing elements from. It will still remove all elements from that position onwards. The next example offers more control by limiting how many items are removed. The last example offers the most control and shows that you can specify a position in an array, and then remove a certain number of elements at that position *and* replace them with new elements from a specified list. The list in this example is two strings ('D' and 'E'), but could have also been an array.

Note that while you could use `splice()` to delete all of the elements from an array, it is usually easier to assign the array an empty list, which we can denote by using an empty pair of parentheses:

```
@array = ();
```

Note that this destroys the *contents* of the array, but not the array itself. You can still add new elements to the array without having to re-establish the array name by using the my keyword.

Problem 4.15.1 Create two arrays; the first array should have an odd number of elements, and the second can have any number (though don't get too carried away!). We'll let you use your imagination in deciding what these arrays are named and what they contain.

Use the `splice()` function to replace the middle element of array 1 with all of the elements of array 2. For added bonus points,[49] try to make your script work out what the middle position of array 1 is going to be – assume that you don't know how long the array is, only that it contains an odd number of elements.

---

[49] You did know that we award bonus points right?

The @ARGV array

There is no argument about how useful this is.

We have covered a lot of the basics of Perl, but so far all of our scripts have only dealt with variables that have been defined *within* the script. From a programming perspective this is generally considered to be a bad thing. Technically, we say that in these scripts all of the variables are *hard coded*. This means that every time you run the script you will get the same result,[50] and if you want a different result then you have to edit the script. This is not desirable, and in some cases (e.g., you don't have permission to edit the script), it will not even be possible.

Most of the time when we run scripts we ideally want to specify some *external* data that the script should use. For example, let's assume we want a script to multiply two numbers together. It would be great if we could just specify some *command-line arguments* when we run the program, and let those arguments become the numbers to be multiplied:

```
$ multiply.pl 6 9
```

This is actually very easy to do in Perl because all of the items you specify after the script name get placed into a special array.[51] And that array is called ... drum roll please ... the @ARGV array. It's called @ARGV because it's a vector (array) that contains command-line arguments. We've had to wait until we introduced arrays before we could mention the @ARGV array, but expect to see @ARGV used a lot more from this point onwards. Let's see some examples of how you can use it in your scripts.

Example 4.16.1 This script makes a simple calculator that uses the exponentiation operator to calculate X to the power of Y. You must specify both numbers as command-line arguments when you run the script.

```
1.   #!/usr/bin/perl
2.   # pow.pl
3.   use strict; use warnings;
4.
5.   print $ARGV[0] ** $ARGV[1], "\n";
```

Understanding the script
Essentially, this is just a one-line script. If you provide two numbers when you run the program, then both of these will automatically be available via the @ARGV array. Just like regular arrays, we can access individual elements of @ARGV by using the array subscript notation. Line 5 uses the exponentiation operator (**) with the two values in @ARGV and prints the result. If you didn't specify two numbers (separated by one or more spaces) after the script name, you will see a 'Use of uninitialized value' error.

---

[50] Unless of course, your script uses the rand() function, which is exactly why we have used it so much in the earlier scripts.

[51] There are many special variables in Perl and these are technically called 'predefined names.' We will see some more of these later in the book.

Renaming elements stored in @ARGV

A bad habit to get into is to keep on using the @ARGV name throughout your script. In a short script, you may easily remember that $ARGV[1] corresponds to the year you were born, and that $ARGV[4] contains your mother's maiden name. However, as your scripts get longer it will become harder to keep track of which element of @ARGV contains which piece of data. Therefore you should adopt the following 'good working practice':

> *Always assign the contents of @ARGV to another array*
> *or set of variables.*

In many cases, this should be the first thing your script does. @ARGV is not a meaningful name for your data,[52] so if $ARGV[1] is to contain your favorite flavor of milkshake, then $flavor is a better choice of name. Sometimes you will want to reassign @ARGV to a number of different variables and sometimes you will just want to assign all of @ARGV to a single array. This will largely depend on whether your command-line arguments represent different types of thing (e.g., color, height, name), or whether they are all of the same type (e.g., a list of species names).

Example 4.16.2 In this example we will use @ARGV to capture someone's name and year of birth in order to determine whether they were born in a leap year. After you type the script, run it as follows:

```
$ leapyear.pl "Andy Warhol" 1928
```

```
1.   #!/usr/bin/perl
2.   # leapyear.pl
3.   use strict; use warnings;
4.
5.   my ($name, $year) = @ARGV;
6.
7.   print "Hello $name. You were ";
8.
9.   if ((($year % 4 == 0) and ($year % 100 != 0)) or ($year % 400 == 0)){
10.      print "born in a leap year.\n";
11.  }
12.  else{
13.      print "not born in a leap year.\n";
14.  }
```

---

[52] No doubt, there is someone in the world who works on something that uses ARGV as an abbreviation for their data (the Average Ratio of Goats to Vets?). If you are that person, then we apologize for any confusion you experience when using @ARGV and we grant you, and only you, special dispensation to continue using @ARGV as an array name throughout your script.

Understanding the script

Line 5 takes the two elements specified on the command line and assigns these to a list containing $name and $year.

Line 7 starts a print statement, but notice that there is no newline character included; this statement will be concluded when we know whether the year was a leap year or not.

The if statement on line 9 then does the math (with the modulo operator) to calculate whether $year is a leap year. This logic can be stated 'If the year is divisible by 4 *and* the year is *not* divisible by 100, *or* the year is divisible by 400, then it is a leap year.'

Checking what's in @ARGV

In the previous example, we used 'Andy Warhol' as the name to be captured by $name. It was important to place this within quotes so Perl recognized it as a single argument. What would happen, though, if we mistakenly run the script like so:

```
$ leapyear.pl Andy Warhol 1928
```

In this case @ARGV gets three items, which is not what we want. Perl assumes that command-line arguments will be separated by one or more space characters, *unless* they are enclosed within quotes (single or double).[53]

It becomes very important to get into the habit of checking that @ARGV contains what you expect it to contain. The first thing to check is that there is the right number of things in @ARGV; then you can usually also check whether each item fits certain criteria. If you are capturing someone's age, then this should be a number between 0 and ~115. As the next example shows, it is very common to use the die() function when checking the contents of @ARGV.

Example 4.16.3 This script takes a pair of latitude and longitude coordinates and then attempts to determine whether those coordinates lie roughly within the state of Kansas, USA.

```
1.   #!/usr/bin/perl
2.   # not_in_kansas_anymore.pl
3.   use strict; use warnings;
4.
5.   my ($latitude, $longitude) = @ARGV;
6.   die "Need latitude and longitude coordinates\n" if (@ARGV != 2);
7.
8.   # check latitude and longitude
9.   if ($latitude < -90 or $latitude > 90){
```

[53] Strictly speaking this is an issue with Unix rather than with Perl, and the same logic applies if you want to run any Unix command. Unix will use space characters as the delimiter between command-line arguments, *unless* they are placed within quote characters.

```
10.       die "Latitude should be between -90 and +90\n";
11.  }
12.  if ($longitude < -180 or $longitude > 180){
13.       die "Longitude should be between -180 and +180\n";
14.  }
15.
16.  # are we in Kansas?
17.  if (($latitude > 36 and $latitude < 41) and
18.       ($longitude > -103 and $longitude < -94)){
19.       print "We are (probably) in Kansas\n";
20.  } else {
21.       print "We are not in Kansas anymore\n";
22.  }
```

Understanding the script

On line 5 we capture two command-line arguments and place them in two variables that we are going to assume contain a latitude and longitude coordinate.

For this script to work we need two, and only two, arguments. We test whether there are two arguments by checking the length of @ARGV on line 6. This is another example of evaluating an array in scalar context – we ask whether an array (@ARGV) is not equal to 2. This only makes sense if we are treating the array as a number, and so Perl does the magic for us and calculates the size of the array. We stop the script with a die() function *if* the length of the array is not equal to two.

Any time you write a script that needs a certain number of arguments, you should always check that you have the right number. There is no point proceeding if an argument might be missing.

Lines 9–14 then check whether the values are in a sensible range to be considered valid latitude and longitude coordinates. If they are not, then the script issues a die statement which lets the user know what range of values are expected.

If we reach line 17 then we know we must have two values which are suitable as a pair of latitude/longitude coordinates. We then have a complex if statement which tests whether those coordinates lie within a particular rectangle of land.[54] Note that we split the if statement over two lines, and use whitespace to make it easier to read.

Problem 4.16.1 Write a script that just counts how many things you specify on the command line. The script should insist that you specify at least three command-line arguments.

---

[54] Apologies to any Kansans who take offense at having their state described as being rectangular in shape. If you could just straighten out the Missouri river, it would help our example a lot.

This chapter is definitely the best place to learn about definition

Scalar variables can be declared but not defined. Consider the following code:

```
my $number;
print "$number\n";
```

In this example the variable $number is declared on one line and we then try to print its value, even though it doesn't contain anything. In this situation we say that $number is *undefined*. It remains undefined until we assign it a value, at which point it is *defined*. If you try printing an undefined value you will see an error such as:

```
Use of uninitialized value in concatenation (.) or string ...
```

Undefined values are not necessarily bad, but you generally want to know if a variable contains an undefined value (especially if you were expecting it to be defined).

How to tell whether something is defined

You will frequently write scripts that get their input data from text files which might contain millions of values. It will be impossible to manually check these files to ensure that every value is present. Sometimes an error in your input file (e.g., an accidentally added blank line) will cause your variables to not contain any value. To determine if a variable is defined, you can use the defined() function. This simply returns true if the argument is defined and false if the argument is undefined. If you want your program to quit if it finds an unexpected undefined variable, you can use a statement like this:[55]

```
die "error" if not defined $variable;
```

Parentheses are optional, and you can also use the other form of the not operator if you wish. This means that the following code is functionally equivalent to the above example:

```
die "error" if (!defined ($variable));
```

Making a variable undefined

If you want to intentionally make a scalar variable undefined,[56] you can use undef in either a function or assignment context:

```
my $var1 = "defined";
my $var2 = "defined";
undef($var1);        # as function
$var2 = undef;       # as assignment
print $var1;         # reports uninitialized value
print $var2;         # reports uninitialized value
```

---

[55] Your error message will hopefully be more informative than this.
[56] There are occasions when you will want to 'reset' the value of a variable.

You can also undef an array and it will destroy its contents. It does not, however, turn the array into an undefined value. Only scalars can have undefined values.

Undefined values are an essential debugging tool

The combination of use warnings and undefined values makes debugging much easier. Consider the following erroneous code, which is meant to simulate flipping a coin.

```perl
my $coin;
if     (rand() < 0.5)    {$coin = "heads"}
elsif (rand() >= 0.5)  {$coin = "tails"}
print  "$coin\n";
```

There is nothing wrong with the syntax of this code. The error is a logical one. The rand() function is called twice, and each time it gets a different value. If the first random number generated is 0.71 and the second number is 0.12, neither of the conditions will have been true. Therefore, sometimes $coin remains undefined. Being able to identify errors in logic is really important, and undefined values are incredibly useful for debugging. Not many programming languages have this feature, so rejoice that you are learning Perl.

Special cases for undefined values

In general, evaluating undefined variables causes a warning message to appear. There are two exceptions to this rule:

(1) Undefined values are logically false.
(2) Undefined values can be incremented.

Exception 1 means you can use undefined values in conditional statements. Exception 2 means you can treat undefined values as having a value of zero or an empty string. The following code demonstrates both these properties and does not produce any warnings:

Example 4.17.1

```perl
1.    #!/usr/bin/perl
2.    # undef_ok.pl
3.    use strict; use warnings;
4.
5.    my ($var1, $var2);
6.    if (not $var1 or $var2) {
7.        $var1++;
8.        $var2 .= "foo";
9.        print "$var1 $var2\n";
10.  }
```

Understanding the script

Line 5 declares two variables without defining their contents. If you tried to `print` these, Perl would report warning messages.

Line 6 shows that you can perform logical tests with undefined values. They always evaluate to false (which means that the `if` statement on this line will evaluate as true).

Line 7 increments `$var1`. Since this is a numeric context, zero is the initial value, and after the statement `$var1` has the value 1.

Line 8 concatenates 'foo' onto `$var2`. This is string context, so the initial value of `$var2` is an empty string.

Line 9 prints the variables which results in '1 foo'.

This section will be very useful … sort of

As in real life, lists are great, but sorted lists are even better. Imagine how tedious it would be to look through a telephone book if it wasn't sorted. Sorting lists and arrays imposes order on your data, and it can also make it easier to perform certain mathematical operations. For example, if you sort an array of numbers, then you know that you can access the minimum value at array index 0 and the maximum value at array index −1 (which corresponds to the last element in an array). Perl has an incredibly flexible sorting function, but it can seem a little complicated, so you may want to come back and read this part again later.

Example 4.18.1 Create the following program. We will be modifying it several times. Can you predict what the sorted output will look like? Do you think the numbers should be sorted before or after the strings?

```
1.   #!/usr/bin/perl
2.   use strict; use warnings;
3.
4.   my @data = qw(3.14 1000 21 red blue yellow);
5.   print "before: @data\n";
6.   @data = sort @data;
7.   print "after: @data\n";
```

Understanding the script
Line 4 creates an array that contains both numbers and strings.

Line 6 does the actual sorting using the appropriately named sort() function. Note that we assign the sorted output back to the same array, but we could have also assigned it to a new array (e.g., @sorted_array).

The output from the program should look like this:

```
before: 3.14 1000 21 red blue yellow
after: 1000 21 3.14 blue red yellow
```

This might not be what you were expecting. The three strings were sorted correctly, but the numbers appear to be out of order. This is because all of the elements were sorted alphabetically by their ASCII values (see Chapter 4.9). In ASCII, 1 is less then 2, which is less than 3. Alphabetically, 1000 is less than 21. If we want to sort numerically, then we have to change the default sorting behavior of Perl.

Sorting numbers vs. sorting strings
A sorting function must compare one value to another. In Perl, there are two operators that allow you to compare pairs of values:

<=> compares numbers;
cmp compares strings.

These operators return −1, 0, or +1, depending on whether the left-hand operand is less than, equal to, or greater than the right-hand operand.

```
"a" cmp "b";    # -1 because a is less than b alphabetically
"b" cmp "a";    # +1
"a" cmp "a";    # 0
1   <=>  2;     # -1 because 1 is less than 2 numerically
2   <=>  1;     # +1
1   <=>  1;     # 0
"a" <=> "b";    # 0 because strings are numerically zero
100 cmp  2 ;    # -1 because 1 is less than 2 alphabetically
```

To change the sorting behavior in Perl, you have to give the sort() function an alternate syntax, which can look a little strange. Perl's sort() function reserves two variables called $a and $b for sorting.[57] When we use the default sorting behavior we don't see these special variables, but the following two lines of code are equivalent:

```
@data = sort @data;              # basic alphanumeric sort
@data = sort {$a cmp $b} @data;  # same, but with all the details revealed
```

The comparison routine is the part that happens inside the curly brackets after the sort. Modify line 6 of your code from Example 4.18.1 to try out each of the following two sorting assignments in turn:

```
@data = sort {$a cmp $b} @data; # alphabetic
@data = sort {$a <=> $b} @data; # numeric
```

You should find that the second example, sorting numerically, produces warnings. The numbers are correctly sorted, but the strings are not:

```
before: 3.14 1000 21 red blue yellow
Argument "red" isn't numeric in sort at sort.pl line 6.
Argument "blue" isn't numeric in sort at sort.pl line 6.
Argument "yellow" isn't numeric in sort at sort.pl line 6.
after: red blue yellow 3.14 21 1000
```

This warning message is fairly self-explanatory; you should sort numerically only if you are sorting numbers, not strings. If you have a mixture of strings and numbers, you can do the following to sort them:

```
@data = sort {$a <=> $b or $a cmp $b} @data;
```

This uses the Boolean or operator to chain the two comparisons. If the numeric sort results in a zero, as would happen with a string, the alphabetic comparison is used instead. Unfortunately, there are still warnings because we are comparing strings as numbers. We can quiet these by temporarily turning off the warnings pragma. This

---

[57] For this reason, please don't name your variables $a or $b. Note that you also don't need to declare them with the my operator.

is done by placing 'no warnings;' before the sort and 'use warnings;' after the sort:

```
no warnings;
@data = sort {$a <=> $b or $a cmp $b} @data;
use warnings;
```

You can put almost anything you want between the curly braces in order to sort data in more imaginative ways. Let's try sorting by the length of each string (in the case of numbers, the length of the string representation of the number).

```
@data = sort {length($a) <=> length($b)} @data;
```

The output is as follows:

```
before: 3.14 1000 21 red blue yellow
after: 21 red 3.14 1000 blue yellow
```

Reverse-sorting
Sometimes you will want to perform a reverse-sort. This is easily done by just switching the position of $a and $b when using either of the comparison operators:

```
@data = sort {$b cmp $a} @data; # reverse alphabetic sort
@data = sort {$b <=> $a} @data; # reverse numeric sort
```

Alternatively, Perl also has a reverse() function which, unsurprisingly, can reverse the order of an array. This function also works on strings.

Example 4.18.2 This short script shows how you can use the reverse() function in combination with sort().

```
1.   #!/usr/bin/perl
2.   # reverse.pl
3.   use strict; use warnings;
4.
5.   my @array = qw(b c h d e g a f); # unsorted
6.   my $string = "reversed";
7.
8.   my @reversed_array = reverse(sort @array);
9.   my $reversed_string = reverse($string);
10.  print "$reversed_string: @reversed_array\n";
```

Understanding the script
After declaring an array and a string on lines 5 and 6, we use the reverse() function to take the sorted array and then assign the reversed output to a new array.

Line 9 shows how reverse can also be used on a string in exactly the same way.

Problem 4.18.1 Write a script where you input a series of numbers on the command line. The script should discard the highest and lowest values and print out what is remains. You might want to revisit the 'Array manipulation' chapter for this part. Your script should also check that you have at least three input numbers and act appropriately if there are not enough.

You might need to read this section over and over again

Loops are one of the most important constructs in programming. Once you have mastered loops, you can do some really useful programming. Loops allow us to do things like count from 1 to 100, cycle through each element in an array, or even process every line in an input file. In general, loops allow us to iterate through a list of numbers or a list of things and do one or more actions in each iteration of the loop. There are three main loops that are used in programming, the for loop, the foreach loop, and the while loop; this chapter will focus on the first of these three looping mechanisms.

The for loop

The for loop generally iterates over a series of integers, usually from zero to some other number. You can think of the integer as a 'loop counter' which keeps track of how many times you have been through the loop (just like a lap counter during a car race). The for loop has three important components that are specified each time we create a loop:

(1) Initialization – provide some starting value for the loop counter.
(2) Validation – provide a condition for when the loop should end.
(3) Update – specify how the loop counter should be changed in each loop cycle.

If we return to the car race analogy, we can imagine a car having to drive 10 laps around a circular track. At the start of the race the car has not completed any laps, so the loop counter would be initialized at zero. The race is clearly over when the counter reaches 10 and each lap of the track updates the counter by 1.

Example 4.19.1 Create and run the following program that produces a simple for loop which will just count from zero to nine.[58] This is a very short script because all our loop is going to do is print out the count for each iteration of the loop.

```
1.      #!/usr/bin/perl
2.      # loop.pl
3.      use strict; use warnings;
4.
5.      for (my $i = 0; $i < 10; $i++) {
6.          print "$i\n";
7.      }
```

---

[58] We could count from 1–10, but we are sticking with the programming concept that arrays start at position 0 rather than 1. This will become more useful later, when we start looping through arrays.

Understanding the script

On line 5 we set up the loop by specifying the three required components (initialization, validation, and update). These components are always placed inside parentheses, and are separated by semicolons. In this loop we first declare a new variable (my $i) to act as our loop counter. It is a convention in programming to use $i as a loop variable name because of the use of i as a counter in mathematical notation; e.g.:

$$\sum_{i=0}^{n} x_i$$

You could name your loop counter anything you wanted to, but we suggest that for now you use $i. Let's look at the three main components of our loop:

```
$i = 0      # initialization: start our loop with $i equal to zero
$i < 10     # validation: keeps the loop going as long as $i is less than 10
$i++        # update: $i is incremented by 1 during each loop iteration
```

The contents of a loop are placed in a block of code just like when we have an if statement. This means that the code has to be enclosed within a pair of curly braces and the code is typically indented in the same way.

In this loop there is only one line of code that does anything. This is the print statement on line 6. It is critical that you realize that this line of code is executed *ten times* in total. Each time Perl evaluates this line of code the value of $i will be different because of the update step of the loop.

Changing how the loop updates

The 'update' component of the loop should describe a way of modifying the value of the loop counter variable, otherwise the loop will never end. Consider the following 'infinite loop':

```
for (my $i = 0; $i < 10; 1) {
    print "$i\n";
}
```

In this code, $i starts at zero but is never modified by the update condition (which is just set to be a true value). This means the code enters the loop and is never able to leave because the validation condition ($i < 10) will never be met. If you try running this code, your terminal window will become full zero values and you will need to press control + c to interrupt the script (this sends the Unix 'interrupt' signal and is a useful way of stopping scripts that run out of control).

We often want to count upwards with intervals of 1. However, the for loop in Perl lets you count upwards or downwards.

Example 4.19.2 Let's make a loop to count down for a rocket launch and add suitable print statements depending on the value of the countdown.

```
1.   #!/usr/bin/perl
2.   # countdown.pl
3.   use strict; use warnings;
4.
5.   for (my $i = 10; $i >= 0; $i--) {
6.       print "$i) ";
7.       print "Ignition sequence started"    if ($i == 9);
8.       print "We have ignition"             if ($i == 3);
9.       print "LIFT OFF!"                     if ($i == 0);
10.      print "\n";
11.  }
```

Understanding the script

Line 5 reverses the behavior of the three loop conditions. We now initialize $i to 10 and count *down* using the decrement operator ($i--). We continue to stay in the loop as long as $i is greater than or equal to zero.

Inside the loop we have two print statements that always occur (lines 6 and 10) and three print statements that are conditional on the value of $i.

As well as counting up and down in increments of 1, we can change the interval to be anything we want:

Example 4.19.3 This script uses numbers specified on the command line to provide all the three values used by the loop (start, end, and increment).

```
1.   #!/usr/bin/perl
2.   # you_decide.pl
3.   use strict; use warnings;
4.
5.   my $usage = "usage: you_decide.pl <start> <end> <increment>\n";
6.   die $usage unless @ARGV == 3;
7.
8.   my ($start, $end, $increment) = @ARGV;
9.   die "Start value should be less than end\n" if ($start >= $end);
10.
11.  for (my $i = $start; $i <= $end; $i += $increment) {
12.       print "$i\n";
13.  }
```

Understanding the script

Lines 5 and 6 check that there are three values specified on the command line. Note that we store the usage message in its own variable.

It makes sense to provide at least one other check to ensure that the start value is less than the end value; this is done on line 9. You could also add a check to ensure that the increment value was in a suitable range (relative to the start and end values).

The for loop on line 11 uses all three of the user-specified variables to control the loop. If we wanted to streamline the code a little we could replace the above for loop with the following:

```perl
for (my $i = $start; $i <= $end; $i += $increment) {print "$i\n"}
```

In this example we can reduce the whole loop down to a single line of code (meaning we don't need a semicolon at the end of the block). If you make a for loop which only does one thing, it is sometimes cleaner to write the loop in this way.

Looping through arrays

One of the most common operations you will do as a programmer is to loop through the contents of an array. Let's do that now.

Example 4.19.4 To make it more interesting, this script will loop over two arrays simultaneously.

```perl
1.   #!/usr/bin/perl
2.   # loops.pl
3.   use strict; use warnings;
4.
5.   my @animals = qw(cat dog cow);
6.   my @sounds  = qw(meow woof moo);
7.
8.   for (my $i = 0; $i < @animals; $i++) {
9.       print "$i) $animals[$i] $sounds[$i]\n";
10.  }
```

Understanding the script

Lines 5 and 6 declare two different arrays, but both have three elements. As you can see, the elements in the @sounds array pair up with the elements in the @animals array.

The for loop on line 8 starts at 0, which is where all arrays start, and continues as long as the loop variable $i is less than the length of the array. Notice that we don't store the array length anywhere. Instead, we can get Perl to deduce the length by making it evaluate it in scalar context. It is very common to see this syntax, e.g., $variable < @array, used in the validation component of for loops.

The important point about the loop itself is that the loop counter ($i) is also used as the array index position for both arrays. You will see this behavior used a lot in for loops!

This script works only because both arrays are the same length and the information in both arrays is connected. The script would not be of any use if one array was longer than the other.

Problem 4.19.1 Write a program that takes a series of numbers specified on the command line and then:

(1) sorts them into ascending numerical order;
(2) loops through the sorted values and prints each value alongside a running total.

For each thing in this section, you should spend a while learning it

The foreach loop

The for loop is great for looping through lists when the specific (numeric) position in the list is important. But sometimes you just want to iterate through a list and you may not care what position you are at, or even how long the list is. In these situations you can use a foreach loop which allows you to iterate through the contents of an array without using a loop counter variable. The basic syntax of a foreach loop looks like this:

```
foreach my $thing (@array) {
    # code using $thing goes here
}
```

Rather than use a loop counter variable, the foreach loop uses a temporary variable that receives a copy of each element in the array you are looping through. It is common, though not essential, to name the temporary variable after the singular form of the array name:[59]

```
foreach my $protein  (@proteins)  { ...
foreach my $result   (@results)   { ...
foreach my $species  (@species)   { ...
```

Let's see the foreach loop in action:

Example 4.20.1 This is a script that takes a list of words and then counts which of those words are longer than an arbitrary length.

```
1.   #!/usr/bin/perl
2.   # long_words.pl
3.   use strict; use warnings;
4.
5.   my @words = qw(cat catastrophe dog doggedness fox foxtrotting);
6.   my @long_words;
7.
8.   foreach my $word (@words) {
9.       print "$word\n";
10.      if (length($word) >= 10) {
11.          push(@long_words, $word);
12.      }
13.  }
14.
15.  print "There were ", scalar(@long_words), " long words:\n";
16.  print "@long_words\n";
```

---

[59] Alternatively, you may want to name your arrays in the singular rather than plural. $result[0] sounds like it contains a single item, while $results[0] sounds like it has multiple items. If an array is going to be indexed numerically, we prefer the singular. If it is accessed in a foreach context, plural reads better.

Understanding the script

Line 5 populates the array that we will be looping through (@words), and line 6 declares the array that will contain only the long words.

On line 8 we have the foreach loop and we use $word as the temporary variable. With each iteration of the loop $word will contain 'cat' then 'catastrophe' then 'dog' and so on. Remember, these are just *copies* of the data in the array.

A simple if statement on line 10 checks to see whether the word is over a certain length. If it is, then it gets pushed to the end of the @long_words array, and these words are then reprinted on line 15.

One important point when using the foreach loop is that you should never use a foreach loop to modify the array you are looping through. Imagine if line 11 in the above example looked like this:

```
11.     push(@words, $word);
```

Rather than pushing $word on to the @long_words array, this line of code is pushing $word on to the array that is the subject of the loop itself. This would mean that @words would keep on growing in size, and the loop would never end.

Looping using the range operator

Note that you do not have to specify an array to loop through with the foreach operator. All that is needed is something that contains (or returns) a list of items. You can even write the list yourself, and sometimes this is easier than storing those items in an array. You could loop through the three domains of life like so:

```
foreach my $domain ('archaea', 'bacteria', 'eukaryotes') {
    # do something with $domain here
}
```

Perl also provides something called the *range* operator, which is represented by two dots (..). This operator provides an easy way of specifying a range of consecutive numbers or letters and it is often used in conjunction with foreach loops. Here is another way you can loop through numbers 1 to 10:

```
foreach my $i (1 .. 10) {
    # do something with $i
}
```

Perl uses the range operator to fill in the missing values between 1 and 10. This style of loop is much more similar to the for loops that we saw earlier. If you want to be a lazy typist (potentially an admirable quality if you are concerned about RSI), you can even use for rather than foreach. This is because foreach is just a synonym of for. Here's a short loop that cycles through each letter in the alphabet:

```
for my $letter ('a' .. 'z') {print "$letter\n"}
```

This is a foreach loop, though it is using the for synonym instead. One way of thinking about this is that the range operator is indirectly specifying all of the three conditions needed for a `for` loop. In the above example, the initialization component would be that the loop starts at 'a'; the validation component tells the loop to keep going until it reaches 'z'; and the update component will ensure that the loop covers all 26 letters of the alphabet.

The `while` loop

The `while` loop continues to iterate as long as some *condition* is met, where the condition is some notion of *true* or *false*. The 'condition' part of a `while` loop can be as simple or as complex as you want it to be. The basic syntax of a `while` loop looks like this:

```
while (condition) {
        # code to execute while condition remains true
}
```

It's very similar to how the `if` statement works, with the difference that there should be some code inside the `while{...}` block that will eventually cause the condition to be false (thereby exiting the loop). Here is an example of a very simple `while` loop which keeps doubling a number until some limit is reached:

```
my $x = 1;
while ($x < 1000) {
        print "$x\n";
        $x += $x;
}
```

In this example the code will continue to loop *while* the value of $x is less than 1000, and $x is doubled in each iteration of the loop. It is important that the test condition will be testing something that is going to change; but it is possible to write code that contains a test condition that will never be met, or that will always be met, such as:

```
while (0) {
        print "this statement is never executed because 0 is false\n";
}
while (1) {
        print "this statement loops forever\n";
}
```

The first `while` loop will never print anything at all because a zero value is always treated as false by Perl; the loop will run only while the value of zero evaluates to true, which is never going to happen.

The second loop will start but never end because the test condition ('while 1 is true') is always true. In fact, anything which isn't a zero or the null string ("") will always evaluate as true. We often use `while` loops to loop through an array. Let's see how we can do that.

Example 4.20.2 This script contains a whi<span></span>le loop that will cycle through each number
in a list that you specify on the command line. It will then report whether the number is
odd or even:

```perl
1.    #!/usr/bin/perl
2.    # odd_or_even.pl
3.    use strict; use warnings;
4.
5.    die "Specify at least 5 numbers\n" if (@ARGV < 5);
6.    my @numbers = @ARGV;
7.
8.    while (@numbers) {
9.        my $number = shift(@numbers);
10.
11.       if ($number % 2 == 0) {
12.            print "$number is even\n";
13.       } else {
14.            print "$number is odd\n";
15.       }
16.   }
```

Understanding the script

In each iteration through the loop, the array @numbers is shortened by removing one
item from the front of the list (using the shift function). The condition of the while
loop remains true as long as there is something in the @numbers array. Therefore, the
loop ends when the length of the array is 0 (empty).

In this example, the array is emptied by the shift() function as we iterate through the
while loop. This is often not what you want to happen (i.e., you want to keep the array
unmodified), so most of the time you will loop through arrays with for or foreach.

The do **loop**

The do loop is a variation of the while loop, and is sometimes known as a do-while
loop; do loops are not used as often as the other loops. Unlike the while loop, it always
executes at least once. The basic syntax of the loop is:

```perl
do {
    # code to be performed
} while (test condition);
```

Even if the test condition always evaluates as false, the code inside the do block will
still execute once.

When to use each type of loop?

There will be situations where you can use different types of loop structure to achieve exactly the same goal for a program. Conversely, there are times when only one type of loop will do. It might not always be clear to you how to make the correct choice, but with practice it becomes more obvious. Sometimes you should first ask yourself 'Do I need to keep track of the position in a loop?' If you don't need to keep track of the position then maybe a foreach or while loop is more appropriate. If the value of the loop counter is important, then you might need to use a for loop. Experiment with different loop structures to see what works and what doesn't.

Problem 4.20.1 To ensure you understand how to use each of the three main types of loop, write a script that uses for, foreach, and while loops to loop backwards through a list of three items.

If this does not make a lasting impression, redo this chapter before the next one

By default, loops continue to iterate until they reach their end point, but there are times when you will want a little more control. This can be achieved by use of the next, redo, and last keywords. The next keyword immediately restarts the loop *and* advances the loop counter variable. The redo keyword restarts the loop also, but does not advance the loop variable. The last keyword terminates the entire loop. If we return to our car race analogy of Chapter 4.19, the next keyword has the effect of teleporting the car *forwards* from wherever it is to the start line, advancing the lap counter in the process. In contrast, the redo keyword teleports the car *backwards* to the start line meaning the car has to do the lap again. The last keyword has the effect of raising the black flag to a car and forcing it to exit the race immediately. The rest of this chapter will show these keywords in use.

next

Sometimes we want to ignore certain values from a list or array if they don't meet certain criteria. We have already seen this in action by the use of certain logical operators (if, unless, etc.). Imagine we wanted to print out the squares of only those numbers from a list that are even. We could do this like so:

```
foreach my $number (1 .. 20) {
    if ($number % 2 == 0) {
        my $square = $number ** 2;
        print "$number squared is $square\n";
    }
}
```

However, this code means that we have to create an if block which adds an extra level of indenting to our code. This is fine, but it makes our code a little messier than it otherwise could be. Now let's see how the same code could be rewritten by including a next keyword.

```
foreach my $number (1 .. 20) {
    next unless ($number % 2 == 0);
    my $square = $number ** 2;
    print "$number squared is $square\n";
}
```

We now use next as part of an unless test to effectively skip to the next number if the current number is odd. You can read this as 'go to the next number unless the current number is even.' Using next like this means we avoid a second level of indenting in the code. It is important to appreciate that these loops still count from 1 to 20.

It is very common to see next used at the start of a loop to skip over any undesirable values. Imagine you had to calculate the square root of each number in an array of numbers and you wanted to add a line of code that would skip any negative numbers.

Your loop might therefore contain the following code that could check for any negative numbers:

```perl
next if ($number < 0);
```

redo

It is quite common to see the next keyword used in Perl scripts, but it is not quite so common to see the redo keyword, even though they behave in a very similar way. As mentioned earlier, the important point about redo is that it *doesn't* advance the loop counter variable. This means that if used incorrectly, you can get stuck in endless loops.

Example 4.21.1 This simple program tries to (randomly) choose three colors from a list of six colors, but it also ensures that one of those colors isn't pink.

```perl
1.    #!/usr/bin/perl
2.    # avoid_pink.pl
3.    use strict; use warnings;
4.
5.    my @colors = qw(white orange pink blonde brown blue);
6.
7.    for (my $i = 1; $i <= 3; $i++){
8.        my $random_color = $colors[rand(@colors)];
9.        redo if ($random_color eq "pink");
10.       print "$i $random_color\n";
11.   }
```

Understanding the script

Line 5 creates an array of six different colors. We then have a for loop (lines 7–11) which will count up to three.

Inside the for loop, we first choose a random element from the @colors array and assign that to a separate variable (line 8). If the chosen color is pink we invoke the redo keyword (line 9) and we re-run that iteration of the loop without iterating the loop counter ($i).

If the randomly selected color is anything other than pink then the if statement on line 9 will evaluate as false and we proceed to line 10 where the color is printed.

This script will only ever print three non-pink colors.[60] We use the redo keyword because any time we choose pink we want to make the script 'pick again.' If we used the next keyword instead, the script could potentially produce no output at all (e.g., if it chose pink three times in a row).

---

[60] Obviously we could just exclude 'pink' as a color from the original array, but we could imagine scenarios where we don't know what the original colors are going to be – maybe a user specifies the colors on the command line.

last

The final way of controlling a loop is simply to quit the loop before it finishes. To do this we use the `last` keyword. Any time Perl comes across this keyword it will immediately stop the loop at that point. Like the `next` and `redo` keywords it is almost always used as part of a logical test.

Example 4.21.2 Here is a program that takes one million guesses at what a very simple (three letter) password might be. If it correctly deduces what the password is, it stops guessing.

```perl
1.   #!/usr/bin/perl
2.   # password_cracker.pl
3.   use strict; use warnings;
4.
5.   my $password = "cat";
6.   my @alphabet = ('a' .. 'z');
7.
8.   for (my $i = 1; $i <= 1000000; $i++){
9.           my $first = $alphabet[rand(26)];
10.          my $second = $alphabet[rand(26)];
11.          my $third = $alphabet[rand(26)];
12.          my $guess = $first . $second . $third;
13.
14.          if ($guess eq $password){
15.              print "Attempt $i: the password is $guess!\n";
16.              last;
17.          }
18.  }
```

Understanding the script

We create a password on line 5 and then use the range operator on line 6 to populate an array with all of the letters of the alphabet. We then setup a `for` loop which will count up to one million.

Inside the `for` loop we use the `rand()` function three times in order to randomly select three letters from the alphabet array (lines 9–11). We then use the concatenation operator to combine them into a single string which is placed in a variable `$guess`.

An `if` statement on lines 14–17 then checks to see whether the current value of `$guess` matches the password. If it does then we print the answer, along with the number of guesses it has taken, and use the `last` keyword to stop the script.

In the preceding example we see the `last` keyword appear all by itself on a line of code. However, it is still part of a logical test (the surrounding `if` statement); it is also common to see `last` used as part of the alternative `if` syntax:

```
last if ($condition == $target_value);
```

Here is a more complex example that illustrates using redo and last together.

Example 4.21.3 This script finds any prime numbers that occur between 100 and 200.

```
1.   #!/usr/bin/perl
2.   # primes.pl
3.   use strict; use warnings;
4.
5.   my $n = 0;
6.   while (1) {
7.       $n++;
8.       redo if ($n < 100); # keep skipping numbers until $n = 100
9.       last if ($n > 200); # breaks out of while loop
10.
11.      my $prime = 1; # assumed true
12.      for (my $i = 2; $i < $n; $i++) {
13.          if ($n % $i == 0) {
14.              $prime = 0; # now known to be false
15.              last; # breaks out of for loop
16.          }
17.      }
18.
19.      print "$n\n" if $prime;
20.  }
```

Understanding the script

Line 6 introduces a while loop, but the test condition for this loop is set to be always true. This tells us that the loop will never end unless we deliberately break out of it using the last keyword.

Line 8 contains a redo. This short-circuits the while loop as long as $n is less than 100. You could have also used next here, because there is no loop counter variable in a while loop. Essentially, lines 7 and 8 of the loop will happen 99 times, but the loop doesn't go any further. Only on the 100th iteration will line 8 fail to evaluate as true.

Line 9 uses the last function to terminate the while loop, effectively ending the program, if $n is greater than 200.

Lines 11–17 determine if a number is prime. These lines will only be executed while $n is greater than 99 and less than 201. This method starts off assuming $n is prime (line 11). It then uses a for loop to check all the numbers between 2 and $n-1 to determine if $i is a factor of $n. If $i is a factor of $n (line 13), then there is no point in calculating any further because $n is not prime. So $prime is set to false (line 14) and the for loop is terminated using last (line 15).

Loop labels

Note that when you use last, it applies only to the *current* block of code that you are in. If we have multiple blocks of code (as in the previous example), you need to be sure which block will be exited:

```perl
for (my $i = 0; $i < 10; $i++){
    for (my $j = 0; $j < 10; $j++){
        last if ($j > $i);
        print "$i vs $j\n";
    }
}
```

In this example we have two nested for loops which both count from 0 to 9. The inner loop contains a last keyword to stop the loop if the inner counter ($j) is greater than the outer counter ($i). This use of last will only exit the inner loop and not the outer loop. If you want to exit the outer loop, you can use loop labels. A label is generally written in upper-case characters (to avoid conflicting with keywords) and followed by a colon. Here is another version of the nested loop which uses a label for each loop (OUTER: and INNER:).

```perl
OUTER: for (my $i = 0; $i < 10; $i++) {
    INNER: for (my $j = 0; $j < 10; $j++) {
        next OUTER if $j > $i;
        print "$i vs $j\n";
    }
}
```

In this example we now call next OUTER inside the inner for loop. This reads as 'Jump to the next iteration of the loop named OUTER.' Loop labels can sometimes make your code more readable. But they can make your code confusing if you use them too often. When in doubt, leave them out.

Labels can be placed anywhere in your program. Every 1000 programs or so, you might find the need to employ the dreaded goto.[61]

```perl
goto BLOCK_OF_DOOM if $something_terrible;
BLOCK_OF_DOOM: {
    # do something really important
}
```

---

[61] The goto keyword is rarely used in Perl. Try to keep it that way. It does what you might expect: execution goes immediately to the label.

The ins and outs of getting data in and out

So far our programs have only managed to include external data by using command-line arguments when we run a script. This is a workable solution for some simple scripts, but generally this is not a very common way to receive data. It is much more common to have external data in a *file*, which is then read by a program. Fortunately, reading files in Perl is incredibly simple.

Reading from a file specified on the command line

The easiest way to read a file with Perl is to specify the file name after the script (so it is added to the @ARGV array); this allows us to use a new operator which looks like <>. This is Perl's file operator; it is used to read lines of text from files and it also keeps track of how many lines it has seen.

Example 4.22.1 Create the program below, which will count the number of characters and lines in a specified file. When you run it, include the name of a valid text file on the command line *after* the program name:

```
$ line_count.pl some_text_file.txt
```

The valid text file can even be one of the Perl scripts you have already written; just make sure to specify the full path to the text file (if it is not in the current directory).

```
1.   #!/usr/bin/perl
2.   # line_count.pl
3.   use strict; use warnings;
4.
5.   die "Usage: line_count.pl <file>\n" if (@ARGV != 1);
6.   print "Processing file $ARGV[0]\n";
7.
8.   my $lines = 0;
9.   my $characters = 0;
10.
11.  while (my $line = <>) {
12.      $lines++;
13.      $characters += length($line);
14.  }
15.  print "$lines\t$characters\n";
```

Understanding the script

Lines 5 and 6 check that we have only one thing specified on the command line, and then print out the name of the file.

On lines 8 and 9 we introduce two variables which will count the number of lines and characters in the file. We also initialize both of these to contain a value of zero.

Line 11 then introduces a `while` loop, but the condition of the loop might seem strange to you as it contains something that we haven't seen before. This is the `<>` file operator. By default, this reads *one line at a time* from the file specified by `$ARGV[0]`. In this example we assign the contents of each line from the input file to a variable (`$line`).[62]

The `while` loop will continue to iterate as long as there are new lines to be read from the input file. This means that on line 12, the value stored in `$lines` is incremented by one for each iteration of the loop; it will end up containing a count of all the lines in the file. The `$characters` variable on line 13 adds the length of the current line to itself, such that it will end up counting all characters in the input file. Finally, after exiting the `while` loop, we print out the final values of `$lines` and `$characters`.

If you use `<>` to read files that are specified on the command line, then it is important that you specify an actual file that can be read. If you specify a file that doesn't exist, you will see an error, but it does not stop the script:

```
$ line_count.pl bad.txt
Processing file bad.txt
Can't open bad.txt: No such file or directory at line_count.pl line 11.
0 0
```

Reading multiple files and reading multiple lines

The script you just wrote would be more powerful if it could calculate the line and character count of multiple files at once. Well, luckily you can do this with just a very minor change to your script.[63] After modifying your script, you just need to specify multiple files on the command line when you run the script:

```
$ line_count.pl file1.txt file2.txt file3.txt
```

This works because the file operator (`<>`) is really looping through *everything* in the @ARGV array. If you wanted to, you could even use Unix wildcard characters to search all text files in the current directory:

```
$ line_count.pl *.txt
```

The file operator is one of those things that can be used in different ways depending on the context. In the earlier example, we assigned the output from each use of the `<>` operator to a scalar variable. If we wanted to read just *one* line of the file we could do this by using the file operator outside of the `while` loop:

```
my $line = <>;
```

---

[62] Remember that `$line` and `$lines` are completely separate entities. They just happen to share a similar name.
[63] In the earlier example, line 5 requires that only one command-line argument is specified. Increase this value to be able to process more files.

However, if we instead assign the <> operator to an *array*, then *all* of the file is read:[64]

```perl
my @file = <>;
print "$file[0]"; # print the first line of the file
```

Note that it is also possible to read a line from a file without doing anything to it. The following code is perfectly valid and would just read one line from a file (and throw it away):

```perl
<>;
```

You might wonder why you would want to do this. Well, sometimes you might need to ignore just the first line of a file. The above code would read the first line of a file and any subsequent use of the file operator would read from line 2 onwards. This is because the file operator keeps track of how many lines it has processed. This means you can even process a file like this:

```perl
<>;                         # skip 1st line of file
my $second_line = <>;       # keep 2nd line in variable
<>;                         # skip 3rd line
my $fourth_line = <>;       # keep 4th line in new variable
```

The default variable $_

It's time to learn some Perl magic. To do this we are going to write a very short script.

Example 4.22.2 The following script will do nothing more than print out the contents of whatever file you specify when you run it. As with the previous script, make sure you specify a suitable text file.

```perl
1.  #!/usr/bin/perl
2.  # echo.pl
3.  use strict; use warnings;
4.
5.  while (<>) {
6.      print;
7.  }
```

Understanding the script

You might be forgiven for thinking we have left something out from this script. But if you run it you may be surprised to see that it is working. How is this possible? We don't specify a variable to store the output of the <> file operator and we don't seem to include anything for the print statement to print. What's going on?

---

[64] You may want to be careful if you do this. Imagine if your file has one million lines; you might run out of memory if you try to read the entire file in one go like this.

The solution to this mystery lies with something called the default variable $_.
This is one of several 'magic' variables in Perl. Depending on the situation, Perl will automatically assign some data to this variable. In this script, $_ contains each line of the input file. Although you don't see it, Perl is actually performing the following operation:

```
5.    while ($_ = <>) {
```

The file operator should always assign the data that it reads from a file to a variable. However, if you don't include a variable and you are accessing the file operator as part of a loop, then Perl will automatically assign it to $_ for you.

The default variable is used by many of Perl's functions and operators. This is also what is happening on line 6. Because we don't specify anything for the print() function to print, Perl uses the $_ variable instead. Line 6 is exactly the same as if we had typed:

```
6.    print $_;
```

You may initially find the $_ default variable a strange thing, but you should get used to using it because it is commonly used in lots of Perl programs. As you learn more about Perl, you'll discover that there are a lot of places where if you don't specify a variable, then Perl will use $_ instead. This sometimes has the advantage of greatly simplifying your code. The disadvantage of this is that sometimes code can be so over-simplified it might, paradoxically, be harder for you to understand. Here is one more quick example to show some other uses of the default variable:

```
foreach ("Keith Bradnam", "Ian Korf"){
    my ($first, $last) = split;
    print "First name: $first\n";

}
```

The output from this code would be:

```
First name: Keith
First name: Ian
```

This foreach loop would normally assign each item in the list to a temporary variable, but in this example we don't include any variable at all. Can you guess what happens to each item? That's right, they get assigned to $_. The split() function in the loop normally requires a variable and a pattern that can be used to split the variable. However, if you specify neither of these things, then Perl assumes you want to use whitespace as the pattern, and $_ as the variable.

Confusing? Yes, a little. But you do get used to it. For now, feel free to give names to all of your variables instead of using $_. By the way, in addition to $_, there are a large number of other special variables in Perl with equally strange symbols. We will show you several more of them as we progress through this book.

Reading from standard input

In addition to reading data from files, and specifying arguments on the command line, there is one other way to input data into a program. On Unix systems, *standard input* usually refers to the data that is sent to a program.[65] When you are typing commands in a Unix terminal, standard input is provided by the text you type on your computer keyboard. We can make our Perl programs interactive by allowing them to receive input while the program is running. This is done by using a special Perl filehandle called STDIN.[66] The following code will require a user to type some input:

```
print "What is your name?\n";
my $name = <STDIN>;
print "Hello $name\n";
```

If you include code like this in your scripts, then every time Perl sees <STDIN>, it will wait until you enter some text. As soon as you press enter on your keyboard, Perl will take all of the text you have typed and, in this case, place it in a variable.

The <STDIN> filehandle can also be abbreviated to just <>, but as you might remember, this is also used to process any files that are specified on the command line. This dual-identity of the file operator can get confusing! Imagine you had the following code:

```
#!/usr/bin/perl
# stdin.pl
use strict; use warnings;
my $input = <>;
print "$input\n";
```

If you run the script without specifying any command-line arguments, then the script will expect you to type some input. However, if you specify the name of a file on the command line, then <> will read one line of input from that file.

In general we would advise you to not rely too much on using <STDIN> to make your scripts interactive. If your scripts have to stop to accept input from a user then they will not be as fast as scripts where all the input data is specified in advance (from files or from the command line). It also means that your scripts are more difficult to put into pipelines[67] with other programs.

---

[65] We explore the concepts of standard input and output in more detail in Chapter 5.3.
[66] This is the common name for standard input on all Unix computers.
[67] Pipelines are properly introduced in Chapter 5.7.

Otherwise known as the ins and outs of ins and outs

The previous section showed us how we can use <> to read files that are specified on the command line. However, if we want to open multiple files, then using <> is not always a good solution as our scripts might not need to read all the files at once. Using <> also assumes that we want to treat all of the files in the same way, but clearly we don't always want to do this. For example, one file might contain a list of email addresses and the other file a list of postal addresses. You wouldn't want to process both files with the same code. The <> operator also doesn't allow us to write new files or append to existing files. The solution to all of these problems is to use the open() function. This can open a file for reading *or* writing, but not both at the same time.

The open() function

The open() function can seem a little confusing because there are several different ways it can be used. Let's start off with learning how to use it in the most traditional way. Any time you want to use the open() function to open a file you'll need to specify three things:

(1) the name of a file to open
(2) something called a *filehandle*
(3) something called the *mode*

Filehandles are a type of special variable used when performing file operations. Once you associate a file with a filehandle you no longer need to use the actual file name. It is a convention to always use upper-case characters for filehandle names.

The *mode* controls whether you are reading from, or writing to, the file in question. The actual value for the mode borrows from how we access files in Unix.[68] I.e.,

```
<    # read from a file
>    # write to a file
>>   # append to an existing file
```

Okay, enough preamble! Time to see how we can open a file for reading.

Example 4.23.1 Create the following program. This will read the contents of a file you specify on the command line and then just print the same information to the screen but in upper-case characters.[69]

```
1.   #!/usr/bin/perl
2.   # shout.pl
3.   use strict; use warnings;
4.
5.   my ($file) = @ARGV;
```

---

[68] We cover this in more detail in Chapter 5.5.
[69] Please don't ever use this program for any reason other than learning Perl. It's impolite to shout.

```
6.
7.   open(IN, "< $file");
8.
9.   while (my $line = <IN>) {
10.      $line = uc($line);
11.      print "$line";
12. }
13. close(IN);
```

Understanding the script

Line 5 takes the name of the file you specify on the command line and assigns it to an appropriately named variable.

Line 7 then uses the open() function to open the file for reading. Notice the arguments that are used: IN is the name of the filehandle; '<' is the mode, telling us that we are reading from the file; and $file contains the name of the file. In this example, the mode and the file name are enclosed between double quotes (so the variable interpolates). The filehandle should be named something suitable, e.g., IN, INPUT, DATA, etc. It's common to see filehandles named after the type of file they are reading, e.g., CSV, FASTA, GFF.[70]

Line 9 should look a little familiar. Instead of using <> by itself, we now use our named filehandle inside the brackets. Only the file associated with IN will be read. This whole line reads as 'while we can read a line from whatever file is associated with the IN filehandle and store that line in $line, perform the following code....'

Notice that when we print out our modified line of output (line 11), we are deliberately not including a newline character. That's because each line of input that was read from $file already ended with a newline.[71] So $line already contains a newline at the end. We'll explain a better way of dealing with this later.

Finally, notice that there is a close() function on line 13. Every time you create a filehandle with open(), you should get into the habit of closing the filehandle with close(). Ideally, you should do this as soon as you have finished working with the file.

Another way of specifying the read/write mode

In the above example we specified the mode in conjunction with the file name. Perl mimics the behavior of Unix, where you can add as much whitespace as you like before or after the file name, including none at all. If you'd prefer, you can even separate out the mode and make that a separate argument. Furthermore, if you are opening a file for

---

[70] Note that Perl will expect you to use the name of a valid file that exists in the filesystem. That may, or may not, be the case and later on we will show you how to check whether the named file actually exists *before* you try opening it.

[71] If you have a text file that contains multiple lines, then each line ends with a newline character, even though you don't see it. It also counts towards the length of characters on each line.

reading, the '<' is actually optional. Finally, you can also specify a file by using its name directly rather than providing a variable name. All of this means that if you have a file called myfile.txt, and this name is stored in a variable $file, then all of the following do exactly the same thing:

```
open(IN, "< $file ");          # spaces either side of $file
open(IN, "<$file");            # no spaces
open(IN, "<myfile.txt");       # no spaces, using actual file name
open(IN, "myfile.txt");        # no '<', still opens file for reading
open(IN, "<", $file);          # three-argument form
open(IN, "<", "myfile.txt");   # three-argument form, using actual file name
```

For now, we would suggest choosing just one of these methods and sticking with it. But read the following section first....

A better way of specifying a filehandle

For a large part of Perl's existence you *had* to specify a filehandle, as detailed above, when opening a file. However, since version 5.6.0 of Perl,[72] a slightly different method became available, and this is a method we prefer.[73] But because you might see a lot of scripts using the older method, we felt we should show you that first.

The new method uses something called *indirect* filehandles and the basic difference is that you can now use a regular scalar variable as the name of a filehandle:

```
open(IN, "< $file");       # older method
open(my $in, "< $file );   # newer, indirect filehandle method
```

You still have to declare the variable with the my keyword, but after that you use the indirect filehandle just as if it was a regular filehandle:

```
while (my $line = <$in>){
     print "$line";
}
close($in);
```

There are some other subtle differences in using this form of filehandle, but as we prefer it we will use it in all future examples.

Using the file operator to read lines from a file

In the last chapter we explained that *any time* you use the file operator <> and assign to a scalar variable, you will read one line[74] from the file, and Perl will also remember the

---

[72] Version 5.6.0 was released in 2000, though many people will have learned the older method and sometimes it's hard to teach old dogs new tricks.
[73] The newer method means one less thing to explain. Filehandles are just scalars and you don't need to learn to use special upper-case filehandle names.
[74] It is possible to change this default behavior to make the file operator read multiple lines at a time, but for 99% of what you will probably be doing, reading one line at a time will be exactly what you want.

current position in that file. The same is true when using named filehandles. Consider the following code, which assumes you have already opened a filehandle called $in:

```
my $first_line = <$in>;
<$in>;
while (my $line = <$in>) {...}
```

The first line in this example will read one line from the input file, which will be the first line of the file. The second line above reads the second line of the input file, but doesn't do anything with it. When we enter the while loop, the filehandle will be pointing to the third line in the file.

Checking that your file has opened successfully

You should realize that if you try opening any file there is always the possibility that you can't open it or write to it. Most likely, this will be because you have typed the file name incorrectly. It could also be because you don't have the correct Unix file permissions to read from, or write to, that file. It is important that you *always* check that files have been opened correctly. There is no point proceeding with a script that attempts to loop through one million lines from your input file, if the file name is incorrect.

The open() function, like many Perl functions, returns a true or false value to indicate whether it was successful or unsuccessful. This is not a value that we need to store in a variable; we can instead just test the entire open() operation:

```
open(my $in, "< $file") or die "Can't open $file\n";
```

What is happening here? We've seen the die() command before, but we haven't seen it used like this! This statement reads as 'Try opening the specified file, but if it can't be opened, then stop the script.' The 'or' part of this statement is only evaluated if the open() is unsuccessful. If it is successful then the open() function returns 'true,' which means the 'or' part doesn't need to evaluate what comes next. If you find this confusing, don't worry about it for now. This syntax is commonly used for lots of Perl functions where if the function fails, then you want to do something (usually die). Just remember that you should:

> *Always test that a filehandle is created successfully*

We might go a little further and add that you should always print out a suitably useful die() message. You might write a script that opens a dozen different files. If every error message just said:

```
Fail!
```

then that doesn't tell you *which* file failed to open. Another of Perl's special variables is the $! variable. This variable is only used when things fail, and it (hopefully) contains a useful error message about the nature of the failure. This means that you can include the following in your die statement when opening a file:

```
my $file = "myfile.txt";
open(my $in, "<$file") or die "Can't open $file. $!";
```

This would print the following error message (if the file can't be opened)[75]:

```
Can't open myfile.txt. No such file or directory
```

Writing output

The ability to read files is useful, but even more useful is the ability to create new output files or append to existing files. Luckily we don't have to learn much more in order to be able to do this.

Example 4.23.2 This script will read a text file, reverse the contents of each line, and then print the modified output to a new file. After you run this script, check your directory for a new file called <file>.rev where <file> is the name you specify on the command line when you run the script.

```
1.   #!/usr/bin/perl
2.   # reversamatic.pl
3.   use strict; use warnings;
4.
5.   die "Please specify a suitable text file\n" if (@ARGV != 1);
6.   my ($infile) = @ARGV;
7.
8.   my $outfile = "$infile.rev";
9.
10.  open(my $in,  "<$infile") or die "error reading $infile. $!";
11.  open(my $out, ">$outfile") or die "error creating $outfile. $!";
12.
13.  while (<$in>) {
14.      chomp;
15.      my $rev = reverse $_;
16.      print $out "$rev\n";
17.  }
18.
19.  close $in;
20.  close $out;
```

Understanding the script

Lines 5 and 6 check that we have a file specified on the command line and then assign the name of that file to a variable.

---

[75] Note that you don't need to include a newline character when printing $!; it will automatically include one for you.

Line 8 creates a suitable output file name by appending .'rev' to whatever the input file name is. It is a common practice to create output file names that are based on the input file name.

Lines 10 and 11 then create filehandles for both the input and output files. Notice that the output filehandle uses the '>' mode to symbolize that we are writing to a file. The act of creating a filehandle for writing to a file is enough to actually create the file itself, though at this point it will be empty. We use the 'or die...' syntax to check that both filehandles were opened successfully.

When we set up the while loop on line 13, notice that we don't assign the file operator to any variable. This should inform you that we are using the special default variable $_. Each line that is read from the input file will be stored in $_.

Line 14 introduces the chomp() function. This function simply removes a newline character from the end of a variable (but only if one is present). Any line that you read from an input file will normally end with a \n character. We often forget about this character, but it's there. It is quite common to chomp your $_. Normally, you specify a variable that you want to chomp, but if you don't specify any, then Perl assumes you want to chomp $_.

One of the most important lines in this script is line 16. We have modified each line of input, and stored the reversed string in $rev. Now we want to print this output to whatever filehandle is specified by $out. All we need to do is include the filehandle as part of the print function.

Lines 19 and 20 close the two filehandles. You should always get into the habit of making sure that every open() function has a matching close() function.

When writing to files using the '>' mode, you should know that if you mistakenly try to write to a file that already exists, you will overwrite the file. This will happen as *soon* as you open the filehandle and all data in that file will be erased. You have been warned!

Until this chapter, we have only ever used the print() function without specifying any filehandle at all, and we have seen that this will send output to the terminal. However, we can now reveal that every time you print to the screen, Perl is in fact using a special filehandle called STDOUT. This refers to something called standard output, which we cover in more detail in the Advanced Unix part of the book (see Chapter 5.5). Printing to STDOUT is the default behavior when no filehandle is specified. For now you just need to know that the following two lines of code behave identically, though for convenience we just use the former syntax and not the latter:

```
print        "Some output\n";   # no filehandle specified
print STDOUT "Some output\n";   # now with STDOUT explicitly defined
```

It's very important that you understand how to use filehandles for both reading files and writing to files. If you are unclear, you should review the contents of this section before proceeding to the next section.

Problem 4.23.1 The one thing we haven't explained in this section is how to append to an existing file. To do this, you just need to set the mode to '>>'. Write a script that attaches your name to the end of any specified file. Check that this script works by appending to an existing file. Be careful not to use the '>' mode by mistake as you will overwrite the original file.

How can we explain this without making a hash of it?

So far we have seen how we can store data in scalars and arrays. With these two data types we have been able to write many useful programs. However, there are some situations where these two data types do not easily help us solve problems. Let's take a look at an example which illustrates the limitations of arrays.

---

*Example 4.24.1*This script reports the population for a few specified countries. To look up the population for China, you would run the script like so:

```
$ population.pl China
```

```
1.   #!/usr/bin/perl
2.   # population.pl
3.   use strict; use warnings;
4.
5.   my ($search) = @ARGV;
6.   die "Usage population.pl <country_name>\n" unless (@ARGV == 1);
7.
8.   my @country     = qw(Australia  China    France   Russia);
9.   my @population   = qw(21.4       1325.6   62.0     141.8);
10.
11.  for (my $i = 0; $i < @country; $i++){
12.      if ($search eq $country[$i]) {
13.          print "$population[$i] million people\n";
14.      }
15. }
```

Understanding the script
Line 5 retrieves the country from the command line and stores it in a variable called $search. Line 6 ensures that one thing, and one thing only, is specified on the command line.

Lines 8–9 use two arrays to store the names and populations of four different countries. Note the use of whitespace to align the name of the country with its respective population.

To find and report the population for a country, the for loop on lines 11–15 compares each country name with the search string. When the string matches, the population is printed.

In the above example we have two sets of information that are clearly related. Each country has a population, and furthermore each country can only ever have one value for its population. Our script is able to keep track of the association between the two data sets because they are in the same order in both arrays. Whenever we want to get information about a country we know that the name of the country and its population will be at the *same* position in both arrays.

Using two arrays like this is not a very good solution for storing the association between the country's name and its population. For instance, if we wanted to delete a pair of values, we would have to use the `splice()` function on *both* arrays. As the data sets become larger then this strategy also becomes very computationally inefficient. You have to look through half the records on average before you find the one you are looking for.

Let us now see a better way of managing this type of information, a way that allows us to more easily keep the association between pairs of related items. It happens to be very computationally efficient too.

What are hashes?

In addition to scalars and arrays, Perl allows for a third data type called *hashes*.[76] It is very common to use the analogy of a dictionary when explaining hashes.[77] Entries in dictionaries consist of two related concepts: A word that you want to look up and an associated definition for that word. You will never see a dictionary that contains the same look-up word twice. There is no point because look-up words are clearly meant to be unique. But the definitions associated with those words do not necessarily have to be unique because a word can have multiple definitions.[78]

Conversely, we also know that two different words can both have exactly the same meaning. If we think back to our previous example, we can imagine that it's possible (if a little unlikely) for two countries to have exactly the same population.

How do hashes work in Perl?

Just like our dictionary example, hashes in Perl associate pairs of items. Strictly speaking, we say that a hash contains *keys* and *values*. A key is like a look-up word in a dictionary, and the value is like a definition. One small deviation from our dictionary analogy is that a regular hash in Perl only has one value for each key. However, it is possible – and often desirable – to associate many values to one key. That will be covered in Part 6 of this book. Once you create a key–value pair in a hash, you can always access the value if you know the key. Hashes should ideally be used for storing pairs of things that have some sort of meaningful connection. Here are some possible key–value pairs that we might consider storing in a Perl hash.

---

[76] In older versions of Perl these were also known as 'associative arrays.'

[77] This might even be a legal requirement; if it is not, then maybe it should be.

[78] The dictionary on our Macs defines 'hash' as either a 'meat-based dish, usually served with potatoes' or a 'mixture of incongruous things; a mess.' Sadly it lacks the definition of a hash as a data type in the Perl programming language. Bad dictionary!

Key	Value
Social security number	Name
Zip/postal code	State/county
Computer serial number	Make and model of computer
Running distance	World record at that distance

These examples all use keys that will contain unique values – there is only one zip code with the value '95616.' Try not to fall into the trap of *assuming* that a key is going to be unique. For example, certain actors' names like 'Harrison Ford' are so famous that we only ever think of one person, but according to IMDB there are two famous actors with that name.[79]

Hash syntax

Just as we use dollar signs to denote scalar variables ($variable) and 'at' symbols to denote arrays (@array), we use yet another character to represent hashes: the percent sign (%hash). We only use the percent sign when we are referring to the *entire* hash. When we refer to one thing in the hash we use a dollar sign and a pair of curly braces. Consider the following code:

```
my %zip_code;
$zip_code{'95616'} = "California";
print "$zip_code{'95616'}\n";
```

Here we declare a new hash (%zip_code) and then add one key–value pair to the hash. When we deal with individual items in the hash, the key is always placed after the hash name in curly braces and we prefix the whole thing with a dollar sign. This is logically consistent with how we deal with single items in arrays, which also use dollar signs. The final line above would print the value associated with the key '95616', and so would print 'California'.

Note that a hash cannot contain multiple entries that use the same key:

```
my %favorites;

$favorites{'food'}   = "cheeseburgers";
$favorites{'drink'}  = "tea";
$favorites{'color'}  = "green";
$favorites{'food'}   = "pizza";
```

In this example, we add a number of different items to a hash (%favorites). After declaring the hash, each line of code would increase the size of the hash by adding one new key–value pair. However, the fourth hash assignment uses the same key as an earlier assignment. As there can only be one hash entry that uses the 'food' key, the previous value ('cheeseburgers') will be overwritten and the hash would end up containing three key–value pairs.

---

[79] The first Harrison Ford actor was a leading man in the silent film era and starred in 87 films.

Some people find hashes confusing. One way to think about them is that they are arrays, but the indices are strings rather than integers. Consider the following two similar statements. Both print a single value from an array or hash. The number 0 is an index in the array just as the string 'key' is an index in the hash:

```
print $array[0];
print $hash{'key'};
```

As a convenience, the string inside the curly brackets does not need to be quoted. All of these are equivalent.

```
$hash{'key'}    = 1;      # single quoted string
$hash{"key"}    = 1;      # double quoted string
$hash{key}      = 1;      # unquoted string

my $string      = 'key';
$hash{$string}  = 1;      # variable containing string
```

Defining a hash

We often want to add multiple key–value pairs to a hash when it is first declared. Perl offers a few different ways of doing this:

```
# method 1
my %starwars = ("Yoda", "good", "Vader", "bad", "Han", "cool");

# method 2
my %startrek = ("Vulcans",   "good",
                "Borg",      "bad",
                "Klingons",  "depends");

# method 3
my %comics = (Spiderman   => "hero",
              Venom       => "villain",
              Punisher    => "unsure");
```

Methods 1 and 2 are identical except they differ in the use of whitespace. Most people would probably say that method 2 looks cleaner and makes more sense. In both cases the keys of the hash are immediately followed by their respective values. Method 3 looks a little different as it introduces something that looks like an arrow (=>) but behaves just like the commas in method 2, with the added bonus that we don't have to quote the hash keys. You can use quotes if you want to, but they are generally unnecessary.[80]

Let's revisit the previous program that used a pair of arrays to associate countries and populations. This time we will use a hash, and you will see how much better it is.

---

[80] Except when your hash keys contain whitespace characters, reserved keywords, or function names, in which case you need to use quotes.

Example 4.24.2

```
1.    #!/usr/bin/perl
2.    # population.pl
3.    use strict; use warnings;
4.
5.    my ($search) = @ARGV;
6.    die "Usage population.pl <country_name>\n" unless (@ARGV == 1);
7.
8.    my %population  = (
9.        Australia   => 21.4,
10.       China       => 1325.6,
11.       France      => 62.0,
12.       Russia      => 141.8,
13.   );
14.
15.   print "$population{$search} million people\n";
```

Understanding the script

Lines 5–6 are the same as in Example 4.24.1, but lines 8–13 introduce new code which declares a hash with key–value pairs for country and population. This means that we use one data structure in place of the two arrays that we used before.

Line 15 retrieves the population associated with the specified country in one step. In the previous array example we had to loop through the array of countries to find the name, and then report the population at this position in the array. The hash version is much cleaner and much more efficient.

Naming hashes

If you work with a lot of hashes, it can sometimes help to make the hash name explain something about the data it contains. Hashes typically link pairs of connected data, e.g., the identifier for a DNA sequence and the GC% content of that sequence; or the name of a politician and the number of votes they received. Based on these two examples, which of the following hash names make the most sense?

```
%seq
%sequences;
%sequence_details;
%sequence2gc;
%sequence_to_gc;

%vote;
%names;
%name2votes;
%name_to_votes;
```

Choose suitable names for your hashes. You should, of course, be choosing suitable names for *all* of your Perl variables.

Problem 4.24.1 Write a script that creates two hashes. The first hash should contain some key–value pairs that are combinations of countries and the internet two-letter code for that country (e.g., France = 'fr', China = 'cn', etc.).[81] These are good examples of keys and values that are both unique. The second hash should reverse the order of keys and values, e.g., the key will be the two-letter code and the value will be the name of the country. The final part of your script will take an argument specified on the command line and check that value in both hashes to see if it exists and then print out the corresponding value. If you specify 'China' on the command line, then the script should print out 'cn'; if you specify 'cn' your script should print out 'China'.

To solve this last part you will need to use the `exists()` function. This function checks whether a specified hash key exists; you can use it like this:

```
print "$key exists\n" if exists($hash{$key});
```

---

[81] If you're interested, these abbreviations are called 'country code top-level domains' or ccTLDs. If you are not interested then you don't need to read this note.

The key point about this chapter is the value it will provide

Perl has many functions that work with hashes, and this chapter will explore several of them and also show you ways of looping through and sorting hashes. Before we start exploring the wondrous ways in which we can work with hashes, let's first check that you can tell the difference between the following:

1. `$thing`
2. `@thing`
3. `$thing[0]`
4. `$thing[$thing]`
5. `%thing`
6. `$thing{'thing'}`
7. `$thing{$thing}`

The point of this exercise is to remind you that it's possible to end up with a Perl script that reuses the same names for scalars, arrays, and hashes, and each of these can be a different thing altogether. The answers to the above list are:

(1) A scalar variable called `$thing`.
(2) An array called `@thing` (no relation to `$thing`).
(3) The first element of the array called `@thing`.
(4) The `$thing`-most element of the array `@thing` (`$thing` is a number).
(5) A hash called `%thing` (no relation to `$thing` or `@thing`).
(6) A key called 'thing' in the hash called `%thing`.
(7) A key in the `%thing` hash which will be equal to the value of `$thing`.

If you are still confused by some of these differences, you should go back and revisit the previous chapters and make sure you are comfortable with the differences between scalars, arrays, and hashes.

Adding and removing key–value pairs

In the last chapter we saw that we can create hashes to represent useful lookups such as countries and their corresponding populations. What if we want to add some new data to an existing hash? This is incredibly simple; just assign a new key–value pair.

```
my %sound;                 # empty hash
$sound{'dog'} = 'woof';    # now contains a key and value
$sound{'cat'} = 'meow';    # now contains two keys and values
```

If the key already exists, it will be replaced with the new value.

```
$sound{'cat'} = 'mew';    # redefined 'cat' key
```

To remove a key–value pair, you use the aptly named `delete` keyword:

```
delete $sound{'cat'};
```

Note that if you wanted to delete *all* of the key–value pairs from a hash, simply assign an empty list to the hash:

```
%sound = ();
```

Determining whether a hash contains a specified key

To test if a particular key exists in a hash, you use the `exists()` function. This returns 'true' if the key is present in the hash and 'false' otherwise. The `exists()` function does not care if the value is defined or not, just if the key is there:

```
print "Exists\n" if exists $hash{$key};
```

If you want to determine if a key has a defined value, you can use the `defined()` function. This returns true if the key exists *and* has a defined value. If you want to test if a key exists and has a *true* value,[82] first test if the key is defined (or exists), and then test if it has a true value:

```
if (defined $hash{$key} and $hash{$key}) ...
```

If you do this in the reverse order (testing for truth before testing for existence), you may get a 'use of uninitialized value' warning message, which we never want to see in any of our programs.

keys() **and** values()

One of the most common things to do with a hash is to extract a list of all of the keys or all of the values. To do this we can use the appropriately named `keys()` and `values()` functions:

```
my @keys     = keys %hash;
my @values   = values %hash;
my $keys     = keys %hash;    # scalar context, returns the number of keys
my $values   = values %hash;  # as above
```

The first two examples use the `keys` and `values` functions in array context; both operations extract a list of the keys or values and assign those to an array. However, as the last two examples show, you can also use these functions in scalar context to return the number of keys or values.

Looping through hashes

Unlike arrays, whose elements are in a defined order starting at 0, there is no obvious order to the elements of a hash. If you execute the following code, there is no guarantee

---

[82] Remember that testing for logical 'truth' is not the same thing as testing whether something is defined. If something is (logically) true then it must also be defined, but the opposite situation is not always true. E.g., if $x contains '0' (zero) or the empty string ("") then it is defined, but not true.

that the keys or values will come out in alphabetical order, the order you added them, or any other way that makes sense to you.[83]

```
foreach my $key (keys %hash) {print "$key\n"}
```

When you loop over the keys of the hash, the order will often not be important. But if it is important, you can sort the keys (see below) or define an additional array that holds the keys in the order you added them (see farther below). Now let's see an example of looping over some keys in a typical context.

Example 4.25.1 This script will simply take a list of all words that you input on the command line and then count the occurrence of each word using a hash. It will then report them in the default order. When you run this script make sure you duplicate some words on the command line – e.g.:

```
$ word_count.pl apple banana apple cherry banana cherry cherry
```

```
1.   #!/usr/bin/perl
2.   # word_count.pl
3.   use strict; use warnings;
4.
5.   # count words on the command line
6.   my %count;
7.   foreach my $word (@ARGV) {$count{$word}++}
8.
9.   # loop through hash, printing word counts
10.  foreach my $word (keys %count) {
11.      print "$word\t$count{$word}\n";
12.  }
```

Understanding the script

Line 6 defines a hash that will contain the counts of words. Line 7 loops over the words stored in the @ARGV array and increments the count of each word that it comes across. It is almost magical in its power.

The first word of our example input is 'apple'. In the first iteration of the loop, `$count{apple}` does not exist. Perl creates the key and gives it an undefined value. The ++ operator increases the undefined value from 0 to 1. The end result is that `$count{apple}` contains the value of 1. The next word is 'banana', and after two iterations of the loop, the count of 'apple' and 'banana' are both 1. On the third iteration, 'apple' is incremented to 2. This continues until all the words are counted. If the loop syntax of line 7 seems too concise, consider this unnecessarily verbose alternative:

---

[83] Unless you are a computer, in which case, you should be reading our other book, *01100101 01111000 01110100 01100101 01110010 01101101 01101001 01101110 01100001 01110100 01100101*.

```
foreach my $word (@ARGV) {
    if (not exists $count{$word})  {$count{$word} = 1}
    else                           {$count{$word}++}
}
```

Lines 10–12 loop over all the keys of the hash %count and report the counts for each word. The order of the keys might not be what you expect. Try using several different combinations of words and observe how the output changes.

Looping through hashes in sorted order

If you want to report the words in a specific order, the easiest solution is to sort the keys. To perform the default alphabetical sort, change line 10 as follows:

```
10. foreach my $word (sort keys %count) {
```

In this example, the keys() function happens first, and the list of keys is then passed to the sort() function. Sometimes you may prefer to sort a hash by its values instead of the keys. For example, rather than printing the words in alphabetical order, you might want to print the most common words first. The solution is simple, but a little confusing at first. Rather than comparing the keys themselves, you want to compare the values those keys have. Instead of comparing $a and $b you compare $count{$a} and $count{$b}. The following statement will sort in descending order by the value of each key–value pair:

```
10. foreach my $word (sort {$count{$b} <=> $count{$a}} keys %count) {
```

See Chapter 4.18 for more information on sorting.

Looping through hashes in a predefined order

If you want to loop through the keys in the order you placed them into the hash, you have to keep a separate array containing the keys. Every time you add a key to the hash, you have to remember to push the key on to an array as well:

```
my @keys_in_order;
my %hash;
$hash{'thing'} = 1;
push @keys_in_order, 'thing';

$hash{$key} = $val;
push @keys_in_order, $key;
```

This strategy has some difficulties, however. If a key is reassigned or deleted, the array could get out of sync with the hash.

Looping through a hash with `while` and `each`

The canonical way to loop through a hash is with a `foreach` loop, but it is occasionally more succinct to use `each()` in a `while` loop. Compare these two loops that perform the same operation.

```
while (my ($key, $value) = each %hash_with_long_name) {
    print "$key, $value\n";
}
foreach my $key (keys %hash_with_long_name) {
    print "$key, $hash_with_long_name{$key}\n";
}
```

The `each()` function returns a key–value pair every time it is called until it has exhausted all the elements of the hash, at which point it returns false. It does not destroy the hash. Like the `keys()` and `values()` functions, there is no order to how `each()` retrieves information from a hash.

Quick review

Function	Meaning
each %hash	Returns a key–value pair from the hash
keys %hash	Returns an array of keys
values %hash	Returns an array of values
exists $hash{key}	Returns true if the key exists
delete $hash{key}	Removes the key and value from the hash

There's nothing 'regular' about them

We first saw the matching and substitution operators in Chapter 4.11: The former lets you see whether a variable or string contains some specified text, the latter lets you substitute one string for another. When we used these operators before, we were only searching for fixed strings and the only flexibility we had was whether to include the 'ignore case' option. These operators are actually much more powerful than they first appeared because we can use them to search for *patterns* instead of fixed strings.

Why do we need to search for patterns?

Sometimes we don't know exactly what text we are trying to match, but we know something about what *possible* variations could be considered a valid match. For instance, if you were trying to find the word 'realize' in an email message, you might also want to search for the British-spelling variant 'realise.' We could probably think of many situations where being able to (computationally) deal with alternative spellings would be useful.[84]

Aside from spelling problems, pattern-matching is an important issue for many computer-related fields. Consider the problem of trying to extract email or web site addresses from a piece of text. These are examples where we want to capture strings that are highly variable but which are usually constrained by some invariant text: A web site name should start with `http://` and end with `.com`, `.net`, `.gov`, etc. Knowing that all web site names are anchored in this way means it should be possible to make a pattern that would match any valid URL.

Let's imagine a simple scenario and see how we might tackle it *without* using patterns. Assume we want to search a file of recipes for spicy dishes that involve chili peppers. The problem is that 'chili' is not the only common spelling of this plant. One solution would be:

```
if ($text =~ m/chili/    or $text =~ m/chilli/ or
    $text =~ m/chillie/  or $text =~ m/chile/){
    print "Chili pepper\n";
}
```

In this example we allow ourselves four alternative spellings (using the or logical operator) in order to be able to match all of the valid spelling alternatives. Although this works, it is not a generalized solution – if there were ten different spellings of 'chili,' the code would have to be changed again and would start to get very long and unmanageable. Because of the similarities between the different spellings, you might think there should be an easier way to capture all of the variants. Well, luckily there is.

---

[84] E.g., if you were looking for a particular Asian grilled lamb dish then you might be looking for a: kebab, kebap, kabab, kebob, kabob, kibob, kebhav, or kephav. These are all valid English spellings in different countries.

Regular expressions to the rescue

Matching flexible patterns is possible if we use something called *regular expressions* (also commonly referred to as regexes or regexps). A regex is *one* defined pattern that can be interpreted to potentially match *many* different strings.[85] Regexes are not unique to Perl; they are used by several Unix commands (notably `grep`), and are implemented by many programming languages as well as many advanced text editors.[86] The power of regexes is that they allow you to define patterns which can be as broad or as specific as you wish. For instance, you could create a regex which would match strings that contain the name *Ann*, *Anne*, or *Annie*. However, you could also create a regex that describes a pattern such as:

> any words that are 3–6 letters long, but contain two consecutive vowels and must be the last word of a sentence, and not preceded by a hyphen.

You may not have realized it, but we have already been using regexes. Consider the following line of code:

```
print "Match\n" if "Ian and Keith" =~ m/Keith/;
```

Here we are using the matching operator to see whether a string ('Ian and Keith') contains the pattern 'Keith'. In this case, the pattern 'Keith' is a regex, it just happens to be a regex that is only trying to match one thing. As we will see shortly, the real power of regexes happens when they are used to match complex patterns, but we often want to match fixed strings, and that's fine too.

When you are writing your own scripts you should always try to make them as flexible as possible (see Chapter 7.4 on Abstraction for more details). Using regexes to find and match patterns rather than fixed strings is one way in which you can do this.

---

[85] Of course, a pattern may also match just a single string, or even no strings at all.

[86] There are a few subtle differences in how different programming languages and Unix use regular expressions, but don't be too concerned by that at the moment.

^If you are s[kc]eptical, you (probab|definite)ly won't enjoy this chapter!{1,3}$

Regular expressions (regexes) rely on using many standard keyboard characters, which are imbued with 'special powers.' These *metacharacters* can be a little confusing because they appear similar to characters that often have a completely different meaning when you use them outside of the binding operator (=~). This chapter will cover many of the metacharacters that are common to regexes on many different computer systems. Because there are a lot of metacharacters, this chapter will include many short code examples, rather than complete scripts.

Alternation

You can easily create a pattern to match more than one string by simply separating all of the possibilities with the pipe character: |.[87] Here are some simple examples:

```
print "Ancient element\n"  if ($substance    =~ m/earth|air|fire|water/i);
print "Stop codon\n"       if ($seq          =~ m/TAA|TAG|TGA/i);
print "Author's name\n"    if ($text         =~ m/Ian|Keith/i);
```

Note that in these examples, *everything* between the two forward-slashes of the matching operator is the pattern that we are trying to match. So the last example has *one* pattern that specifies 'Ian' or 'Keith'; it is not two separate patterns. It is common to use alternation as part of a larger pattern, in which case you also need to use parentheses, which act as *grouping* metacharacters:

```
if ($word   =~ m/fire (alarm|engine)/) {...} # 'fire alarm' or 'fire engine'
if ($sport  =~ m/(basket|foot)ball/) {...}    # basketball or football
if ($name   =~ m/Ste(v|ph)en/) {...}          # Steven or Stephen
```

Using alternation gives us the ability to more easily solve the chili pepper problem from the last chapter:

```
# method 1
print "Chili pepper\n" if ($text =~ m/chili|chilli|chillie|chile/i);
# method 2
print "Chili pepper\n" if ($text =~ m/chil(i|li|lie|e)/i);
```

The second method works because all of the possible spellings start with 'chil'.

Note that you can also use the alternation metacharacter but only provide one possibility. If you do this, you are effectively implying a second option of *not* matching the pattern:

```
if ($text =~ m/foot(ball|)/) {...}          # matches football or foot
```

Furthermore, you can also nest groups inside each other.

```
if ($text =~ m/foot(ball(s|)|)/) {...}    # footballs, football, or foot
```

---

[87] The pipe character is one of those characters that often lives in very different positions on your keyboard, depending on what keyboard you own. Have a good look, it should be there somewhere.

Character classes

In our chili example all possible spelling variants end with 'i' or 'e' so we could possibly simplify our pattern if we could specify that the last character had to end with just one of these two letters. Fortunately, regexes allow us to define a *character class*, which is a set of allowed characters. The basic syntax for using a character class is to put all of the possible characters inside square brackets. Here are some examples:

```
if ($word =~ m/reali[sz]e/) {...}      # matches British and US spellings
if ($hero =~ m/spider[ -]man/) {...} # allow for space or hyphen as separator
if ($year =~ m/198[048]/) {...}        # matches Olympic years in the 1980s
```

You can put as many characters as you want inside the square brackets, and can also include non-alphanumeric characters. Note that character classes are only ever specifying *one* character to match. In the final example above, we specify a pattern that matches four characters, the last of which must be one of three possible digits.

It is very common to see character classes in solving various DNA/protein sequence matching problems. A simple character class allows you to specify an 'unknown' DNA base where it must be one of four possibilities:[88]

```
my $dna = "ACGATGAGCCAGTG";
print "DNA contains proline codon\n" if ($dna =~ m/CC[ACGT]/);
```

Note that we could also use character classes in addition to the alternation metacharacter to solve our chili spelling problem:

```
# method 3
print "Chili pepper\n" if ($text =~ m/chil([ei]|l(i|ie))/i);
```

Here we make use of a character class and two levels of grouping. This probably introduces more complexity than is needed as it makes the regex very hard to read. This can be a common problem and some regexes can quickly grow to fill an entire line of code.

Negated character classes

Sometimes we want part of a pattern to match anything as long as it *isn't* from a specific range of characters.[89] We can do this with *negated character classes* which involve putting a caret symbol (^) inside the square brackets of the character class. If you wanted to test a string to make sure it *only* contained vowels, you could do this:

```
print "Vowels only please\n" if ($string =~ m/[^aeiou]/);
```

---

[88] If you wanted to include nucleotide ambiguity codes: N, R, Y, etc., then you could also put them inside the character class.

[89] This should not be confused with the 'not match' operator ($!), which acts on a whole pattern and not specific parts of a pattern.

Negated character classes can be used to help search for simple DNA motifs, i.e., report if a sequence contains a subsequence such as 'T, followed by A or C, followed by *anything but* G,[90] and ending with C':

```
if ($dna =~ m/T[AC][^G]C/i) {...}
```

Character ranges

Another useful option when specifying a character class is to use a dash to specify a *range* of characters or numbers:

```
print "Password contain numbers!\n" if ($password =~ m/[0-9]/);
print "PIN contains letters!\n"     if ($pin      =~ m/[a-z]/i);
print "Name contains A-K\n"         if ($name     =~ m/[A-K]);
```

Note that in the second example we include the ignore-case option, otherwise this pattern would only match lower-case letters. A better check for a suitable PIN would be to use a negated character class to check that it only matches non-numbers :

```
print "Not a PIN!\n"                if ($pin      =~ m/[^0-9]/);
```

You can use character ranges with the transliteration operator to easily convert strings from lower- to upper-case (or vice versa):[91]

```
$text =~ tr/[a-z]/[A-Z]/;
```

Anchors

Most of the patterns we used in the last chapter were quite permissive, meaning they might match more things that we want to. Let's consider our 'chili' matching example again:

```
print "Chili pepper\n" if ($text =~ m/chili|chilli|chillie|chile/i);
```

This code works, but it would also return a match if $text was any of the following:

chilliest
visitchile.com
trochile[92]

Clearly we would not want our script to match these words. Ideally, we want to be able to *anchor* our patterns and specify that a pattern must match at the start and/ or end of a string. To ensure that a pattern matches from the beginning of a string, we can use the caret (^) symbol. This can be confusing because, as we saw earlier, the

---

[90] Matching 'anything but G' would also include matching non-DNA characters such as 'Z,' '3,' '@,' etc. But hopefully any DNA-related script would also check that all characters are valid on another line of code.

[91] Of course, this is still not as easy as using the lc() and uc() functions.

[92] Trochile is the suborder of birds that includes hummingbirds, but you knew that, right?

caret symbol is also used as the negation metacharacter within a character class. Some examples are:

```
$text =~ m/^hat/;      # matches hat, hate, hatch, etc., but not chat or what
$text =~ m/^ /;        # matches as long as $text starts with a space
$text =~ m/^[^0-9]/;   # matches as long as $text starts with a non-digit
```

The last example uses the caret symbol as an anchor *and* negation character class meta-character. If you want a pattern to only match if it occurs at the *end* of a string, you can use the $ character, and this can also be combined with the ^ anchor.

```
$text =~ m/old$/;      # matches old, bold, gold, etc., but not older
$text =~ m/^unique$/;  # will only match 'unique'
$text =~ m/^$/;        # will match the empty string
```

The second pattern in the above example uses both anchors, which is a common prac-tice and ensures that the pattern has to match the entire string. The last example looks like it wouldn't match anything, but it matches the empty string ''. This can be very useful for matching blank lines in a file, and you can use this with the next function to skip any blank lines you come across while looping through a file:

```
while (my $line = <$file>) {
    next if ($line =~ m/^$/); # skips to next line in file
}
```

Note that for the purposes of matching the 'end' of a string, newline characters do not count. This means that the following code will match:

```
print "Match!\n" if ("The end\n" =~ m/end$/);
```

The dot metacharacter

Outside of a regex, the dot symbol acts as Perl's concatenation operator. However, *inside* a regex the dot metacharacter has a very simple meaning: *match any single character.*[93] This means you can use the dot metacharacter as a wildcard to match anything at all:

```
$text =~ m/.at/;       # matches bat, cat, hat, 4at, #at, etc.
$text =~ m/^...$/;      # matches anything with exactly three characters
```

You should only really use the dot metacharacter if you know that a character in your pattern can be *anything* at all (including spaces and all forms of punctuation). Usually, this is not the case and we often know that we want a letter, number, or range of certain punctuation characters in a particular pattern.

How to match special characters

Consider the following code; will the print statement be executed?

[93] One exception to this is that a dot metacharacter will not match a newline.

```
my $text = "A common plant used in research is A. thaliana";
print "Text contains pattern\n" if ($text =~ m/A./i);
```

This code will print the specified statement, but can you see what part of $text the pattern is matching? It actually matches in six different places. The dot in the pattern is acting as a metacharacter and not as a regular period. This means it matches anywhere in $text where there is a letter 'A' (upper- or lower-case) followed by any single character. From Perl's point of view, as soon as it finds any match it will execute the print statement, and so the pattern is actually matching the very first two characters of $text (the letter 'A' and the space character).

This may not have been what we wanted, so how could we make the dot character in the pattern not act as a metacharacter? The answer is to prefix the dot with a backslash, which acts as an 'escape character'; we first saw this in Chapter 4.10. You can do this for any of the metacharacters that are used in regexes:

```
$text =~ m/Mr\./;      # matches 'Mr.', no special meaning to '.'
$text =~ m/a\|b/;      # matches the string 'a|b'
$text =~ m/\(|\)/;     # matches '(' or ')'
$text =~ m/\\/;        # matches '\'
```

Note that in the final example we are trying to match a backslash character, which requires us to escape the character with yet another backslash character.[94]

Matching things that repeat

We have already seen a lot of regular expression metacharacters, but in some ways we have barely begun to get at their real power. Let's imagine that we were looping over the lines of a long file and we wanted to spot lines that might contain telephone numbers. In the United States, telephone numbers have a standard format consisting of a three-digit area code followed by a three-digit exchange code, and ending with a four-digit subscriber number. However, people will often use different formatting when writing these numbers:

```
5551234567        # no spaces
555 1234567       # one space for area code
555 123 4567      # spaces for area and exchange codes
(555) 123 4567    # parentheses around area code
(555) 123–4567    # parentheses and hyphen
555–123–4567      # just hyphens
```

There are probably other permutations that we could come up with as well. So, how could we begin to make a regex that could capture all of these? Let's start by tackling the easiest situation: a number with no spaces:

```
$number =~ m/[0-9][0-9][0-9][0-9][0-9][0-9][0-9][0-9][0-9][0-9]/;
```

---

[94] When you start matching slash characters (forwards or backwards), it is sometimes easier on the eye to choose a different delimiter for the matching operator, rather than use the default forward-slash character.

It seems wasteful to have to reproduce the same pattern over and over again. Fortunately, there are a series of *quantifier* metacharacters which will help us make it much easier to specify parts of a regex that we want to repeat. If we know that we want something to occur exactly *n* times, then we can just append {n} after any part of a regular expression pattern:

```
$text =~ m/a{5}/;        # matches aaaaa
$text =~ m/Yes!{3}/;     # matches Yes!!!
$text =~ m/.{10}/;       # matches any 10-character string
$text =~ m/A{2}|B{3}/;   # matches two As or three Bs
$text =~ m/[a-z]{4}/;    # matches any four (lower-case) letters
$text =~ m/(X|Y){5}/;    # matches five Xs or five Ys
```

The important point to note about using this quantifier is that it only applies to the preceding *character*, not the entire pattern. The exceptions to this are when you use a quantifier after a character class or grouped pattern. In the last example above, the {5} quantifier applies to all of the (X|Y) pattern and not just to the last closing parentheses. We can now easily use this quantifier to more easily solve our telephone number problem:

```
$number =~ m/[0-9]{10}/;
```

However, this only matches *exactly* ten digits. If $number contained more or fewer digits, it wouldn't match. So how could we also make our regex also match UK telephone numbers, which are 11 digits? The easy solution is that the {n} quantifier can also be used to specify a minimum *and* maximum value. If you want to match a pattern that occurs between *m* and *n* times, the syntax is simply {m,n}. You can also use this quantifier to specify a minimum without specifying a maximum, with the syntax {m,}. Here are some brief examples:

```
$text =~ m/A{2}/;     # match exactly two As
$text =~ m/A{2,}/;    # match two or more As
$text =~ m/A{2,4}/;   # match between two and four As, inclusive
```

There are three more quantifiers that are very commonly used in regular expressions, but which can be difficult to understand:

```
+    # match 1 or more times
?    # match 0 or 1 time
*    # match 0 or more times
```

The last two quantifiers can be the most confusing because the concept of matching something *zero times* is not always easy to grasp. Consider the following regex:

```
$text =~ m/A*/;
```

This reads as 'match if $text contains zero or more copies of the letter A.' This means it will *always* match because every possible string either contains the letter A or it doesn't. Therefore this quantifier should always be used as part of some larger pattern:

```
$text =~ m/[a-z]+ *[0-9]+/;
```

This pattern simply matches a word (one or more letters) followed by any number of spaces (including none at all) and then followed by one or more numbers.

We can now use all of the regexes we have seen so far in order to tackle the telephone number problem. First, let's ignore the issues of parentheses that might occur in phone numbers, and see if we can have one regex that captures all other possibilities:

```
$number =~ m/^[0-9]{3}[ -]?[0-9]{3}[ -]?[0-9]{4}$/;
```

This regex will match any phone number which has the following structure:

(1) Starts with three digits: `^[0-9]{3}`
(2) Zero or one space/hyphen characters: `[ -]?`
(3) Any three digits: `[0-9]{3}`
(4) Zero or one space/hyphen characters: `[ -]?`
(5) Ends with four digits: `[0-9]{4}$`

Note the use of both anchor characters (^ and $) to ensure that $number would only consist of this pattern.

It is not too much more work to extend this pattern to also allow for parentheses around the first three digits:

```
$number =~ m/^([0-9]{3}|\([0-9]{3}\))[ -]?[0-9]{3}[ -]?[0-9]{4}$/;
```

The first part of this regex has to use backslashes to escape the special meaning of the parentheses characters. You can read the first part as 'match three digits (`[0-9]{3}`) *or* match three digits that are flanked by parentheses (`\([0-9]{3}\)`).' As you can see, regexes can quickly become long and confusing.

Advanced metacharacters

All of the metacharacters we have seen so far are fairly standard among Unix and many different programming languages. There are also some other metacharacters used by Perl and some other languages which both simplify some existing patterns and also make it easier to make complex patterns. These metacharacters all include a leading backslash character and they all replace existing patterns that you might otherwise use. Here are the new metacharacters, followed by the normal patterns they equate to:

```
\w    [a-zA-Z0-9_]    match any word character
\d    [0-9]           match any digit
\s    [ \t\n\r\f]     match any whitespace character
\W    [^\w]           match any non-word character
\D    [^0-9]          match any non-digit
\S    [^\s]           match any non-whitespace character
```

Notice the word metacharacter (\w) will match both lower- and upper-case letters, numbers, and also an underscore character. The whitespace character includes a regular space, a tab character, a newline, and also some things that we've not even talked about before. The last three metacharacters are all negated versions of the first three metacharacters.[95] We can use these new metacharacters to do things such as:

```
$text =~ m/\w+/;           # match a single word
$text =~ m/\w+\s+\w+/;      # match two words separated by some spaces
$text =~ m/\w+\W+\w+/;      # match two words separated by some non-words
$text =~ m/^\S/;            # match if $text doesn't start with whitespace
$text =~ m/(\d|\s)/;        # match if $text contains digits or whitespace
$text =~ m/[\d\s]/;         # same as previous example
$text =~ m/\S+\s{2}\S+/;    # match two spaces between non-space characters
```

Note that the last example could be used to check a piece of text to see if you have accidentally included two space characters between words. By using \S+ rather than \w+, it doesn't matter if a word is followed by a period, comma, etc. It can take time to learn all of these metacharacters, but it is worth the effort.

Problem 4.27.1 All of the short code examples so far have only used the matching operator. However, we can also use regular expressions in conjunction with the substitution operator. Imagine you had the following information in a file:

```
1 Spider-Man
2 Mr. Fantastic
3 Superman
4 Iron Man
5 Captain America
6 The Flash
7 Batman
```

You want to write a script to process this file and look for all the superheroes who have 'man' as part of their name. You then want to remove the leading number and space character that starts each line, and then print out just the hero's name.

Problem 4.27.2 Write a script that would match only those lines in the following text that contain US state abbreviations (two letters) and zip codes (five digits):

```
CA 95616
95618
CA 90210
TX
DC 20500
```

Assume that this text is inside a file you need to process. The script should print some suitable output for those lines that contain both fields, but also warn if either field is missing. A good solution will also print some suitable message if neither of these fields are present.

---

[95] It takes a while to start thinking in terms of 'non-words' and 'non-digits,' but they are very useful.

Let's go to (W|w)ork!

The previous chapter introduced what might have seemed like hundreds of different regular expression (regex) metacharacters, but it didn't really cover many of the powerful ways in which you can use regexes in your scripts. This chapter will go over some of the ways you can put regexes to good use. We could devote many more chapters to regexes (and we will devote at least one more in Part 6), but this chapter should be enough to arm you with all of the basics.

Using regular expressions with the split function

The binding operator (=~) is not the only place where we can use regexes. Another common use is in conjunction with the split() function. We have already seen this function before (in Chapter 4.15) and we learned how we could turn a scalar variable into an array by splitting it at a fixed character:

```
my @array = split(/","/, $string); # split on commas
```

But sometimes we will have data which is not so neatly separated. What if we have a line of text with six text fields that are separated by variable amounts of whitespace:

```
my $string = "First_name Second_name Age    DOB Height    Weight";
my @array = split(/" "/, $string); # split on space character
```

Using this code would result in @array containing many empty elements because of the multiple space characters between words in the input string. Of course, it is not always easy to tell whether something consists of multiple spaces or a mixture of spaces and tab characters. The better solution is to use a regex to split the string:

```
my @array = split(/\s+/, $string); # split on whitespace character(s)
```

This pattern (\s+) will use 'one or more whitespace characters' as the delimiter when splitting $string. It therefore doesn't matter if the fields are separated by spaces or tabs, and this method will ensure that @array ends up with just six elements.

If you use split() without specifying the pattern or the variable, the pattern is assumed to be whitespace and $_ is used as the variable. In addition, leading whitespace is removed from the line. The following code snippet will print the first and third columns of a space-delimited file.

```
while (<>) {
    my @array = split;
    print $array[0], "\t", $array[2], "\n";
}
```

Using variables inside regular expressions

So far we have used various patterns to match against strings, but these patterns have also been text strings. We can also use variables inside the matching operator and Perl will use the contents of that variable as the pattern.

Example 4.28.1 This simple script takes some input from the command line and tests whether that input matches the string 'REGULAR EXPRESSIONS'. You should try running this script with the following commands:

```
$ pattern_test.pl "REG"
$ pattern_test.pl "LAR\sEXP"
$ pattern_test.pl "^\w{7}\s\w{11}$"
```

```
1.   #!/usr/bin/perl
2.   # pattern_test.pl
3.   use strict; use warnings;
4.
5.   die "Please provide a pattern\n" if (@ARGV != 1);
6.
7.   my $pattern = $ARGV[0];
8.   my $text = "REGULAR EXPRESSIONS";
9.
10.  if ($text =~ m/$pattern/){
11.      print "\'$pattern\' matches \'$text\'\n"
12.  } else {
13.      print "\'$pattern\' doesn't match \'$text\'\n"
14.  }
```

Understanding the script

Line 7 takes the input string and assigns it to $pattern. This variable is then used as a pattern to compare to $text (line 10). A simple if-else statement prints one of two things depending on whether the pattern matches the text.

You should hopefully notice that all of the example commands work. This means you can include any regex metacharacter inside $pattern and it is understood by the matching operator.

Note that you could also use multiple variables to specify different parts of a regex. The following code would match if the $text variable contained either Batman, Batwoman, Catman, or Catwoman.

```
my ($x, $y, $z) = ("bc", "man", "woman");
print "Match\n" if $text =~ m/[$x]at($y|$z)/i;
```

How to capture parts of a matching pattern

It is not always enough to know that some text matches a particular pattern. We often want to extract just the specific parts that match. Fortunately, this is very easy to do in Perl. We have already seen the grouping metacharacters (), which are used in conjunction with the alternation metacharacter |. These grouping metacharacters can also be used to surround parts of a pattern that we want to be able to extract. If a pattern matches a string, then the grouped parts of the pattern are stored in some special variables. The first group gets saved in a variable called $1, the second group gets stored in $2, the third group in $3, and so on. Imagine that we want to capture two names from a string and put those in separate variables:

```
"Batman vs Catwoman" =~ m/(\w+) vs (\w+)/;
my $hero    = $1;
my $villain = $2;
```

In this example we make groups from the two parts of the pattern that specify words. These matches are automatically transferred to the special variables $1 and $2, which we immediately assign to some new variables. It is common to reassign the special variables as soon as possible because the contents of $1 might be replaced if the matching operator is used again:

```
"Weather is sunny"  =~ m/Weather is (\w+)/;   $1 contains 'sunny'
"Weather is cloudy" =~ m/Weather is (\w+)/;   $1 now contains 'cloudy'
```

Example 4.28.2 Let's imagine we have a file ('dates.txt') which contains dates as part of some longer sentences.[96] It might look something like this:

```
I was born on 03/01/1982
On 11/23/2010 it will be my thirtieth birthday
I am going to be 21 this year
I'm looking forward to starting my new job on 06/15/10
Something big is going to happen on 8/8 this year
10/29/64 was the day that we were married
```

Notice that the third line does not contain any date at all, and the fifth line contains an incomplete date (no year is specified). Also, some dates use four digits for the year, and others use two. The following script will process this file and print out the day, month, and year for any line that contains a complete date. To run this script, you'll need to recreate the dates.txt file and specify this when you run the script:

```
$ find_dates.pl dates.txt
```

```
1.   #!/usr/bin/perl
2.   # find_dates.pl
3.   use strict; use warnings;
4.
5.   die "Please provide an input file\n" if (@ARGV != 1);
6.
7.   while (my $line = <>) {
8.       next unless $line =~ m#\d{1,2}/\d{1,2}/\d{2,4}#;
```

---

[96] We will use the US month/day/year convention.

```
9.
10.        $line =~ m#(\d{1,2})/(\d{1,2})/(\d{2,4})#;
11.        my ($month, $day, $year) = ($1, $2, $3);
12.        print "Month = $month, Day = $day, Year = $year\n";
13.  }
```

Understanding the script

Line 7 introduces a while loop that will loop through all of the lines in the specified file.

The first thing we do inside the loop is check whether $line contains a valid date. To do this we use a regex \d{1,2}/\d{1,2}/\d{2,4} to match the date. This is not a perfect solution as the year component will also match a three-digit year.[97] Because our regex expression contains forward-slashes, we use a different delimiter (#) for the matching operator.

Line 8 checks for a match to our pattern as part of a next unless construct. The next command will skip to the next input line *unless* $line contains a date pattern. We could have also written this with reversed logic, i.e., perform the next command if $line *doesn't* match a date:

```
next if $line !~ m#\d{1,2}/\d{1,2}/\d{2,4}#;
```

When we reach line 10 we know that $line must contain a date, but we don't yet know where it is. The regex on line 10 is the same as the one on line 8, but now we place grouping metacharacters around the parts of the pattern which will become the month, day, and year.

Line 11 uses list context to reassign the matching patterns from the special variables ($1, $2, and $3) to new variable names.

Note that, in practice, we would not need to include line 10 at all, as we could instead include the grouping metacharacters as part of line 8:

```
8.     next unless $line =~ m#(\d{1,2})/(\d{1,2})/(\d{2,4})#;
```

This line of code would perform the twin functions of (1) checking to see whether there is a match, and (2) capturing specific parts of the input to $1, $2, etc.

Using the special variables ($1, $2, etc.) in substitutions

We often have text from which we want to remove characters from either end. For example, maybe the text contains quotation marks that we want to delete, or maybe we want to strip the 'www' and '.com' parts from a web address. This becomes very easy to do when using regexes with the substitution operator. We can use the grouping operators to capture the part we want to keep and then just use $1 as the replacement part of the substitution operator:

```
my $url = "www.perlforever.com";
$url =~ s/www\.(\w+)\.com/$1/; # $url now contains 'perlforever'
```

---

[97] We could also make a pattern for a two- *or* four-digit year with (\d{2}|\d{4}).

Looping through multiple matches

We can use the grouping metacharacters to capture multiple parts of a matching string, but this is only useful if you know exactly how many matches to expect. If you don't know how many matches will be in a string, then we need to do something different. Imagine we have a string that contains an unknown number of email addresses that we wish to capture:[98]

```
my $text = 'Email me at captain.kirk@starfleet.com or jim_kirk@gmail.com';
```

We could make a regular expression to capture a valid email address,[99] but if we then try to use this with the matching operator, it will only ever match the first email address:

```
if ($text =~ m/([\w\.]+@[\w\.]+\.[a-z]{2,4})/i) {
    my $email = $1; # $email gets captain.kirk@starfleet.com
}
```

To solve this problem we need to be able to loop through all possible matches in our input string. This can be achieved by using a while loop in conjunction with the global match option (/g), which we first saw in Chapter 4.11:

```
while ($text =~ m/([\w\.]+@[\w\.]+\.[a-z]{2,4})/gi) {
    my $email = $1;
}
```

This loop reads as 'while we can keep on finding matches to the pattern in $text, put each match in turn into $1.' It is important to include the global match option after the pattern, otherwise the loop will never end.

Capturing all matches in one go

Rather than looping through all possible matches in a string, we can also assign the result of the matching operator to an array, and each array element will contain each of the successive matches. You still need to use the global match option:

```
my @email = ($text =~ m/[\w\.]+@[\w\.]+\.[a-z]{2,4}/gi);
# $email[0] gets captain.kirk@starfleet.com
# $email[1] gets jim_kirk@gmail.com
```

Capturing patterns like this in *list context* can also be useful when you know exactly how many matches there will be, as you can avoid the need to use $1, $2, etc:

```
my $text = "The meeting will be at 4:30 pm";
my ($hour, $minute) = ($text =~ m/\s+(\d+):(\d+)\s+[ap]m/);
```

---

[98] The string is deliberately in single quotes so that we don't need to escape the @ sign with a backlash. Otherwise, Perl would think we were including some array names, and would try to replace them with their contents.

[99] This is not a straightforward problem, and the regular expression we use here is a very simplified pattern that will not match all email addresses.

Note that this syntax works even if you wanted to capture just one pattern from a string:

```
my ($day) = ($text =~ m/\s+(\w+day)/);
```

Summary

Regular expressions are an incredibly useful part of Perl and are frequently used in areas such as bioinformatics. Try not to be overwhelmed by all of the different metacharacters we have introduced, but remember if you are ready to learn even more, then see Chapter 6.1, which covers some advanced regular expression topics.

Problem 4.28.1 The DNA sequences of protein-coding genes begin with a start codon (ATG), then have a number of three-nucleotide (nt) codons before ending with a stop codon (TGA, TAA, or TAG). Create a regex that will test whether an unknown variable ($seq) contains a valid coding sequence. Award yourself bonus points if your regex copes with the following criteria:

(1) $seq can be upper or lower case (this varies a lot in bioinformatics!).
(2) $seq might also contain N characters (unknown bases) in addition to A, C, G, and T.
(3) $seq is at least 100 nt long (most coding genes are longer than this).
(4) $seq has a length that is a multiple of three nt, i.e., $seq consists of codons.

Perl is like the Force: it has a light side, a dark side, and it holds the Universe together[100]

One of the things Perl excels at is being the glue between different programs. Consider the following scenario. Your boss emails you several .doc files and instructs you to perform the following procedures on each file: (1) count the number of words; (2) determine if the word 'patent' appears in the text; (3) convert the .doc file to a text file, making sure that all characters are upper case. If you were only emailed one file, you would probably open the document in a word processing program to complete the tasks. But what would you do if there were thousands of files? If you're a Perl programmer, you write a few lines of code to glue a couple of programs together and then go and have lunch while the program runs.

Capturing program output with the backticks operator

Perl's backticks operator ` is a very convenient way to run commands and capture the output directly. The backtick looks like an apostrophe but is actually a different character and will be hiding somewhere on your keyboard.[101] It's another one of those characters that should always be used in pairs. Whatever you put between a pair of backticks will be executed in the Unix shell, and the output will be returned to you in an array or scalar, depending on how you asked for it.

Example 4.29.1 This example runs the Unix `ls` command twice using the backticks operator and simply prints out the name of each file that `ls` finds.

```
1.   #!/usr/bin/perl
2.   # backtick.pl
3.   use strict; use warnings;
4.
5.   my @files = `ls`; # output of command returned as an array
6.   foreach my $file (@files) {print $file}
7.
8.   my $files = `ls`; # output of command returned as a string
9.   print $files;
```

Understanding the script
On line 5, we place the `ls` command inside a pair of backticks. The first thing Perl will do is run the command. Because the backticks are on the right-hand side of an assignment, Perl will then assign whatever output is provided by the command to the named variable (in this case an array). Each element of the @files array will contain a file name followed by a newline.

---

[100] Originally, this was a comment about duct tape, but it is equally true of Perl.
[101] The location depends on your keyboard layout, but try the upper-left corner to begin with.

Lines 8–9 are similar to lines 5–6, but the output from the backticks operator is assigned to a single scalar variable rather than to an array. Note that $files contains all of the newline characters from the output of the ls command.

Note that you can also use Perl's qx// operator as an alternative to using backticks. They behave in an identical manner, but the qx operator allows you to choose any pair of delimiters to surround the command you want to execute – the following lines of code are all equivalent to each other:

```perl
my @files = `ls $dir`;
my @files = qx/ls $dir/;
my @files = qx(ls $dir);
my @files = qx'ls $dir';
```

The last example uses single-quote characters as the delimiter, which means the command does not undergo variable interpolation.

How to check whether external commands ran successfully

If you capture program output using the backticks operator, there is always the possibility the program didn't run properly. This may cause your captured output to consist of an error message, or maybe be completely blank. It is a good idea to test whether programs ran properly or not; we can do this by inspecting the value of yet another of Perl's special variables, $?. This variable stores the *exit status* from any program you try running. If a program runs successfully then the value of $? will be zero.[102]

Example 4.29.2 This simple script uses backticks to run the Unix which command in order to determine the location of a program specified on the command line.

```perl
1.   #!/usr/bin/perl
2.   # backtick_error_test.pl
3.   use strict; use warnings;
4.
5.   die "Please specify the name of a Unix command\n" if (@ARGV != 1);
6.   my $program = shift;
7.
8.   my $location = `which $program`;
9.
10.  if ($?) {
11.      print "Error running \'which $program\' command\n";
12.  } else {
13.      print "$program is located at $location";
14.  }
```

---

[102] Of course, just because a program runs successfully, it might still not contain the output you want.

Understanding the script

As long as one, and only one, thing is specified on the command line, it gets passed to a variable $program using the shift function (lines 5–6).

On line 8, we use backticks to run the Unix which command and the results are saved to a variable. We then test the status of $? using an if-else block which either prints an error message or the location of the specified command.

Running other programs with the system() function

The backticks operator should mostly be used when you want to capture the output from a command. Sometimes you might want your Perl script to run other commands (including other Perl scripts) which don't return any output. In these situations you should use the system function. Like the backticks operator, system runs a command in the shell and sets $? to zero if it succeeds, or a non-zero error code if it fails.

Example 4.29.3 Here is a short script that just runs the Unix clear command to clear the screen. As this command produces no output, we will use system to run it.[103]

```
1.   #!/usr/bin/perl
2.   # clearscreen.pl
3.   use strict; use warnings;
4.
5.   system("clear");
6.   die "error running clear\n" if ($?);
```

Understanding the script

On line 5 we run the clear command via the system function and then we test the exit status on line 6 by inspecting $?, and issue a die command if the clear command did not run correctly.[104]

In practice there are easier ways of testing the exit status of a system function call than the way we just showed you. If the system function worked without any problem it will return a zero. This can be tested immediately as part of a logical test using the or operator. Remember that when we use the or operator, we provide a list of possible conditions to test. Perl checks each condition *in turn* and as soon as any condition returns true, Perl knows that it doesn't need to check the remaining conditions. Previously we have used the or operator as part of an if statement, but let's see how it can be used on its own with the system function:

```
system("$command") == 0 or die "Error running $command\n";
```

---

[103] This would be a somewhat pointless script if all it is going to do is run the clear command. But for the sake of teaching you some Perl, we will lower our standards and be a little bit pointless.
[104] Having a die command as the last line of a script is also a little bit pointless. Perhaps we should rename this script to pointless.pl.

If the `system` function returns zero, then the '`== 0`' part of the code will evaluate as true and therefore Perl doesn't need to evaluate the rest of the `or` condition. This means that nothing else happens, which is what you want. Alternatively, if `system` fails, it will contain a non-zero exit status. This is not equal to zero and so Perl proceeds to evaluate the next part of the `or` condition. In this case, the `die` function must be run before Perl can evaluate the truth of the entire statement, and running the `die` function means the script will end.[105]

The `system` function is useful when you want to run a command and you don't need to capture the output directly into your script. If you do need the output, you can redirect the results of the `system` call to a file and then read the output from a file following an `open` statement.

You can also use the `system()` function by passing it a list of items, where the first item would specify the program you want to run, and the subsequent items would become arguments to that program:

```
system("ls -l", $file) == 0 or die "Can't run ls -l against $file\n";
```

Using pipes to send and receive output from a program

Previously we saw that the `open` function allows one to read from, or write to, files. You can also use the `open` function to read from, or write to, programs. This is done by appending or prepending a pipe character to the command to be run as part of the `open` function.[106] The syntax can look very similar to how you open a file for reading:

```
open(my $fh, "file");    # open the file called 'file' for reading
open(my $fh, "prog |"); # run the command 'prog' and capture output
open(my $fh, "| prog"); # send input to the command called 'prog'
```

Example 4.29.4 This script uses two pipes: one to read output from the `ls` command, and one to send input to the `wc` command.

```
1.   #!/usr/bin/perl
2.   # pipes.pl
3.   use strict; use warnings;
4.
5.   open(my $in, "ls |") or die "Can't open pipe: $!";
6.   while (<$in>) {print}
7.   close $in;
8.
9.   open(my $out, "| wc") or die "Can't open pipe: $!";
10.  print $out "this sentence has 1 line, 10 words, and 51 letters\n";
11.  close $out;
```

---

[105] Alternatively, use the syntax: `system("$command") && die "Can't run $command\n";`
[106] The syntax borrows a lot from how we use pipelines in Unix (see Chapter 5.7).

Understanding the script

Lines 5–7 show how you can read from a process. Line 5 associates the $in filehandle with the ls command. Note the pipe symbol that *follows* the command name. Once the filehandle is connected to ls, each time the filehandle $in is accessed, *one line* of output from the ls program will be read.

Line 6 uses a while loop to read one line at a time from the output of the ls command. Each line is stored in $_ and then printed. Line 7 closes the pipe.[107]

Lines 9–11 show how you can write to a process, in this case the wc command. Here, the pipe symbol is placed *before* the program name, just as you would do on a Unix command line (see Chapter 5.7).

Line 10 prints to the filehandle specified by $out, which effectively means that the print command is being used to send input to the wc program. The wc command will only print its output when the filehandle is closed or when the script is terminated. Note that this line does not print the text 'this sentence has 1 line, 10 words, and 51 letters' to the screen. This string is only ever used as input to the wc command.

Being able to read from and write to programs is very powerful. You can automate a lot of tasks this way. What if you want to read *and* write to a process? You might try to do something like this:

```
open(my $inout, "| program |") or die; # sorry, does not work
```

Unfortunately, filehandles only support reading or writing to a process, but not both. If you want to read and write to the same process, you have to use the Perl modules IPC::Open2 or IPC::Open3.[108]

---

[107] You don't have to close your pipes necessarily, because they will automatically close when the variable goes out of scope. But it's never a bad idea.

[108] Perl modules are pre-packaged parcels of Perl code that are used to achieve a specific goal or tackle a certain problem. An online repository of free Perl modules is available at http://cpan.org. Chapter 6.2 of this book contains a brief introduction on how to use modules.

Subroutines can get you out of a dysfunctional relationship

Writing efficient code

Good Perl programs should not try to reinvent the wheel, and the environmentally friendly mantra of 'Reduce, Reuse, Recycle' also works well as a manifesto for programming. But what should you do if two (or more) parts of your Perl script need to do exactly the same thing? Maybe you have a script that needs to process two files and count how many lines match a certain pattern in both files. Part of your script might end up looking like this:

```perl
my ($file_A, $file_B, $pattern) = @ARGV;

# file 1
my $line_count1 = 0;
open(my $input_A, "$file_A") or die "Can't open $file_A $!";
while (<$input_A>) {
    chomp;
    $line_count1++ if m/$pattern/;
}
close($input_A);
print "$line_count1 lines matched $pattern\n";

# file 2
my $line_count2 = 0;
open(my $input_B, "$file_B") or die "Can't open $file_B $!";
while (<$input_B>) {
    chomp;
    $line_count2++ if m/$pattern/;
}
close($input_B);
print "$line_count2 lines matched $pattern\n";
```

Hopefully you will have noticed the tremendous amount of redundancy in this code. The two halves of the script are essentially identical, except that they are working with different files, and use different counter variables. It would be much more efficient if we could parcel up all of the lines of code that are doing the same thing, and treat them almost as a separate script. Then we could just send it the name of a file and a pattern as input, and get back the line count as output. Well, what we have just described is something called a *subroutine*, which are used by all programming languages, including Perl.

The syntax of a subroutine

To make a basic subroutine we just need to create a block of code and give it a name using the special sub keyword. We then call that code from somewhere else in the script using the name we have given it. The simplest subroutines can be used to achieve some repetitive tasks, such as printing a set of error messages and exiting a script:

```perl
sub print_error {
    print "You did something wrong\n";
    print "We cannot be held accountable for your stupidity\n";
    print "Goodbye\n";
    exit(0);
}
```

This code describes a subroutine called `print_error`. It is declared with the `sub` keyword and then all of the code inside the subroutine is indented within a block. Subroutines can be defined anywhere in a program, but it is common to put them at the end of a Perl script. In other languages you might find them at the beginning.[109] You can do it either way, but try to be consistent. We can run the code contained within a subroutine just by specifying the name of it from somewhere else in the script. Sometimes you might want to call a subroutine *if* some condition has been met, but you can also just call it on its own:

```perl
# subroutine call with no conditionals
print_error();

# subroutine calls based on conditional statements
print_error() if ($bad_thing_happened);

unless ($result eq "good") {
    print_error();
}
```

We can imagine this code being used in a script that has to check for errors in many different places. The advantage of using a subroutine for this is that if you needed to change the error message, you only have to change the code in *one* place.

Note that when calling subroutines we generally include parentheses after the subroutine name.[110] Because of this, you might be thinking, 'Hmm, those subroutine calls look an awful lot like Perl functions.' Well, that's because they are! Writing a subroutine is just the same as writing your own function, and in this sense the words 'subroutine' and 'function' are pretty much interchangeable.

Passing data to subroutines

Subroutines become much more powerful when you pass data to them and you can pass one or more items to a subroutine in a number of ways. Items to be passed should be specified between the parentheses that follow the subroutine name:

---

[109] Subroutine definitions are global, so your script will know where to find them even if the code that uses a subroutine occurs earlier in the script than the subroutine itself.

[110] But also see the last section of this chapter.

```
fruits("apples");          # passes one string to the 'fruits' subroutine
add_them_up(1, 5, 4);      # passes three numbers
mean_size($small, $big);   # passes two variables
math("mean", @numbers);    # pass a string and an array
how_many_words(@words);    # passes the contents of the @words array
count(split /\s+/, $text); # passes the results of the split() function
```

No matter *how* you specify the parameters to be passed, they are *always* passed as a single list of scalars. In the fourth example above, a string and an array are passed. The array will lose its identity and will be added on to the end of a list that starts with the specified string.

Upon arrival in a subroutine the contents of the list that has been passed are immediately placed in a special Perl variable called @_.[111] You can access this array like any normal array – e.g., $_[0] will contain the name of the first item you pass to the subroutine. However, just like when you work with @ARGV, it is often best to first reassign the contents of @_ to new variables that have more meaningful names. Let's imagine we have a subroutine called process_fruit() and we want to pass three items to it:

```
process_fruit("apple", 5, "green");
```

Now we will consider four different ways we could potentially deal with these items once they 'arrive' inside the subroutine:

```
# method 1, list assignment
sub process_fruit {
    my ($fruit, $quantity, $color) = @_;
}

# method 2, array assignment
sub process_fruit {
    my @fruit_details = @_;
}

# method 3, scalar and array assignment
sub process_fruit {
    my ($fruit, @details) = @_;
}

# method 4, using shift
sub process_fruit {
    my $fruit = shift;
    my $quantity = shift;
    my $color = shift;
}
```

---

[111] Apart from also having an underscore, this special array has nothing to do with the special variable $_. Except that they are both 'special' of course!

The first method just reassigns the @_ array to a new list of variables. This is useful if the things being sent represent different categories of data. Alternatively, if you are passing a list of the same type of things, it is often better to assign the data to an array (method 2). For example, if your subroutine was calculating some property from a set of numbers, you would probably want to keep those numbers in a single array. Method 3 shows that you can still keep some items as single scalars, and some items as an array. The last method exploits a special property of the shift() function. Inside a subroutine, each invocation of shift will remove an item from the @_ array – i.e., shift is just a shortcut for shift(@_). It is important to note that all of these methods pass *copies* of the original data. You can modify data in a subroutine and it won't affect the original version of that data.[112]

Alternative syntaxes when calling subroutines

So far we have been calling all of our subroutines in exactly the same way, placing parentheses after the subroutine name, and optionally specifying a list inside the parentheses. This is our preferred syntax, and it is also the same syntax used by most programming languages. We suggest you do not deviate from this practice. However, you will see some alternative syntaxes. Let's look at all the ways that we could potentially call the print_error subroutine we saw earlier.

```
1.    print_error();        # preferred, with no parameters
2.    print_error(2, 3);    # preferred, with two parameters
3.    print_error;          # omitting trailing parentheses
4.    print_error 2, 3;     # two parameters, no parentheses
5.    &print_error;         # prefixed by an ampersand character
6.    main::print_error;    # prefixed by package name
7.    &::print_error 2, 3;  # oh, give me a break!
8.    $obj->print_error();  # OOP syntax
```

Lines 1 and 2 demonstrate the preferred syntax. Please use this always.

Lines 3 and 4 omit the trailing parentheses. It is common to omit parentheses for Perl built-in functions like print. You can omit parentheses for your own subroutines if you declare your subroutines before you call them (i.e., put them at the start of your program). You can also omit parentheses if you *prototype* your functions.[113]

Line 5 prefixes an ampersand before the function name. The ampersand was actually a required part of the Perl syntax at one time. Now it is only required for function references, which is not a topic we cover in this book.

---

[112] There is a way for a subroutine to process the original version of data and not a copy, but we won't be able to explain this until we have introduced the subject of *references* in Chapter 6.3. This means that if your script has a large array that takes up 1 GB of memory, and you then pass that array to a subroutine, then your script will need an extra 1 GB of memory.

[113] Prototyping a subroutine tells it what kinds of arguments it takes. This is generally necessary in compiled programming languages. In Perl, it is unnecessary and provides only a minor convenience. We have therefore omitted its description.

Line 6 adds the package name to the prefix. If this statement makes little sense, don't worry about it. It's not so important right now. But consider reading Chapters 4.7 and 7.4.

Line 7 includes the ampersand, defaults the package name to main, and includes arguments without parentheses. All legal, but we'll beat you over the head if we catch you doing it.

Line 8 shows the object-oriented syntax. See Chapter 6.8 for more information.

One good (re)turn deserves another

Sometimes it is enough to just send data *to* a subroutine, but usually we also want to *return* a value (or values). This is particularly true for subroutines which process data in order to produce a result from a calculation. Returning data from a subroutine can be done by using the aptly named `return` keyword. You can return multiple values in a return statement, or even none. Sometimes we just return 1 or 0 to indicate success or failure. Here is an example of a subroutine that returns the product of two numbers:

```
my $result = product(4, 5);

sub product {
    my ($x, $y) = @_;
    my $product = $x * $y;
    return $product;
}
```

This code returns the contents of the variable `$product` from the subroutine and this is immediately assigned to the `$result` variable. If you wanted to be concise, you could return the result of a calculation directly, without first storing it in a separate value:[114]

```
sub calculation {
    my ($x, $y) = @_;
    return $x * $y;
}
```

In most cases we will return values from subroutines, which are immediately assigned to new variables. However, if you don't need to store the returned value, you can still do things like print it:

```
# print but don't store returned value
print product(4, 5),"\n";
```

You should also be aware that if you return a value but don't do anything with it, then Perl will not complain or produce any error message. The following code example is valid, if not a little pointless:

```
# don't do anything with returned value
product(4, 5);
```

You might find this next part strange, but you can also return values from a subroutine without using the `return` operator. Consider the following code:

---

[114] If you wanted to be *really* concise, this whole subroutine could be further reduced to:
```
sub product { return $_[0] * $_[1] }
```

```perl
print "1 + 2 = ", add(1, 2), "\n";

sub add {
    my ($x, $y) = @_;
    $x + $y;
}
```

This code will result in the string '1 + 2 = 3' being printed, even though we didn't seem to return anything from the subroutine. This is because in the absence of any return operator, Perl will always return *the last expression evaluated*. Note that the last expression to be evaluated may not always be the last line of the subroutine.

Returning multiple values from a subroutine

Just as you can send multiple items to a subroutine, you can also return multiple items. Returned items are always passed as a single list of scalars – i.e., in the same way that parameters are passed to a subroutine. Let's consider how we might deal with the return values from some fictional (and unseen) subroutines:

```perl
my @numbers      = (1, 2, 5, 8, 10);

my $mean         = mean(@numbers);          # send array, assign to scalar
my ($min, $max)  = min_and_max(@numbers);   # send array, assign to list
my @odd_numbers  = find_odd(@numbers);      # send array, assign to array
my $odd_mean     = mean(find_odd(@numbers));
```

In the last example, we send an array to one subroutine, find_odd(), and then the returned values are immediately sent to another subroutine, mean(), the result of which is assigned to a single variable ($odd_mean).

Using multiple return statements

You can place multiple return statements within a subroutine and it is often desirable to do so. Usually this will be done as part of a set of conditional statements, e.g., *if A then return X, else if B then return Y*. When Perl reaches any return statement it will exit the subroutine at that point.

In Chapter 4.16 we showed some code that calculated whether any given year was a leap year or not; let's revisit that code and place it in a subroutine:

```perl
sub leap_year {
    my ($year) = @_;
    if ((($year % 4 == 0) and ($year % 100 != 0)) or ($year % 400 == 0)) {
        return 1;
    } else {
        return 0;
    }
}
```

This subroutine is passed a single variable (hopefully a four-digit year) and it just returns '1' if it is a leap year or '0' if it isn't. We could alternatively return text such as 'This is a leap year,' but in some ways it is better to keep the subroutine simple and let the calling code decide what to do. Returning true/false values such as 1 or 0 keeps our code as flexible as possible and allows us to access the subroutine in different ways:

```
# basic if-else
if (leap_year($year)) {
    print "Leap year\n";
} else {
    print "Not a leap year\n";
}

# only print if true
print "Leap year\n" if leap_year($year);

# print unless true
print "Not a leap year\n" unless leap_year($year);
```

Writing a custom sorting function

In Chapter 4.18 we learned to use the sort function and we also learned how we could modify the default behavior of this function by including a short block of code after the function name:

```
@data = sort {$a <=> $b or $a cmp $b} @data;
```

Subroutines offer another way to write even more complex sort functions, or just allow us to hide the complexity of the sorting code.

Example 4.31.1 This script creates a subroutine to perform the same alphanumeric sort as just shown.

```
1.   #!/usr/bin/perl
2.   use strict; use warnings;
3.
4.   my @data = qw(3.14 1000 21 red blue yellow);
5.   print "before: @data\n";
6.   @data = sort alphanumeric @data;
7.   print "after: @data\n";
8.
9.   sub alphanumeric {
10.      no warnings;
11.      my $result = $a <=> $b or $a cmp $b;
12.      use warnings;
13.      return $result;
14.  }
```

Understanding the script

On line 6, the contents of the @data array are sent to the alphanumeric subroutine. Whatever this subroutine returns is stored back to @data, overwriting the original contents of the array.

To quiet the warning messages associated with comparing strings as numbers, we need to temporarily turn off the warnings pragma on line 10. The comparison is performed on line 11 and stored in a new variable. The result is not immediately returned because we need to turn warnings back on at line 12. Note that $a and $b were not declared as lexical variables because these are special variables defined by Perl.

The output is free of warning messages with both strings and numbers sorted properly.

```
before: 3.14 1000 21 red blue yellow
after: red blue yellow 3.14 21 1000
```

You can make up many kinds of sorting function. These do not need to use the <=> or cmp operators. As long as your function returns −1, 0, or +1, it will work with Perl's sort.

When to use subroutines?

Subroutines should be used any time, perhaps even *every* time, you find yourself writing the same lines of code in multiple places. Even if that repeated code is only a few lines long, there are benefits to be gained by placing it in a subroutine. Imagine you have two lines of code that are reused in ten different places in your script. If you decide to update the functionality of that code (or fix a bug), you would have to edit your script in ten different places. If you put the code in a subroutine, you would only have to make one change.

Subroutines can also help improve the readability of your code. Rather than see all of the details of how you calculate some mathematical function, it might be cleaner to keep that code in a subroutine, which keeps it hidden from the main body of the code.

Once you develop your own useful functions by using subroutines, you will find that you want to use them again and again. One way to reuse code is to simply copy-and-paste your functions from one program to another. However, an ever better way is to place that subroutine in a function *library*. This will be explored in Chapter 6.2.

Part 4 summary 4.32

if ($current_Perl_knowledge < $target_Perl_knowledge) {redo} else {next}

If you have made it this far and have worked through all of the Perl chapters, then we have to congratulate you. Well done! We have covered a lot of material and hopefully you will have understood much of it. We also hope that you can already see practical uses and applications for the Perl you have learned. This part of the book was designed to equip you with the basic set of skills you will need to accomplish most problems that you might want to tackle. But don't let the word 'basic' fool you, the skills we have covered are both diverse and powerful. You can now write scripts to perform mathematical operations; process text files and do complex pattern matching; and loop through lists and arrays in myriad ways.

You may want to stop now and take a breather. You've certainly earned it! Take a short vacation maybe, but whatever you do, don't leave it too long before you return to Perl. Like learning any language (including French, Spanish, etc.) memories will fade with time if you don't keep with it. The best way to really understand Perl is to keep on using it and maybe even start to consider what lies ahead. And what does lie ahead? Well, next is a return to Unix to learn some more advanced concepts. However, if you are hungry to learn even more Perl, then feel free to jump ahead to Part 6 'Advanced Perl' to see more of the powerful ways you can put Perl to work for you. The distinction between what we have called 'Essential' and 'Advanced' Perl is very arbitrary. The topics in Part 6 may allow you to tackle new problems, but in some cases they may just help you tackle existing problems in a more efficient manner.

Advanced Unix 5

Where you will become BFF with GFF

The average life scientist will spend a lot of time working with data. Increasingly, this data will exist in the form of large data sets that will have been downloaded or extracted from one of the many large biological databases that are accessible on the internet. Such *in silico* data might consist of a small number of very large files, a large number of very small files, or anything in between these extremes. However, in many cases the default file format for those files will be plain text. The actual format of the plain-text file will vary a lot, but the fact that it is plain text means there are many Unix commands that are just waiting to get their hands on your data.

This part of the book will cover a small number of extremely powerful Unix commands that are well suited for slicing and dicing text files. If you are reading this part of the book *after* working through the 'Essential Perl' section then you will spot the similarities between some of these commands and some of the operators in Perl. Conversely, if you have yet to start learning Perl, you will find this section introduces many topics that will be revisited as you learn Perl.

In addition to providing many powerful commands, Unix also lets you combine those commands in myriad combinations that can enable you to tackle complex tasks. In a nutshell, you can set the output from any one Unix command (which goes to the screen by default), to act as input to another Unix command. In this way you can chain together several commands. Such *pipelines* will also be covered in this part of the book.

Example files
To effectively demonstrate the Unix commands that we introduce in this part of the book, we require some example files to work with. These files are freely available from our 'Unix and Perl for Biologists' web site:

http://unixandperl.com.

If you don't have access to these files (or don't want to download them), you will still be able to learn a lot from following the examples in the chapters of this part of the book. We will also show snippets of the example files.

GFF and FASTA file formats
GFF and FASTA are two file formats that are commonly used in bioinformatics for storing genomic sequence information. Some of the examples in this part of the book will make use of some files that are in this format. Wikipedia gives a good overview of both of these file formats, but the following sections provide a brief description.

FASTA file format
It is increasingly common for biologists to have to work with large files that contain computer representations of DNA or protein molecules. These *sequence files* may contain anything from the sequence of a single gene, to the sequence of an entire genome.

Such files can therefore vary in size from a few bytes, up to several gigabytes. The FASTA file format is a commonly used format for representing such sequence data. FASTA files contain one or more sequences and each sequence must have an associated *header line*,[1] which must start with a > character. Apart from that, there are very few restrictions as to how the content of a FASTA file should be structured.[2] The content of each header line should ideally be unique, and headers typically contain some form of sequence identifier that refers to the sequence that follows. Here is a sample from a FASTA file provided by the Saccharomyces Genome Database (SGD):

```
>YAR060C SGDID:S000000086, Chr I from 217148-217483, reverse complement,
  Dubious ORF
ATGTCAAAAACCGATGTCAAAAAATCTCGTGAGGCCTCCGGAATTTTGACGCTGCAAGTC
AATCTACGGGAAAGAAGAAATTTTTTAAACTTAATGCAAAATAAGCTTTTTTCTTGGAAA
ATAAGATTTTCGGCAATAAAAGGTAAATGCAGCCAAAAATCAAAATACTTCAGAAGAAGT
CGTAGCGAGGACTGCTACGTGGAAGCGGATTTGAAGATCCTTTCCAGAACAAGAAGGAGC
CGAAAGCTGCCAGGAACTGTTCCTGATTTTTTAGGAAAACAATTAATAGGTATCTCGTCT
AGCGTAGTATCTCGAGTTTCCAGAAGTTGCAGATAA
>YAR061W SGDID:S000000087, Chr I from 218131-218334, pseudogene
ATGCCTTATCACTATTTATTTTTGGCACTCTTCACCTACCTGGCCACGTCCAATGTTGTT
TCAGGAAGTACACAAGCATGCCTGCCAGTGGGCCCGAGGAAAAATGGGATGAATGTCAAC
TTTTATAAATACTCATTACTGGATTCAACAACGTATTCCTACCCGCAATATATGACTTCT
GGATATGCCTCGAATTGGAATTAG
>YAL030W SNC1 SGDID:S000000028, Chr I from 87287-87388,87502-87753, intron
  sequence removed, Verified ORF
ATGTCGTCATCTACTCCCTTTGACCCTTATGCTCTATCCGAGCACGATGAAGAACGACCC
CAGAATGTACAGTCTAAGTCAAGGACTGCGGAACTACAAGCTGAAATTGATGATACCGTG
GGAATAATGAGAGATAACATAAATAAAGTAGCAGAAAGAGGTGAAAGATTAACGTCCATT
GAAGATAAAGCCGATAACCTAGCGGTCTCAGCCCAAGGCTTTAAGAGGGGTGCCAATAGG
GTCAGAAAAGCCATGTGGTACAAGGATCTAAAAATGAAGATGTGTCTGGCTTTAGTAATC
ATCATATTGCTTGTTGTAATCATCGTCCCCATTGCTGTTCACTTTAGTCGATAG
```

This example shows three sequences, and each is preceded by its own header line.[3] The sequences that follow each header line are all in upper-case characters and each line of sequence contains 60 characters only. In this case, the sequences represent the open reading frames (ORFs) that correspond to some genes that occur on chromosome I of the *Saccharomyces cerevisiae* genome. Each sequence therefore begins with a start codon (ATG) and ends with a valid stop codon. The header line for each ORF contains two different database identifiers, and the third ORF also has a gene designation (SNC1). Additionally, the header lines are being used to store the chromosome coordinates of the ORF along with some other useful information.

---

[1] Also known as the definition or description line.

[2] Please note, however, that some bioinformatics programs or databases will impose their own formatting requirements for a FASTA file (e.g., specifying that sequence characters are only lower case or upper case or requiring a fixed number of sequence characters per line).

[3] The header lines in this example are all contained on a single line, though for display purposes we have had to wrap them across two lines.

Note that FASTA files may contain spaces and/or blank lines among the sequence and some files may even use upper- and lower-case characters to denote different sequence features. It is also possible that the header lines are so packed with information that they become longer than the sequence that follows. The following is another example of a valid, if untidy, FASTA file:

```
>1
ACGTACGT
A C G T A C G T
A  C  G  T  A  C  G  T
>2
gatcaag GTAGGCGATTGCACCAGAGTTTCACGTTAGTTTCAG ttcaccac
```

GFF

GFF (generic feature format) is a text-file format used extensively in bioinformatics. It is used to describe details of genomic features, notably their coordinates in relation to a larger sequence (e.g., a gene, a BAC clone, or an entire chromosome). Historically, GFF was mostly used to describe details of genes (e.g., their exon/intron structure), but is now used to describe just about any feature that can be localized to a region of a sequence. The region in question can range from a single nucleotide (nt) up to the entire length of the sequence.

GFF files always consist of nine tab-delimited columns of information. Each column contains a certain type of information:

(1) Sequence name – this could be the name of a BAC clone or the name/number of a chromosome or chromosome arm.
(2) Source – when most gene information was provided in the form of computer predictions, the source column would typically list the program that was used to predict the gene structure. These days this field is often repurposed for different means, and might contain the name of the database from which the GFF file was produced.
(3) Feature – the feature is the item that the whole line is referring to. It might be a well-known biological entity such as an exon or an intron, or it might refer to more generic elements such as 'repeat_region' or 'similarity.' Typically, this field does not contain any spaces.
(4) Start coordinate of feature.
(5) End coordinate of feature.
(6) Score – this field was commonly used to store details of whatever score a gene prediction program assigned to a feature, i.e., how accurate it believed the feature to be. These days this field is typically left empty, though the file format requires there to be something present, so you will very commonly see a dot character in this column.
(7) Strand information – either a plus or minus symbol to represent which strand of DNA the feature is associated with.

(8) Frame – another field associated with gene structure. Specifically, this refers to the reading frame of the 'exon' or 'CDS' feature and describes the relationship between the first nt of the exon/CDS and the codon that the nucleotide is contained in. Exons/CDSs can either be in frame 0, 1, or 2, which means they begin with the first, second, or third nt of a codon. As this field is meaningless for any feature which isn't an exon or CDS, it is also common to see this field only contain a dot character.

(9) Attribute field – often used to include one or more database identifiers that relate to the feature in question. Some databases add a lot of information to this field which often makes the whole line wrap around your screen.

Here is a very short example of a GFF file, again taken from the SGD database:

```
chrI   SGD   repeat_region   1     62      . - .   ID=TEL01L
chrI   SGD   telomere        1     801     . - .   ID=TEL01L
chrI   SGD   repeat_region   63    336     . - .   ID=TEL01L
chrI   SGD   gene            335   649     . + .   ID=YAL069W
chrI   SGD   CDS             335   649     . + 0   Parent=YAL069
```

Note that features from different lines can overlap in their coordinate space. GFF files will sometimes contain a feature called 'chromosome,' the coordinates of which define the range of all other features. Also note that in the case of the last two lines, there are two different features ('gene' and 'CDS') that occupy the same range of coordinates. Overall, each line in a GFF file should be unique.

As we learn more about the myriad number of features that can describe any given genome, the size of the respective GFF file for that genome continues to grow. It is not uncommon to see GFF files that contain over one million lines. Some genome databases produce separate GFF files for each chromosome, such that the first column is the same for all lines in the file. Other databases will produce a single GFF file for an entire genome. Because each record is contained on one line, GFF files are very suitable for processing by Unix commands and Perl scripts.

Unix users are licensed to (use the) kill (command)

This book will hopefully equip you with the skills to run Unix commands and to write (and run) Perl scripts. These skills are to be treasured; mastery of this knowledge will allow you to write many amazing programs and utilize many powerful Unix commands. However, knowing how to start a program is sometimes less important than knowing how to *stop* it. In computer terminology, programs that are running on your computer (including Unix commands and Perl scripts) are known as *processes*; the ability to stop, pause, and restart such programs is known as *process control*.

Why would we need to stop a program?

In an ideal world we would all write Perl scripts that are perfectly constructed and which always work first time. However, we do not live in that world! Let's consider a problematic Perl script:

Example 5.2.1 The aim of this script is to simply use a for loop to count to 100, but to also print an additional message when the counter reaches a user-specified number. Can you spot what the problem is?

```
1.   #!/usr/bin/perl
2.   # liveforever.pl
3.   use strict; use warnings;
4.
5.   # set target number, check it is less than 100
6.   die "Usage: liveforever.pl <target integer>\n" unless (@ARGV == 1);
7.   my ($target) = @ARGV;
8.   die "Target integer must be less than 100\n" if ($target >= 100);
9.
10.  for (my $i = 0; $i < 100; $i++){
11.
12.      # count up to 100
13.      print "$i\n";
14.
15.      # print special message when $i reaches target number
16.      print "$target was your magic number!\n" if ($i = $target);
17.
18.  }
```

Understanding the script

The error in this script occurs on line 16. The condition for the if statement should only be executed if $i is *equal* to $target. However, the author of this program has made a common mistake of making a variable assignment ($i = $target) rather than testing for equality ($i == $target). The result of this is that each iteration of the for loop will see line 15 evaluated, and every time we get to line 15 we reset $i to the value of whatever $target is.[4] This value will always be less than 100 (the condition of the for loop) and so the loop will run forever.

The result of running the code in the above example will be a script that will never terminate and which will fill up your terminal window with never-ending output. We therefore need to stop the program. But how?

Interrupting a program

The first thing you can do with runaway programs that need to be stopped is to press control + c. This keyboard combination will send the special Unix *interrupt signal* to stop the command. This should stop most basic Perl scripts that are stuck in any sort of endless loop. In fact, this keyboard shortcut works with most Unix commands,[5] and is an essential piece of Unix to know if you ever get 'stuck.' Depending on what it is that you are interrupting, you may or may not see any additional output. In many cases, the command/program will just stop immediately.

Bear in mind that there may be problems that result if you interrupt a program before it has finished – a program may have just made a large temporary file, but was interrupted before it deleted it. Trying to stop a program by sending the interrupt signal can be useful, but you should also appreciate that not all programs can be interrupted in this way.

Why else might you want to stop a program?

Having a Perl script stuck in an endless loop is a problem that is quite easy to spot (assuming you are looking at the terminal). However, sometimes an error in a Perl script might be causing other problems that might not lead to any suspicious output on your screen.

Sometimes you will run a program and not be sure how long it should take to finish. If you sit there waiting for the final output of the program to appear in your terminal window, you might not be able to tell whether the program is still running or has crashed (or got 'stuck'). Errant Perl scripts might cause your computer to run out of memory, or might try to open too many files at once. These types of problem can often be 'felt' as your computer typically becomes very sluggish and unresponsive. A further complication might arise if you are running several (different) Perl scripts and/or Unix commands at once. How can you tell which one is causing the problem?

In these situations, you ideally need to be able to (1) identify which command or program is causing the problem, and (2) stop it. There are several Unix commands that exist exactly for this type of process control, and we will briefly introduce you to three of them.

---

[4] In this case, the variable assignment becomes the test for the if statement. Because you can always assign a value to a variable, it will always evaluate as true.
[5] Try it by running a Unix command that will take a long time (such as a recursive directory listing of the root directory: ls -R /) and then press control + c to check that you can interrupt the command.

The top command

At any given time a Unix computer might be running anywhere between 50 and 500 different processes.[6] All of the Perl scripts we have seen so far will become a single process when they are run. However, more complex programs can be written such that, when running, they consist of multiple processes – a word processor program might initially consist of a single process, but might launch additional processes to handle tools such as the spell-checker. Normally, you don't care about any of these details, but if you are trying to track down an errant Perl process it is useful to know that there are potentially lots of other different processes happening.

The top command allows you to see most of the processes that are running on your computer in real time.[7] Unlike many other Unix commands, the output of the top command can vary a lot depending on the size of your terminal window (more information can be displayed as you make the window wider). Here is some representative output of the top command:

```
Processes: 86 total, 4 running, 82 sleeping, 378 threads          15:43:27
Load Avg: 2.23, 1.99, 1.33 CPU usage: 56.69% user, 26.33% sys, 16.96% idle
SharedLibs: 7188K resident, 6188K data, 0B linkedit.
MemRegions: 14599 total, 590M resident, 24M private, 387M shared.
PhysMem: 856M wired, 1214M active, 431M inactive, 2501M used, 1598M free.
VM: 207G vsize, 1042M framework vsize, 21081612(0) pageins, 605503(0)
pageouts.
Networks: packets: 40418356/25G in, 44593385/39G out.
Disks: 3034501/85G read, 3302451/80G written.
```

PID	COMMAND	%CPU	TIME	#TH	#WQ	#POR	#MREG	RPRVT	RSHRD	RSIZE	VPRVT	VSIZE
29122	top	6.4	00:00.23	1/1	0	24	43	1216K+	264K	1792K+	18M	2379M
29097	System Event	0.0	00:00.23	2	1	66	105	2872K	6124K	8516K	30M	2685M
29095-	iTunes	2.5	00:12.53	17	2	338	600	68M-	82M+	113M	91M-	1128M
29092-	Pages	0.0	00:15.47	4	1	129	501	12M-	97M+	72M	17M-	1099M
29064	bash	0.0	00:00.01	1	0	17	25	284K	748K	924K	17M	2378M
29063	su	0.0	00:00.02	2	1	36	55	2756K	296K	3868K	23M	2391M
29058	bash	0.0	00:00.00	1	0	17	25	260K	748K	964K	9576K	2378M
29057	login	0.0	00:00.01	1	0	22	54	448K	312K	1580K	10M	2379M
29056	perl5.10.0	42.0	04:02.31	1/1	0	14	39	488K	260K	1540K	18M	2380M
29039	bash	0.0	00:00.02	1	0	17	25	260K	748K	972K	9576K	2378M
29038	su	0.0	00:00.02	2	1	36	54	2724K	296K	3836K	22M	2390M
28999	bash	0.0	00:00.01	1	0	17	25	284K	748K	964K	17M	2378M
28998	login	0.0	00:00.01	1	0	22	54	448K	312K	1580K	10M	2379M
28994-	Terminal	107.3	10:53.46	4/1	1	109	152	11M+	33M-	35M	47M+	942M

---

[6] Or thousands. It depends on the computer, how many users are logged in, and what they are doing. After booting up, a typical Mac computer (with a single user logged in) might launch between 50 and 100 processes.

[7] Most computers typically have some graphical program for viewing the same information (e.g., Activity Monitor on a Mac).

Bear in mind that this output will look completely different on your computer (different versions of the `top` command might also display the output in a slightly different way). The first thing to note is that this command will keep on running (and updating the display) until you press 'q' to quit it. The output will change as you launch different applications on your computer, or run different Unix and Perl programs. The first eight lines of output are showing various strands of information about the current state of memory, disk use, and network traffic on the computer. Following that are details of currently running processes. The first four columns of this particular output show us:

(1) PID – process ID (we will refer to this later).
(2) COMMAND – the name of the command/program that is running (might be truncated).
(3) %CPU – the percentage load on the CPU that the command in question accounts for.[8]
(4) TIME – the length of time the process has been running.

The other columns are not so important for the purposes of this chapter.[9] In the list of running processes you can see one entry for the `top` command itself. You might also spot an entry 'perl5.10.0.' This process corresponds to the errant Perl script from Example 5.2.1 that is still running on our computer (though in theory this could be referring to any Perl script that is active). We now know that this process has an identifier (29056) which we shall use later to stop the script. The `top` command makes it easier to spot which processes, if any, are consuming large amounts of CPU and/or memory.

The `ps` command

Sometimes you just want to see details of only those processes that belong to you. To do this we can run the `ps` command (ps = process status), which shows us details of currently running processes. The default output is far less overwhelming when compared to the output of the `top` command. As we still have the `liveforever.pl` script from Example 5.2.1 running on our computer, we need to open a new terminal window before we can run this command:

```
$ ps
PID    TT   STAT    TIME COMMAND
29039 s000   S    0:00.02 bash
29056 s000   R+ 17:36.54 /usr/bin/perl liveforever.pl 24
29064 s001   S    0:00.02 bash
```

You might think that the `ps` command is suggesting that very few processes are currently running, but bear in mind that the default output of `ps` will usually show you only the running processes that are connected to your terminal.[10] So by default, you won't see

---

[8] The values in this column may not sum to 100% if there are multiple processors or processor cores on your computer (as is the case with most modern computers).

[9] The other columns deal with memory usage (among other things). See the man page for more information, which will also show you how to change the display of what information is shown by default.

[10] By 'connected' we mean those processes that were run from this terminal session (as opposed to scripts running in other terminal windows or any programs you ran yesterday in a terminal window which you subsequently closed).

details of any graphical programs that are running on your computer, or any processes generated by any other user. The ps command has a very large number of command-line options which can drastically change both the amount and type of information shown. You should look at the man page for this command, because sometimes only seeing the default output will not be enough. In this example we see five columns of information in the output:

(1) PID – process ID. This is the same information as in the output of the top command.
(2) TT – a confusing title that refers to something known as the *controlling terminal*. Basically, each terminal window you have open has its own identifier. In this case, we have two windows open (s000 and s001).
(3) STAT – the state of the process. R indicates a running process, S indicates a sleeping process (see the man page for more info about this and other states).
(4) TIME – the amount of CPU time accumulated by the process.
(5) COMMAND – details of the command/program that is running.

We usually see one process for the Unix shell that is being run in each terminal window (in this example we are using the default bash shell). This means the ps command typically produces at least one line of output even if you are not running anything. In this example we also see details of our errant Perl script. The main thing to remember from this section is that if you have recently run a Perl script which hasn't finished, you can easily find out the identifier of the corresponding process by using the ps command.

The kill command

The preceding two sections both show you how you can find the process ID of any running process. If that process belongs to you, then you can easily stop it with the Unix kill command. To stop a process, you just need to supply the necessary process ID:

```
$ kill 29056
```

This will not cause any other output to appear in the terminal where you run the kill command. However, in the other terminal window (which still contains our endlessly running Perl script) you would see something like this:

```
24 was your magic number!
25
24 was your magic number!
25
24 was your magic number!
25
Terminated
```

Specifically, the kill command is used to send a 'terminate' signal to a running process. The 'terminate' signal will try to let the running process finish in a clean manner, i.e., closing any file connections the process had opened. Sometimes this is not enough though, as some processes cannot be killed in this manner. A final solution is to send the more deadly 'kill' signal. This forces a process to stop without first cleaning up any file connections and will sometimes work when the standard `kill` command doesn't. To send the 'kill' signal, you just need to add '-9' to the command:[11]

```
$ kill -9 29056
```

This will cause the script to stop with a slightly different output:

```
24 was your magic number!
25
24 was your magic number!
25
24 was your magic number!
25
Killed
```

There is much more you can learn about process control and we encourage you to try to learn more when you get a chance. As you write larger and more complex scripts, it becomes useful to be able to understand more about how those scripts are running as processes. Process control also allows you to pause processes and switch them between the *foreground* and *background*.[12] However, the essential take-home messages of this chapter are:

(1) All programs and commands run as *processes*.
(2) All processes have an identifier which `top` and `ps` can reveal.
(3) The `kill` command needs these identifiers in order to stop processes.

---

[11] There are other signals that you can force the `kill` command to send to a process. E.g., `kill -2` sends the interrupt signal, which is exactly the same signal that is produced by pressing Control + c. See the `kill` man page for more details.

[12] Sending a running process into the background can return control of the terminal in order to issue more commands (the original command continues to run and will still send its output to the same terminal window).

Have haystacks? Want to find needles? Then grep is the tool for you

Many text-based data files use a common syntax of *one record per line*.[13] From a 'learning Unix' point of view, the content and format of those records is less important than the fact that each record is self-contained on one line. This is important because many Unix commands, such as grep, work on a line-by-line basis. Let's imagine we have a very simple data file called 'heroes_and_villains.txt' that looks like this:[14]

Batman	Bruce Wayne	Hero	DC
Catwoman	Selina Kyle	Villain	DC
Doctor Doom	Victor von Doom	Villain	Marvel
Doctor Octopus	Otto Octavius	Villain	Marvel
Invisible Woman	Susan Storm Richards	Hero	Marvel
Iron Man	Tony Stark	Hero	Marvel
Poison Ivy	Pamela Isely	Villain	DC
Rogue	Anna Marie	Hero	Marvel
Spider-Man	Peter Parker	Hero	Marvel
Supergirl	Linda Danvers	Hero	DC
Superman	Clark Kent	Hero	DC
The Riddler	Edward Nigma	Villain	DC
Two-Face	Harvey Dent	Villain	DC
Wonder Woman	Diana Prince	Hero	DC

This data file conforms to the one-record-per-line format and contains a list of comic book heroes and villains, with details of the character's real name, and whether they appear in comics published by DC or Marvel. Now let's imagine that we would like to ask some questions about the data in this file:

(1) Which lines contain the letters 'Man'?
(2) Which lines contain the letters 'Man' *or* 'man'?
(3) Which lines *don't* contain the letters 'Man' or 'man'?
(4) Which lines only contain the *word* 'Man' (i.e., flanked by spaces)?
(5) How many villains are there?
(6) Which specific lines contain two vowels in a row?

For such a small file, we could obviously answer all of these questions by manually searching through the file by eye. But what if this file contained details of every hero and villain *ever created*? Clearly, you would have to resort to using some other tool to get the answers. That's where grep comes in. This Unix command allows you to very quickly search a file (or files) to see which lines match (or don't match) a specified pattern.

[13] As was the case with the GFF files we saw in the last chapter.
[14] This file is also available on our www.unixandperl.com web site (see the previous chapter for more information).

Using grep to quickly search text files for patterns

The basic syntax of the grep command is:

```
grep [options] <pattern> <file>
```

where <pattern> is the pattern of characters you want to search for and <file> is the name of the file (or files) you are searching against. As we shall see later on, the patterns you might use can become quite complex.[15] Like many other Unix commands, grep has many useful command-line options, and we shall see some of these in action in just a moment. Let's work through the questions we asked earlier to see how we can answer them using grep.

Example 5.3.1 Which lines contain the letters 'Man'?

```
$ grep Man heroes_and_villains.txt
Iron Man     Tony Stark      Hero     Marvel
Spider-Man   Peter Parker    Hero     Marvel
```

Explanation

The default implementation of grep is case-sensitive, so searching for 'Man' will not match 'man' (or 'MAN,' 'MaN,' etc.). The default output simply prints the lines which match the pattern. If you want, you can optionally enclose the search term between single or double quotes (this is necessary if you want to search for a string of characters that contain a space).

Example 5.3.2 Which lines contain the letters 'Man' *or* 'man'?

```
$ grep -i Man heroes_and_villains.txt
Batman            Bruce Wayne              Hero      DC
Catwoman          Selina Kyle              Villain   DC
Invisible Woman   Susan Storm Richards     Hero      Marvel
Iron Man          Tony Stark               Hero      Marvel
Spider-Man        Peter Parker             Hero      Marvel
Superman          Clark Kent               Hero      DC
Wonder Woman      Diana Prince             Hero      DC
```

Explanation

By simply specifying the -i command-line option, grep will perform a case-insensitive search and will therefore match the names of an additional set of heroes and villains, as well as the two matches from the last example.

---

[15] If you have already worked through the Perl chapters on regular expressions then you'll see the concept of patterns in Unix is very similar to those that are used by Perl.

Example 5.3.3 Which lines *don't* contain the letters 'Man' *or* 'man'?

```
$ grep -i -v Man heroes_and_villains.txt
Doctor Doom        Victor von Doom   Villain    Marvel
Doctor Octopus     Otto Octavius     Villain    Marvel
Poison Ivy         Pamela Isely      Villain    DC
Rogue              Anna Marie        Hero       Marvel
Supergirl          Linda Danvers     Hero       DC
The Riddler        Edward Nigma      Villain    DC
Two-Face           Harvey Dent       Villain    DC
```

Explanation

The -v option turns the logic of the grep command upside down and instead shows lines that *don't* match the pattern. In this case we combine it with the -i option (we could also specify these options as -iv or -vi).

Example 5.3.4 Which lines only contain the *word* 'Man'?

```
$ grep -w Man heroes_and_villains.txt
Iron Man      Tony Stark    Hero    Marvel
Spider-Man    Peter Parker  Hero    Marvel
```

Explanation

The -w option asks grep to only find the pattern if it occurs as an isolated word, rather than being part of a larger word. This means the pattern has to be flanked by whitespace (tabs and space characters), or other non-word characters such as hyphens in this case.[16] This means that the 'Man' in 'Spider-Man' is considered a match. The -w option also allows the pattern to occur at the start or end of a line. This means that a line in the input file that *only* contained the word 'Man' would also count as a match.

Example 5.3.5 How many villains are there?

```
$ grep -c Villain heroes_and_villains.txt
6
```

Explanation

Rather than showing you lines that match a pattern, the -c option simply shows you a count of *how many* lines match the specified pattern.

---

[16] Non-word characters are those which aren't word characters, which is anything that isn't a letter, digit, or underscore character.

Example 5.3.6 Which specific lines contain two vowels in a row?

```
$ grep -n -i [aeiou][aeiou] heroes_and_villains.txt
2:Catwoman        Selina Kyle      Villain   DC
3:Doctor Doom     Victor von Doom  Villain   Marvel
4:Doctor Octopus  Otto Octavius    Villain   Marvel
7:Poison Ivy      Pamela Isely     Villain   DC
8:Rogue           Anna Marie       Hero      Marvel
12:The Riddler    Edward Nigma     Villain   DC
13:Two-Face       Harvey Dent      Villain   DC
14:Wonder Woman   Diana Prince     Hero      DC
```

Explanation

The −n option forces the inclusion of the matching line numbers at the start of the output. We can look for two vowels in a row by using the syntax [aeiou][aeiou].[17] Of course, the word 'villain' itself contains two adjacent vowels and so all villains will be included in the output.

Searching multiple files at once

There are many commands like grep that can work with one or more files. If we had split our 'heroes_and_villains.txt' file into two separate files ('heroes.txt' and 'villains. txt') then we could use grep to search both of them by specifying *.txt instead of a single file name – for example, if we wanted to find lines that contained the letters 'der' or 'ler' we could do the following:

```
$ grep -i [dl]er *.txt
heroes.txt:Spider-Man      Peter Parker    Hero      Marvel
heroes.txt:Wonder Woman    Diana Prince    Hero      DC
villains.txt:The Riddler   Edward Nigma    Villain   DC
```

When searching multiple files, grep will always include the name of the matching file at the start of the output.[18] Hopefully, you will have started to appreciate that there is a lot of functionality to the grep command. Read its man page to see all of the other wondrous things it can do.

Using grep with FASTA files

As we saw in the previous chapter, the FASTA file format is a commonly used format for representing sequence data. FASTA files can contain one or more sequences, each of which must have an associated *header line* which starts with a > character. Apart

---

[17] Don't worry if this doesn't make sense. This pattern is an example of a *regular expression*, which will be revisited in far greater detail when you cover 'Essential Perl' in Part 4. If you've already read Part 4 and this still doesn't make sense, then perhaps you need to go back and refresh your memory!

[18] If you also use the -n option, then you will see both the name of the file that contains the pattern *plus* the number of the line that contains the pattern in that file.

from that, there are very few restrictions on how the content of a FASTA file should be structured. Unlike GFF files, however, the information for each record is always spread across multiple lines (at least two). This means you should be very careful when using grep to search for specific patterns within a FASTA file. Imagine you had downloaded the sequences of every gene from the chromosome I of the yeast genome from the SGD database[19] and you then wanted to search for those genes that contained the sequence 'CGTATAT'. You might try this:

```
$ grep -n -i CGTATAT yeast_chr1_orfs.fa
185:GAAAAGATTCATCAACTGGCCAAAAATAAACTAGCTTTAGATTCGTATATCAGTGAAGAT
1456:ACTACCGTTAGTGATGACTTTGAAGGGTACGTATATACTTTTGACAACAATCTAAGCCAG
1548:GGGATCCAGTCAGCCGTATATTCATCGTTAAATGGTGTAAATGGTTTAGAGGGATGGAGA
1771:GTGTCACAAGGCTCCGAATCCGTAGTCTCATGGACAACTTTAACACACGTATATTCCATC
1800:CCTCAAGTACCTCCTGAAACGACTATATTTGATTACGTAACGTATATCCTTACTGGCAGG
```

We use the -i option to make our search case-insensitive, and we use -n to show which lines contain the sequence of interest. Hopefully you have realized that the answer is possibly incorrect. This file actually contains seven occurrences of the pattern and not five as shown. The grep command missed two occurrences which are each split across two lines of the file – the end of the first line below ends with six of the seven desired nucleotides, with the last nucleotide in the pattern starting the line that follows:

```
ATGGCACCAAGTATAGCAACGGTAAAGATAGCCAGGGACATGGTTTTGCCATTACGTATA
TTTGTCAATAGAAAGCAGATCCTTCAAACCAATGATAAGACTAGCAATAAGTCGAATGCC
```

The lesson to be learned here is simple:

> *Be very careful when using grep for patterns that might occur across multiple lines*

---

[19] These files are available from either yeastgenome.org or from the web site that accompanies this course: http://unixandperl.com.

Who let the cat out of the bag?

Back in Chapter 3.20 we learned that we can view text files by using the less command. This is a perfect tool for when you might want to navigate to a certain position within a file, or search for specific text. Sometimes, though, we just want to quickly see what is in a file, and don't want to do anything else. The cat command often provides the best way to do this. You just specify the name of a file (or files) after the command name:

```
$ cat yeast_chr1_orfs.fa
>YAL027W SAW1 SGDID:S000000025, Chr I from 94688-95473, Verified ORF
ATGGCACCAAGTATAGCAACGGTAAAGATAGCCAGGGACATGGTTTTGCCATTACGTATA
TTTGTCAATAGAAAGCAGATCCTTCAAACCAATGATAAGACTAGCAATAAGTCGAATGCC
ACTATATTTGAAGCACCATTATTATCAAATAACTCCATAATCTGCTTAAAATCACCAAAT
ACAAGAATATATTTATCGCAACAAGATAAGAAGAATCTTTGTGACGAGATCAAGGAGGAC
CTGTTATTGATTGTTTACGAACTAGCGTCCCCGGAAATCATCAGTTCCGTACTCAGCAAA
ATAAGAGTTGGTCATTCTACTGATTTCCAAATCAACGTTCTGCCCAAACTTTTTGCAGGT
GCCGATACGGATAATGCGGTAACTTCTCACATCCAGTCTGTGACAAGGCTGGCTAAATTC
AAATACAAGTTGCACTACAAACATAAGTGGGAGCTCGACATATTCATCAACAGCATTAAG
AAGATCGCCAATTTAAGGCACTATTTGATGTTTCAAACATTAACATTAAACGGTTTCTCA
TTAAATGCAGGACCCAAAACGTTATTAGCTAGGAAAATAGAAAAACAGCCCCAGGTACCT
AATTTGTTAATAGAAAATGGGGACGCTGATGCCCTGGATACACCGGTGGAAGAGGATATA
AAACCTGTAATAGAATTTATGTACAAGCCTGTTATTAATTTAGGTGAAATTATTGATGTA
CATGTGTTGCATAGGCCTAGAAGACATAAGGTACGTACCCAGTCGAAGCAACCCCAGGAG
GAATGA
>tL(CAA)A
GGTTGTTTGGCCGAGCGGTCTAAGGCGCCTGATTCAAGCTCAGGTATCGTAAGATGCAAG
AGTTCGAATCTCTTAGCAACCA
>YAL028W FRT2 SGDID:S000000026, Chr I from 92901-94487, Verified ORF
ATGCAAAATGCTCAAATAAAGAGCTCTTCTAAAGGCAGCGGAATAGATGGTACAGATCGC
AATAGCAAAGATGGTGTAGAAAAGAGACCCCTGGAAGATGTAAAGCAAATGATTGACGCT
```

In this example we are only showing a small fraction of all of the actual output (this file contains almost 3000 lines of data). So in this sense, the cat command is most useful when you want to view files that will fit within your terminal window. When the cat command finishes, it immediately returns you to the command prompt. So what should you use when you want to quickly look at some files which might be very large? Well, fortunately, there are a couple of ways of dealing with this to make things more manageable (other than sending the Unix interrupt signal, control + c; see Chapter 5.2).

head **and** tail

Two other very useful commands for viewing text files are head and tail. By default these commands just show you the first or last ten lines of a file:

```
$ head yeast_chr1_orfs.fa
>YAL027W SAW1 SGDID:S000000025, Chr I from 94688-95473, Verified ORF
ATGGCACCAAGTATAGCAACGGTAAAGATAGCCAGGGACATGGTTTTGCCATTACGTATA
TTTGTCAATAGAAAGCAGATCCTTCAAACCAATGATAAGACTAGCAATAAGTCGAATGCC
ACTATATTTGAAGCACCATTATTATCAAATAACTCCATAATCTGCTTAAAATCACCAAAT
ACAAGAATATATTTATCGCAACAAGATAAGAAGAATCTTTGTGACGAGATCAAGGAGGAC
CTGTTATTGATTGTTTACGAACTAGCGTCCCCGGAAATCATCAGTTCCGTACTCAGCAAA
ATAAGAGTTGGTCATTCTACTGATTTCCAAATCAACGTTCTGCCCAAACTTTTTGCAGGT
GCCGATACGGATAATGCGGTAACTTCTCACATCCAGTCTGTGACAAGGCTGGCTAAATTC
AAATACAAGTTGCACTACAAACATAAGTGGGAGCTCGACATATTCATCAACAGCATTAAG
AAGATCGCCAATTTAAGGCACTATTTGATGTTTCAAACATTAACATTAAACGGTTTCTCA

$ tail yeast_chr1_orfs.fa
ACCTTAATTGCCATTCCTGGTTCCCTCTTATTTTGGTTAATATTCTTCCCAATTTATGCT
TCTATATTTCCTCATGCTAACATCTCAAGAGAGTATTATGGTGTGGTTAAACACACGTAT
GGATCCGGTGTATTTTGGTTAACTTTGATCGTTTTACCAATTTTTGCACTGGTAAGAGAT
TTTCTATGGAAGTACTATAAAAGAATGTATGAACCAGAAACGTATCATGTTATTCAAGAA
ATGCAGAAATACAATATCAGCGACTCTAGGCCGCACGTTCAGCAATTTCAAAATGCCATC
AGGAAGGTGAGGCAAGTGCAAAGAATGAAAAAACAAAGAGGATTTGCCTTTTCACAGGCT
GAAGAGGGTGGGCAAGAAAAAATTGTCAGAATGTATGATACTACTCAAAAGAGAGGTAAG
TACGGTGAATTACAAGATGCTTCGGCGAACCCCTTTAATGACAATAATGGACTGGGAAGT
AACGACTTTGAAAGTGCAGAACCGTTCATTGAAAATCCATTTGCTGATGGTAATCAAAAT
TCAAATAGATTCAGTTCTTCGAGAGATGATATTTCATTTGATATATGA
```

Often you will look at a file just to see what type of file it is, or you may only be interested in the first few lines. In these situations the head command will save time compared to using less. Both of these commands also support a -n option that allows you to see a user-specified number of lines:

```
$ head -n 1 yeast_chr1_orfs.fa
>YAL027W SAW1 SGDID:S000000025, Chr I from 94688-95473, Verified ORF
```

Redirecting input and output   5.5

You redirect your left arm in, your left arm out ...

So far we have learned that we can type any Unix command, press enter, and then (hopefully) see the results of that command appear on our screen. However, some Unix commands will produce output that we might want to keep in order to do further work with it. Imagine we wanted to find the details of any ORFs on yeast chromosome I that start within the first 999 base pairs of the chromosome. We could use the grep command with the -E option to specify a regular expression that does this:[20]

```
$ grep -E "from [0-9]{1,3}-" yeast_chr1_orfs.fa
>YAL069W SGDID:S000002143, Chr I from 335-649, Dubious ORF
>YAL068W-A SGDID:S000028594, Chr I from 538-792, Dubious ORF
```

If we wanted to store the two-line output from this command, our only solution would be to use copy and paste. It's not a great solution, particularly if the output to be copied doesn't fit on your screen. Luckily the solution to this problem is simple – you just need to *redirect* the output.

Standard output

If you run any Unix command that produces output, then you might always expect to see that output appear in your terminal window. If it doesn't appear there then you wouldn't be able to see it; you might think this would be a very bad thing. However, Unix actually treats the output from programs like a stream of data that can be redirected to different places. The official term for the output from any command is *standard output*[21] and people sometimes speak of standard output being 'attached' to something, e.g., to your computer screen.

   If you want to 'capture' the output from a command, you can easily achieve this by redirecting standard output into a file. To do this, you simply use the redirect operator which is represented by the > character. Just place this character after any command that has output that needs redirecting and then provide a name of a file to write the output to. Note that when you redirect a command's output to a file you won't see any of the command's output in the terminal.[22]

Example 5.5.1 Let's assume you have downloaded the FASTA file that we have already been working with. Let's capture the details of any 'Verified' ORFs:

```
$ pwd
/Users/nigel/Unix_and_Perl

$ ls -l
total 328
```

---

[20] If you've worked through Part 4 then you will hopefully know about regular expressions; if not then maybe have a quick look at Chapter 4.26.
[21] Often abbreviated to stdout.
[22] You may still see some output, e.g., error messages, and we will deal with this in the next chapter.

```
drwxr-xr-x    6 nigel staff      204  Sep 13 17:17   Code
drwxr-xr-x    2 nigel staff       68  Sep 16 16:08   Temp
-rw-r--r--@   1 nigel staff   165971  Sep 16 14:47   yeast_chr1_orfs.fa

$ grep "Verified" yeast_chr1_orfs.fa > verified_orfs.txt

$ ls -l
total 344
drwxr-xr-x    6 nigel staff      204  Sep 13 17:17   Code
drwxr-xr-x    2 nigel staff       68  Sep 16 16:08   Temp
-rw-r--r--    1 nigel staff     6103  Sep 16 16:11   verified_orfs.txt
-rw-r--r--@   1 nigel staff   165971  Sep 16 14:47   yeast_chr1_orfs.fa

$ head verified_orfs.txt
>YAL027W SAW1 SGDID:S000000025, Chr I from 94688-95473, Verified ORF
>YAL028W FRT2 SGDID:S000000026, Chr I from 92901-94487, Verified ORF
>YAL019W FUN30 SGDID:S000000017, Chr I from 114920-118315, Verified ORF
>YAL033W POP5 SGDID:S000000031, Chr I from 82707-83228, Verified ORF
>YAL027W SAW1 SGDID:S000000025, Chr I from 94688-95473, Verified ORF
>YAL028W FRT2 SGDID:S000000026, Chr I from 92901-94487, Verified ORF
>YAL019W FUN30 SGDID:S000000017, Chr I from 114920-118315, Verified ORF
>YAL033W POP5 SGDID:S000000031, Chr I from 82707-83228, Verified ORF
>YAL035W FUN12 SGDID:S000000033, Chr I from 76428-79436, Verified ORF
>YAL038W CDC19 SGDID:S000000036, Chr I from 71787-73289, Verified ORF
```

Explanation

We first confirm that the only file in the directory is the file of yeast ORFs from chromosome I. Then we use grep to find those lines in the input file that contain the word 'Verified'. However, rather than have the output go to the screen, we use the redirect operator to send it to a new file called 'verified_orfs.txt' instead. We then confirm that this file has been produced and we use the head command to look at the first ten lines of this file.

You should be careful when using redirection. If you accidentally redirect a command's output to a file that already exists, then the file will be overwritten with the output from the Unix command. This will happen even if the command in question doesn't produce any output. For example:

```
$ grep "Clark Kent" yeast_chr1_orfs.fa > grep_output.txt
```

If we assume the phrase 'Clark Kent' doesn't occur anywhere in the file of yeast ORF sequences, then we know that the grep command won't actually find anything. However, if there was any existing text in the file 'grep_output.txt' then it will be overwritten with 'nothing,' causing it to become an empty file. You can prevent redirection from overwriting a file by turning on the *noclobber* setting. You might want to consider adding the following to your login script:

```
set -o noclobber   # for bash-type shells
set noclobber      # for c-type shells
```

You should try to ensure that the name of any redirected output file makes sense and informs you about what data it might contain.[23] The simple ability to redirect program output now gives you much more power as you can now *save* the results of your commands.

Appending to an existing file

Sometimes you will want to create a *new* file using the output of a Unix command, but many times you might want to add to an *existing* file instead. You can easily do this with the *append* redirect operator >>. Imagine that you had a text file which you just needed to add one more line to. Rather than using a text editor, you could instead use the echo command and append the output to the file like so:

```
echo "P.S. I'm learning Unix ... it's great!" >> letter_to_grandad.txt
```

Bear in mind that appended text gets added *immediately* after the last character in a file; Unix won't insert a blank line for you.

Using redirects with the cat command

The cat command that was introduced in the previous chapter becomes more powerful when used in conjunction with the redirection operators. For example, imagine you have some text files that you want to combine into one larger text file. You could do this like so:

```
$ cat a.txt > all.txt
$ cat b.txt >> all.txt
$ cat c.txt >> all.txt
$ cat 1.txt >> all.txt
$ cat 2.txt >> all.txt
$ cat 3.txt >> all.txt
```

In this example the contents of each successive file are added to the contents of all.txt, which is created by the first command (the only one to use the standard redirect operator rather than the append operator). Of course, you could also do this in one step, like so:

```
$ cat a.txt b.txt c.txt 1.txt 2.txt 3.txt > all.txt
```

---

[23] The world of Unix practitioners is full of people who stare at their directory listings and ponder what might be contained in files named 'output' or 'file.' Just because you might know *today* that a file called 'stuff' contains some really important research results, you might not remember this in a week's time.

The `cat` command will read each file in turn and redirect all output to a single file. We can further shorten this operation by using a wildcard character:

```
$ cat *.txt > all.txt
```

However, this method does not give you any control as to the order in which files will be combined.[24] They will be combined in an order determined by the ASCII values of their file names, meaning that files that start with numbers would appear in the final output file *before* any files that start with letters.

Standard input

Now that we have grasped the concept of *standard output* and how to redirect it, we can learn about the related, but opposite, data stream *standard input*.[25] This refers to the data, usually text, that are provided as input for a Unix program. Just as there is a default destination for the standard output of commands (the computer screen), so there is a default source of data for standard input – the keyboard.

Standard input can be a little confusing, but just remember that it is only referring to the input used by *commands*. Most Unix commands we have seen so far take their input, if required, from a file that you specify on the command line. Consider the following simple command:

```
$ less some_text_file.txt
```

In this scenario the `less` command sees that 'some_text_file.txt' is specified as an argument to the command. This is separate from standard input and when you run commands like `less`, you are never prompted to type any more input. However, a few Unix programs – such as `tr` – will require some typed input *after* the command has started, and this input will come from standard input. The `tr` command functions just like the transliteration operator in Perl (Chapter 4.12); you can specify a character (or range of characters) that are to be changed into another character (or range of characters). After specifying which characters are to be changed, you press enter and then the command will wait for you to provide some input text that the modifications will affect:

```
$ tr 'A' 'a'
CAT
CaT
BANANA
BaNaNa
```

In this example the user typed 'CAT' and the `tr` command then changed the 'A' to 'a' and printed the output. The user then typed 'BANANA' and the `tr` command changed it to 'BaNaNa'. To get out of the text input mode you can use control + d, which is the Unix keyboard shortcut that ends text entry. This is a little different to how we have been

---

[24] It would also capture other .txt files that you might not want included in all.txt.
[25] Also known as stdin.

using Unix, and in this example the text manipulation would rely solely on the user providing data via standard input.

Redirecting standard input

Just as we can redirect standard output *to* a file, so we can redirect standard input to come *from* a file. To do this we use the < redirect operator. Generally, this is used much less than redirecting standard output, but there are occasions where it is useful or necessary. It makes using the tr command much more straightforward, as rather than having to type the text we want to change, we can just read it from a file:

```
$ tr 'A' 'a' < yeast_chr1_orfs.fa
>YaL027W SaW1 SGDID:S000000025, Chr I from 94688-95473, Verified ORF
aTGGCaCCaaGTaTaGCaaCGGTaaaGaTaGCCaGGGaCaTGGTTTTGCCaTTaCGTaTa
TTTGTCaaTaGaaaGCaGaTCCTTCaaaCCaaTGaTaaGaCTaGCaaTaaGTCGaaTGCC
aCTaTaTTTGaaGCaCCaTTaTTaTCaaaTaaCTCCaTaaTCTGCTTaaaaTCaCCaaaT
aCaaGaaTaTaTTTaTCGCaaCaaGaTaaGaaGaaTCTTTGTGaCGaGaTCaaGGaGGaC
CTGTTaTTGaTTGTTTaCGaaCTaGCGTCCCCGGaaaTCaTCaGTTCCGTaCTCaGCaaa
aTaaGaGTTGGTCaTTCTaCTGaTTTCCaaaTCaaCGTTCTGCCCaaaCTTTTTGCaGGT
GCCGaTaCGGaTaaTGCGGTaaCTTCTCaCaTCCaGTCTGTGaCaaGGCTGGCTaaaTTC
aaaTaCaaGTTGCaCTaCaaaCaTaaGTGGGaGCTCGaCaTaTTCaTCaaCaGCaTTaaG
aaGaTCGCCaaTTTaaGGCaCTaTTTGaTGTTTCaaaCaTTaaCaTTaaaCGGTTTCTCa
```

In this example we perform the same text manipulation as before, but we instruct the tr command to now read from the 'yeast_chr1_orfs.fa' file. Note that we only show a small fraction of the resulting output, but it is obvious that all occurrences of 'A' have been changed to 'a'.

There are very few commands like tr that need to read from standard input in this way. However, you should appreciate that you can redirect standard input for many of the commands we have already seen. For example, the following two commands will achieve the same result, even though they are behaving slightly differently:

```
$ grep "pattern" datafile
$ grep "pattern" < datafile
```

The first example is using the file called 'datafile' as an *argument* to the grep command. The second example is instead telling grep to read the contents of the file as standard input. There are many commands where the two syntaxes can be used interchangeably and the results will be exactly the same.[26] Don't worry about it too much if you find the concept of standard input confusing. In a couple of chapters' time, we will learn another way of providing input to commands which will hopefully be easier to understand.

---

[26] However, there are some command-line options that might not work as expected when reading from redirected standard input. Also, note that redirected standard input can only read from a single file, whereas you might want grep (or other commands) to read multiple files.

Redirecting standard input and output

One final thing to mention is that you can redirect both the input *and* output of a single Unix command. If we revisit the tr command one more time, we can specify that it reads text input from one file and sends the processed output to a new file, all in one go:

```
$ tr 'A' 'a' < yeast_chr1_orfs.fa > processed_dna.fa
```

Once again, we should point out that you won't see this used very much, and if you'd prefer to not know about this then we give you our full consent to immediately forget about this example.

Standard error 5.6

In which you will learn to see the (standard) error of your ways

If you've been thinking 'Wow, I really loved learning about standard input and standard output, I wish there was another Unix data stream that I could study!' then you're in luck. This short chapter will complete the trinity of Unix data streams and briefly teach you about something called *standard error*.[27] This is a secondary form of output that can be produced by any Unix command.

Example 5.6.1 This example should hopefully reveal most of what you need to know about how standard error works. Follow the instructions carefully; there is a deliberate typo included in this example and you need to reproduce the typo:

```
$ echo "Spiderman" > hero
$ echo "Venom" > villain

$ cat hero villain
Spiderman
Venom

$ cat hero villlain
Spiderman
cat: villlain: No such file or directory

$ cat hero villlain > hero_and_villain
cat: villlain: No such file or directory

$ cat hero_and_villain
Spiderman
```

Explanation
We first use the `echo` command to put some text into two separate files and we confirm that this has happened by using the `cat` command to display the contents of both files.

Then we try viewing the contents of both files again, but this time we include an error in one of the file names. The `cat` command shows us the first file ('hero'), but we see a Unix error that there is no file called 'Villlain.'

We next try repeating the command, but this time we redirect standard output into a file. Notice that you still see the same error message appear on the screen though. Finally, we confirm that the new file ('hero_and_villain') only received the contents of the correctly named file.

Many Unix commands will produce error messages, most commonly when used incorrectly. It is important to appreciate that error messages are controlled by a separate data stream (standard error) from the normal output of the command. This is mostly

---

[27] Often referred to as stderr.

a good thing as it means that if you redirect standard output to a file, that file will not contain the error messages.

Redirecting standard error

Most of the time you will probably be happy with the default situation, where standard error is always attached to your screen, even when standard output is redirected to a file. But there are occasions when you might want to capture standard error to a file too, or maybe even turn off standard error altogether. This can all be accomplished by Unix, though the implementation can vary a little between different types of Unix shell.

The basic point to remember is that Unix uses some special 'labels' for these data streams;[28] standard output is labeled data stream '1' and standard error is data stream '2.' For the sake of completeness, we'll also let you know that standard input is data stream '0.' You mostly only need to know about these numbers if you want to redirect standard error. In practice, you just need to use these numbers *before* the redirect operator – let's look at how we can redirect the standard error stream of a command to a file:

```
$ cat hero villlain
Spiderman
cat: villlain: No such file or directory

$ cat hero villlain 2> errors
Spiderman

$ cat errors
cat: villlain: No such file or directory
```

The first cat command is the same as in Example 5.6.1 and both standard output and standard error streams get displayed on the screen. The second cat command uses the 2> syntax to redirect standard error to a file called 'errors.' When using this syntax, the only thing you see on the screen is the standard output (i.e., the contents of the 'hero' file). Note that you must *not* put a space between the 2 and the > when redirecting standard error like this. If you wanted to *append* standard error on to the end of an existing file, then you can do this by using 2>> instead of 2>.

Another common practice is to combine standard output and standard error into one stream. This means that if you want to capture errors from a program, they would end up in the same file that you redirect standard output to. When combining both output streams, you need to use '&>' for the redirect:[29]

```
$ cat hero villlain &> hero_and_villain

$ cat hero_and_villain
Spiderman
cat: villlain: No such file or directory
```

---

[28] More precisely, these 'labels' are known as 'file descriptors,' but that sounded a little boring to us, so we're calling them labels.

[29] Again we warn you that the syntax of data stream redirection can vary slightly between different Unix shells and that there are also different syntaxes for achieving the same thing.

Other combinations of redirecting standard output and standard error are also pos-sible,[30] but it is perhaps more useful to let you know that you can effectively 'turn off' standard error – sometimes you want to see a command's standard output on the screen but not have it interrupted by error messages. The solution is to redirect standard error to a file on the filesystem called the null device: `/dev/null`. This file is common to all Unix systems and acts as a 'black hole' that essentially just discards any data it receives and doesn't produce any data you can read. It might sound a strange concept, but it's useful for when you want to do things like the following:

```
$ cat hero villlain 2> /dev/null
Spiderman
```

This command reads as 'use `cat` to display the contents of the following two files but if there is any error messages on the standard error stream, then redirect those someplace where I won't ever see them.' One final warning concerning the wisdom of hiding the standard error stream should be stated at this point:

Error messages are usually there for a good reason!

---

[30] If you wanted to, you can perform any combination of redirecting the two output streams. I.e., standard output and/or standard error can be redirected to a file and/or to standard output and/or standard error. We'll leave it up to you to find out more about these various possibilities, though you probably won't have much use for most of them. However, it can sometimes be useful to have standard output and standard error both redirected to *different* files.

Being able to redirect the output of a command to a file, and then being able to use that file as input to another command starts to give us the ability to build workflows. For example, we could use grep to extract the header lines from a FASTA file and then use grep again to extract just those lines that *didn't* contain the 'Dubious' keyword. Finally, we could count how many lines we have in the final output by using a program such as the word-count command, wc:

```
$ grep ">" yeast_chr1_orfs.fa > yeast_headers
$ grep -v "Dubious" yeast_headers > not_dubious
$ wc -l not_dubious
     102 not_dubious
```

In this scenario we have to redirect standard output to produce two intermediate output files in order to get to the final answer. We use the -l option of the wc command to just count lines.[31] We might not have any real need to keep these intermediate files around, and if the original file was huge, we could end up with large temporary files as well. Fortunately, Unix has a very powerful solution that allows you to combine multiple commands together without having to write any intermediate output to a file. Unix achieves this by letting you build *pipelines*.

How pipelines work

A pipeline refers to a set of commands that are linked together. You can take the output of any Unix command and tell Unix to then use that output as input for a subsequent command. There is no practical limit to how many commands you can combine in this way, though it is necessary that any command in the pipeline actually produces output and in turn passes it to another command that can receive standard input. The actual magic of joining commands together is done by the pipe character |, which sort of looks like a vertical pipe.[32] This means we can rework the earlier example into a single pipeline, like so:

```
$ grep ">" yeast_chr1_orfs.fa | grep -v "Dubious" | wc -l
     102
```

This pipeline uses two pipes: the output of the first grep command is sent 'down a pipe' as input to the second grep command; in turn, this grep command sends its output to the wc program. Note that the pipeline does not write to any files.

---

[31] By default, the word-count program displays the number of lines, words, and bytes in any input file.

[32] This is one of those characters that lives on different places on different computer keyboards, but you will need to use the 'shift' key in order to produce this character.

Piping output into the `less` command

It is very common to use a pipeline to just send the final output of a command (or set of commands) to a program like `less` so you can have more control over the output. If we wanted to look at some of the initial output of the last example, we could do this:

```
$ grep ">" yeast_chr1_orfs.fa | less
```

The output of the `grep` command is passed to the `less` program. At this point the user would have access to the results and be able to use any of the tools available in the `less` program, e.g., jump to the end of the data or search for a specific pattern. Note that the output is still not stored in any file though. If you work with large data files, then it becomes very useful to always send the output from any text manipulations to `less`.

Using `cat` to start a pipeline

A common use of the `cat` command is to start a pipeline. This allows you to avoid having to redirect standard input with the < operator. We could take the file of yeast ORFs and send it to the `tr` command to modify the DNA sequence:

```
$ cat yeast_chr1_orfs.fa | tr 'ACGT' 'acgt' > modified_dna.fa
```

Using `cat` in this way achieves the same result as if you had instead specified that `tr` should read from redirected standard input. However, the underlying behavior is a little different, as Unix will need to run an additional process (the `cat` command) to complete the task. For practical purposes you won't notice any losses in performance by constructing your pipeline in this way.

Advanced pipeline construction

The ability to combine different Unix commands in a pipeline means that knowledge of just a handful of commands gives you a diverse toolkit to solve different problems. For example, let's consider how we can create a pipeline to perform a series of manipulations to a GFF file (see Chapter 5.1).

Let's imagine we are interested in whether the genomic features that occur at either end of yeast chromosome IV (the longest chromosome in the genome) are very different to each other. We have a single GFF file[33] which contains details of all features across the 16 chromosomes of the yeast genome (nearly 17 000 features in total). About 2000 of these features are on chromosome IV, so we can aim to compare just 20 features (1%) that occur at either end of this chromosome. Before we show you one way of doing this with a Unix pipeline, just imagine how you might attempt to do this without knowledge of Unix. Maybe you would try to write a Perl script? As you will soon see, many problems that might even be hard to achieve in Perl can sometimes be relatively straightforward in Unix. Here is one solution (note that this command wraps on to a second line):

---

[33] You can download this file from our Unix and Perl web site: www.unixandperl.com.

```
$ grep -E "^chrIV" yeast_genome.gff | sort -n -k 4 | head -n 20 | cut -f 3 | sort | uniq
ARS
CDS
binding_site
chromosome
gene
nucleotide_match
repeat_region
telomere
$ grep -E "^chrIV" yeast_genome.gff | sort -r -n -k 4 | head -n 20 | cut -f 3 | sort | uniq
CDS
binding_site
gene
nucleotide_match
repeat_region
telomere
```

In just two Unix pipelines we get all of the information we wanted and we see that the types of feature that occur at either end of yeast chromosome IV are mostly the same. These pipelines introduce some new Unix commands, which also use several command-line options. Here is a brief breakdown of how these pipelines work (remember to view the man pages for these commands to find out more):

(1) Use grep to find lines in the GFF file which start with the pattern 'chrIV';
(2) sort the resulting output with the sort command, but sort numerically (-n) using the values in column 4 (-k 4), which is the start coordinate of each feature;
(3) take only the first 20 lines of resulting output using the head command;
(4) use the cut command to cut[34] out only those values in column 3 (-f 3), which is the GFF 'feature' field;
(5) sort the resulting output alphabetically;
(6) pass the sorted output to the uniq command, which removes redundancies and keeps only unique output.[35]

The second pipeline is identical to the first except that the first sort command additionally uses the -r option to do a reverse-sort, meaning it will get the last features on chromosome IV (those with the highest start coordinates).

How to test pipelines

You may sometimes be tempted to write a pipeline that chains together a dozen or so commands. If you run a complex pipeline without testing it only to find that it doesn't work, then you will have no idea which of the many steps is causing the problem.

---

[34] By default, the cut command assumes that columns are tab-delimited (which is the case with GFF files), but you can specify to use other delimiters too (see the man page).
[35] This step is not strictly necessary because the sort command has a -u option which achieves the same thing.

Therefore, when you construct complex pipelines in this way, it is essential to get into the habit of testing each step. Check that the output of the first command in the pipeline produces what you are expecting (remember, you can always pipe output to the less or head commands). Then add the second command in the pipeline and test again. Repeat for every step until you know what effect every command is having on the final output.

Also, when first testing pipelines you may not need to use the entire input file. If your initial data file is very large, as many GFF and FASTA files are, then it might take several minutes for your pipeline to run. You might only need a very small amount of data to test that your pipeline works. If you had a GFF file which had over one million lines, you could first do something like this:

```
$ head -n 10000 big_file.gff > small_file.gff
```

As long as the smaller file is representative of the bigger file, then you can use it for all of your testing and things will be much quicker. Only when your pipeline is working with your small test file should you consider moving to working with your very large file.

How to optimize pipelines

If you are constructing a pipeline to process a very large file, then sometimes the order of the commands within the pipeline can be very important. Let's imagine a simple example. You want to convert the DNA sequence of a yeast chromosome to lower-case characters and then use grep to show any lines which contain a particular DNA pattern ('atgtag'). We could do this like so:

```
$ cat yeast_chr1_dna.fa | tr 'A-Z' 'a-z' | grep 'atgtag'
```

However, in doing this we are using the tr command to modify *every* line that is in the input file, even though the subsequent grep command might only be keeping a small fraction of those input lines. It would therefore be more efficient to first use grep only to find the lines that we are interested in, and *then* use tr to format the text:[36]

```
$ cat yeast_chr1_dna.fa | grep 'ATGTAG' | tr 'A-Z' 'a-z'
```

In this example, you might not notice much difference in speed between the two methods. This is because our pipeline is short and the yeast chromosome sequence in question is (relatively) small. However, if you run the same pipeline against the entire *human genome* sequence, then the difference between the two methods becomes much more noticeable. On our Mac computer the former method took three minutes to process a file containing the human genome sequence, whereas the latter method took just under one minute.[37] In general you should always try to put commands like grep, which might be filtering the data, earlier in the pipeline. This will reduce the work for any commands that follow.

---

[36] Note that you could also avoid using the cat command in the second example because we could just specify a file name as an argument to grep. This might not be any quicker, but it might be cleaner and easier to follow.

[37] Any bioinformatician who is reading this several years *after* this book's publication might want to take this moment to laugh at how slow our early twenty-first-century computers were.

Advanced text manipulation

In which we come to the end of the line

Unix is a great OS for working with bioinformatics data because such data often exists in the form of structured text files and there are so many useful commands that can be used to slice-and-dice text files. This is the final chapter in this part of the book and it will introduce two other extremely powerful Unix commands that can be used to manipulate text. But first we will take a quick look at an issue that can arise when working with text files produced on a non-Unix OS.

Line endings

When you press the return/enter key on your keyboard you may think that this causes the same effect no matter what computer you are using. The *visible* effects of hitting this key are indeed the same – if you are in a word processor or text editor, then your cursor will move down one line. However, the underlying effects of pressing enter can differ depending on what OS your computer is running.

Technically speaking, pressing enter generates something called a *newline*, which is a special character that is represented internally by either a *line feed* or *carriage return* character, or a combination of both characters. It all depends on your OS. This is something that you normally never need to know about, except if you are a Unix user and you receive a text file that was generated on a non-Unix system. If this happens then you might find that what looked like a multi-line text file now looks like one long line of text with strange ^M characters appearing where you would expect the newlines to be. For example, a text file that looks like this on a non-Unix computer:

```
a,1,alpha
b,2,beta
c,3,gamma
d,4,delta
```

might instead look something like this when you look at it with less:

```
a,1,alpha^Mb,2,beta^Mc,3,gamma^Md,4,delta
```

The strange ^M characters represent carriage returns. Fortunately the carriage return can also be represented by a special escape sequence \r. This means you can 'fix' such files by simply using the tr command to change any occurrence of \r to \n (the special escape sequence that represents a newline character):

```
$ tr '\r' '\n' < old_file > new_file
```

This will run the tr command and make it change any carriage returns in a file to newline characters. This is only a short overview of dealing with computer line endings and we explore a more efficient, Perl-based solution of handling line endings in Chapter 7.8.[38] You just need to remember that if you receive or download a text file

---

[38] See also the Wikipedia page on 'Newline' for a far greater overview of this subject.

that looks kind of strange when you look at it with less, then there may well be an easy fix for it.

The sed command

There is a Unix command that is so powerful and flexible that people have written entire books about it. The sed command can be used to perform a wide variety of text manipulations. If you have already worked through Part 4 of this book then you will quickly see that Perl's substitution operator is based on the sed command. One of the (many) functions of this command is to substitute any one text string for another. This happens on a line-by-line basis.

If you look again at the yeast GFF file that we used previously, you will see that each line starts with 'chr' and then the chromosome number. Let's use the cut command to get just the first three columns of this GFF file and then we can show the first ten lines using head:

```
$ cut -f 1–3 yeast_genome.gff | head
chrI     SGD     chromosome
chrI     SGD     repeat_region
chrI     SGD     telomere
chrI     SGD     repeat_region
chrI     SGD     gene
chrI     SGD     CDS
chrI     SGD     repeat_region
chrI     SGD     nucleotide_match
chrI     SGD     binding_site
chrI     SGD     gene
```

If we wanted to change the 'chr' to instead say 'chromosome,' we could do it like this:

```
$ cat yeast_genome.gff | sed 's/chr/Chromosome /' | cut -f 1–3 | head
Chromosome I    SGD     chromosome
Chromosome I    SGD     repeat_region
Chromosome I    SGD     telomere
Chromosome I    SGD     repeat_region
Chromosome I    SGD     gene
Chromosome I    SGD     CDS
Chromosome I    SGD     repeat_region
Chromosome I    SGD     nucleotide_match
Chromosome I    SGD     binding_site
Chromosome I    SGD     gene
```

The 's' part of the sed command puts it in 'substitute' mode, where you specify one pattern (between the first two forward-slashes) to be replaced by another pattern (specified between the second set of forward-slashes). Note that this doesn't actually change the contents of the file, it just changes the screen output from the previous command in the pipeline.

By default, sed only changes the *first* occurrence of the pattern that it finds. Just like in Perl, we can make it perform all possible replacements on each line by appending a 'global' switch. If we wanted to 'highlight' potential start codons[39] in a DNA sequence, we could make them stand out like so:

```
$ cat yeast_chr1_dna.fa | sed 's/ATG/- ATG -/g' | grep ATG | head
CAACCCACTGCCACTTACCCTACCATTACCCTACCATCCACC- ATG -ACCTACTCACCATAC
TGTTCTTCTACCCACCATATTGAAACGCTAACAA- ATG -ATCGTAAATAACACACACGTGCT
TACCCTACCACTTTATACCACCACCAC- ATG -CCATACTCACCCTCACTTGTATACTGATTT
TACGTACGCACACGG- ATG -CTACAGTATATACCATCTCAAACTTACCCTACTCTCAGATTC
CACTTCACTCC- ATG -GCCCATCTCTCACTGAATCAGTACCAA- ATG -CACTCACATCATT- ATG -
ACATACGTTATACCACTTTTGCACCATATACTTACCACTCCATTTATATACACTT- ATG -TC
AATAATACATAAACATATTGGCTTGTGGTAGCAACACTATC- ATG -GTATCACTAACGTAAA
AGTTCCTCAATATTGCAATTTGCTTGAACGG- ATG -CTATTTCAGAATATTTCGTACTTACA
CAGGCCATACATTAGAATAAT- ATG -TCACATCACTGTCGTAACACTCTTTATTCACCGAGC
AATAATACGGTAGTGGCTCAAACTC- ATG -CGGGTGCT- ATG -ATACAATTATATCTTATTTCC
```

In this example the sed command will process each line in turn, and change all occurrences of 'ATG' to '- ATG -'. Without the 'g' option on the end of the command, sed would only change the first occurrence it finds. Remember, this replacement is occurring on a line-by-line basis. If a potential start codon spans two lines in the file, then sed will not 'see' it. The sed command is worthy of further study, so you are encouraged to look at its man page and/or read up more about it on the internet.

The awk **command**

It is a little unfair to describe awk as just another Unix command. Like Perl, it is a fully fledged scripting language and can be used to write very powerful programs for text manipulation.[40] However, you don't need to learn all of the complexities of AWK (the programming language) when you can achieve many useful things through simple use of awk (the command). There are many web sites devoted to AWK, and so this section will give only a brief glimpse into some of the things you can do with it.

The following examples are all used in conjunction with the yeast_genome.gff file. The very first example will simply print the contents of the file, using AWK's print function. This is not a particularly useful way of using AWK, but as you will see, it becomes easy to add just a few more arguments to this command to do many useful things. The comments before each command indicate what the command will achieve, but you are encouraged to try these yourself:

---

[39] Start codons are represented by the nucleotides ATG and are required at the start of protein-coding genes.
[40] The development of the Perl programming language was partly due to a desire by Larry Wall, the creator of Perl, to overcome some of the limitations of AWK. So without AWK, there may have been no Perl, and hence you would probably be doing something less useful than reading this book right now.

```
# print out contents of input file
$ awk {'print'} yeast_genome.gff

# print out first column of file
$ awk {'print $1'} yeast_genome.gff

# print out column 5 then column 1
$ awk {'print $5, $1'} yeast_genome.gff

# print out start and end coordinates only if start coordinate is > 10,000
$ awk '{if ($4 > 10000) print $4, $5}' yeast_genome.gff

# print out start and end coordinates of features, plus their lengths
$ awk '{print $4, $5, $5 - $4 + 1}' yeast_genome.gff

# same as above ... whitespace is optional
$ awk '{print $4,$5,$5-$4+1}' yeast_genome.gff

# print lines that contain the word 'intron'
$ awk '/intron/ {print}' yeast_genome.gff

# same as above, print is implied
$ awk '/intron/' yeast_genome.gff

# print line numbers of features using NR variable (NR = number of records)
$ awk '{print NR, $3}' yeast_genome.gff

# use 'END' statement to only print the total number of lines in file
$ awk 'END {print NR}' yeast_genome.gff

# calculate sum length of all introns in UTRs and print out total
$ awk '/UTR_intron/ {lengths += ($4-$3+1)} END {print "total length of UTR
    introns = " lengths}' yeast_genome.gff
```

Hopefully this last example begins to show you some of the power of the awk command. In many cases, it can be quicker to extract simple information from text files
using AWK rather than writing a Perl script.

Advanced Perl 6

This chapter will cover a few 'beyond-the-basics' topics that were left out of the earlier chapters on regular expressions[1] (Chapters 4.26–4.28). This chapter still omits many, many aspects of regex use and you should consult the official Perl documentation to find out more.

Greedy quantifiers

Consider the following code. Can you guess what will end up in $match?

```
my $text = "Old password = opensesame, new password = abracadabra";
my ($match) = $text =~ m/password = (.+)\s/;
```

We use the grouping metacharacters to capture the subpattern .+ as part of a larger pattern. The grouped subpattern will match one or more occurrences of any single character. However, we follow the grouped pattern with \s, meaning that the subpattern ends when it matches some whitespace. As the $text string does not end with a whitespace character, this means that $match will never contain 'abracadabra'. So, what will it contain? In this example our regular expression can match one of two different strings:

(1) 'password = opensesame,'
(2) 'password = opensesame, new password ='

The second match could happen because it still satisfies the regular expression pattern of the string 'password =' followed by one or more of any character at all, followed by a whitespace character. So which of the two strings will the regex match? By default, repetition quantifiers are *greedy* and will always try to match as many characters as possible. In this example, the regex will match the longer of the two strings and $match would end up containing 'opensesame, new password ='.

If you try to match multiple groups from a string, and each group uses a greedy repetition quantifier, then the first grouping will be the greediest and subsequent groups will have to fight over what is left:

```
"Peter Parker is Spider-man" =~ m/(.*)er(.*)/;
# $1 gets 'Peter Parker is Spid'
# $2 gets '-man' though this pattern also matches ' Parker is Spider-man'
```

Non-greedy matching

You can override the default greedy behavior by appending a ? after any repetition quantifier. This is known as a *non-greedy* quantifier:

---

[1] Also known as regexes or regexps.

```
my ($match) = "A to Z, a to z" =~ m/(a.*?z)/i;   # $match gets 'A to Z'
my ($match) = "Excited!!!!!" =~ m/(!{3,5}?)/;    # $match gets '!!!'
my ($match) = "Anne" =~ m/(Anne??)/;             # $match gets 'Ann'
```

Note that in the last example the question mark character serves two different functions; first as the repetition metacharacter meaning 'match zero or one times,' and then as the non-greedy quantifier (which essentially forces the preceding metacharacter to match zero times).

Using regular expressions with the default variable $_

If you are using a regex to try to match the contents of the default variable ($_), you can simplify your code because the matching operator will inspect this variable if no other variable is specified:

```
# normal usage
print "Match\n" if ($text =~ m/\w{3,5}$/);

# using default variable
print "Match\n" if m/\w{3,5}$/;
```

Notice that if you are using the default variable, you don't need to include the binding operator (=~). When looping through a file using the <> operator, you can produce some very concise code. Imagine we were processing a file and we wanted to change any line which ended 'today.' to instead end with 'yesterday.':

```
#!/usr/bin/perl
# back_in_time.pl
use strict; use warnings;

while(<>){
    next if m/^$/; # skip blank lines
    s/today\.$/yesterday\./;
    print;
}
```

In this example, the matching and substitution operators and the print function are all working with $_.[2]

Rules for matching

Regexes will always try to match as early as possible in the target string. Depending on the pattern, this can sometimes conflict with some other principles of how regexes try to match. Consider the following:

```
my ($match) = "The fat cat sat on the mat" =~ m/([mscf]at)/;
```

---

[2] If you wanted to remove the newline characters from each input line, you could also use chomp, which is yet another function that will work with the default variable if no other variable is specified.

The character class in this example means that 'mat' should be the first pattern considered by Perl and 'mat' is a valid match to the input string. However, 'mat' matches later
in the string than 'sat', 'cat', or 'fat', so $match will end up containing 'fat'. This means
that even though Perl will normally match the first alternative it can, this is overruled if
a later alternative matches earlier in the string.

Matching word boundaries

There are still some metacharacters we have not seen before. The most common one you
might see is the \b metacharacter, which matches at a *word boundary*. This is defined as
a position *between* a word and non-word character or vice versa.

```perl
my ($match) = "Three, two, one" =~ m/\b(\w{3})\b/; # $match gets 'two'
```

In this example you might think that \b behaves a lot like \W, and you are correct.
However, the non-word part of the \b metacharacter also includes matching the start or
end of a string. Consider the difference between the following:

```perl
my ($match) = "A B C" =~ m/\W(\w)\W/;   # $match gets 'B'
my ($match) = "A B C" =~ m/\b(\w)\b/;   # $match gets 'A'
```

Remember that \b will match any position where you go from a non-word to a word
character (as in the example above), and also when you go from a word to a non-word
character:

```perl
"Spider-man" =~ m/\b-\b/; # matches
```

Because \b matches *between* characters, it is known as a *zero-width assertion*. If you try
to capture the match, the \b does not capture any characters:

```perl
my ($match) = "Spider-man" =~ m/(\b-\b)/; # $match gets '-' and not 'r-m'
```

Like many of the other metacharacters, the \b metacharacter has a negated version, \B.
This matches anything which *isn't* a word boundary.

Tracking the details of a match

When you are looping through a set of matches within one string, you might be interested
in knowing what position in the string you are at with each match. This can be accessed
using the pos() function, which returns the offset of the last match. Here is an example
of how you can use the pos() function; the output of the code is also included:

```perl
my $text = "Bruce Wayne is Batman";
while ($text =~ /(\w+)/g) {
    my $position = pos($text);
    print "Word \'$1\' ends at position $position\n";
}

# output
Word 'Bruce' ends at position 5
```

```
Word 'Wayne' ends at position 11
Word 'is' ends at position 14
Word 'Batman' ends at position 21
```

Even if you don't use the grouping metacharacters, it is still possible to find out what part of a string was matched to a pattern. The matching portion of any regex is stored within another special Perl variable $&. This variable is one of a set of three; the two other variables ($` and $') capture the text to the left and right of the match:

```
"Let's meet on Monday at 4:00 pm" =~ m/\w+day/i;
print "$`\n"; # prints 'Let's meet on '
print "$&\n"; # prints 'Monday'
print "$'\n"; # prints ' at 4:00 pm'
```

Using backreferences

We have already seen how grouped patterns are stored in the special variables $1, $2, etc. These variables can be used *outside* of the regex. However, you can also refer to any matches from grouped patterns *inside* a regex by using something called *backreferences*. These have a syntax of \1, \2, etc.

```
print "Match\n" if ("woof woof" =~ m/(\w{4}) \1/); # matches
```

In this example the \1 backreference is referring to the pattern that gets matched inside the preceding grouping characters. So whatever string is matched by \w{4} will also be used again as \1. Backreferences are therefore well suited for detecting duplicated strings – for example, maybe you wanted to see whether a DNA sequence contains a tandem duplication of any three nucleotides:[3]

```
my ($match) = "ATGCACGCAGCACACCCCTAG" =~ m/([ACGT]{3})\1/;
print "$match\n"; # $match gets 'GCA'
```

Using variables in regular expressions

Consider the following code, which would loop over every line of a file and try to match a regex stored in a variable $pattern. Notice that we have to escape the meaning of the metacharacters in $pattern by using backslashes:

```
my $pattern = "\\b\\w{1,5}\\$"; # find short words at ends of lines
while(my $line = <>){
    chomp($line);
    print "$line contains \'$1\'\n" if ($line =~ m/($pattern)/);
}
```

---

[3] Note that this code will only find the *first* occurrence of any tandem duplication. You would need to loop through the string to find all occurrences.

This code would work as expected, but behind the scenes Perl has to reevaluate the contents of $pattern for each line of the loop. This is because it is possible that the variable could be changed by some other code that occurs inside the while loop. If you know that a variable isn't going to be changed, you can speed things up by adding an /o option to the matching operator. This tells Perl to only evaluate the regex pattern once. The relevant line of code from the above example would now look like this:

```
print "$line contains \'$1\'\n" if ($line =~ m/($pattern)/o);
```

Prohibiting variable interpolation inside regular expressions

Sometimes you might want a regex to search for actual variable names. In these cases you don't want a regex to undergo variable interpolation. A simple way around this is to use single-quote characters as the delimiters for the matching operator:

```
my $pattern = "Some text";
print "Match\n" if ($line =~ m'$pattern');
```

This code will try to match the characters *$pattern* and not the contents of the $pattern variable. The behavior of using single quotes as the delimiters matches the behavior of quoting strings with single characters.

Write once, use many times

At the end of Part 4 of this book, we learned how we could parcel up useful pieces of code into *subroutines*. At some point you will create an amazing subroutine that performs some piece of Perl magic, and you will realize you want to perform the same magic in another program. The obvious solution is to copy-and-paste the subroutine from one program to another. Since subroutines are like mini programs, this often works just fine. But what if you discover an error in the function and now you want to fix all the programs that use it? You'll have to fix each program with the function. Not only is this laborious, but copy-and-paste is one of the more heinous crimes in programming. Thankfully there is a better solution.

Modules

A function library is a file of functions that can be shared among programs. You only have to write the function once, and you can reuse it in as many programs as you want. In Perl-speak, a function library is called a *package* or *module*. Perl module files are saved with the .pm suffix (for Perl module). The first line of a module uses the *package* statement and the last line is simply 1;. All of the functions go between those statements. There is no limit to the number of functions you can place in a library. Enough talk, let's see this in action!

Example 6.2.1 Save the following code as 'Library.pm' in the directory where you store your Perl scripts. This is not a particularly descriptive name, but it will do for now.

```
1.   package Library;
2.   use strict; use warnings;
3.
4.   sub gc {
5.       my ($seq) = @_;
6.
7.       $seq = uc($seq); # convert to upper case to be sure
8.       my $g = $seq =~ tr/G/G/;
9.       my $c = $seq =~ tr/C/C/;
10.      my $gc = ($g + $c) / length($seq);
11.
12.      return $gc;
13.  }
14.
15.  1;
```

Understanding the code

Line 1 contains the name of the library in a package statement. Generally, there is only one package statement per file. Line 2 ensures that the strict and warnings pragmas are used in the library even if they are not in the main program. Lines 4–13 define a function for calculating the GC composition of a nucleotide sequence. A library must 'return true,' and line 15 ensures that this occurs.

The gc() function in 'Library.pm' can now be used in any program you write. The syntax should be familiar: we just use the library.

Example 6.2.2

```
1.   #!/usr/bin/perl
2.   # library_test.pl
3.   use strict; use warnings;
4.   use Library;
5.
6.   my ($seq) = @ARGV;
7.   my $gc = Library::gc($seq);
```

Understanding the script

Line 4 contains the use Library statement. Notice that this is in the header with the other use statements; when Perl sees this line, it will attempt to import the contents of 'Library.pm.' You do not need to include the '.pm' suffix when instructing Perl to use a library.

Once a library is imported, all of its functions are available to the program. Functions are called in the same way we would call a subroutine; the only difference is that we must also prepend the library name as shown on line 7.[4]

Library PATH

From the Unix section of this book, you should be familiar with your $PATH environment variable (see Chapter 3.28). Just like Unix commands, Perl libraries can exist in various places on your computer. In the previous example, we saved the two files Library.pm and library_test.pl in the same directory (we hope you did too). If you didn't save them in the same directory, you would have seen an error that looks something like this:[5]

```
Can't locate Library.pm in @INC (@INC contains: /path1:/path2)
```

Perl looks for libraries in a variety of places, including the current working directory. If you want to see all the places Perl is looking, try printing the special @INC variable.

```
1.   #!/usr/bin/perl
2.   print join("\n", @INC), "\n";
```

---

[4] It is possible to automatically import functions so you don't have to prefix the library name. This is a dangerous practice so we don't advocate it (or even describe it). If you want to live dangerously, look it up yourself.

[5] Your error message will look different. There may be many paths shown with long and potentially strange names.

If you move Library.pm to any of the locations listed in the @INC array, Perl will be able to find it. Some of these locations may be protected, however, so you shouldn't put your libraries in those locations. Instead, you should add your personal library location to @INC. Like other things in Perl, there's more than one way to do it. First, let's move Library.pm to another location so Perl can't find it.

```
$ mkdir lib
$ mv Library.pm lib/
$ perl library_test.pl
```

Now try running library_test.pl again. If you didn't see error messages before, you should see them now. Perl can't find Library.pm anymore. There are several ways to tell Perl where it is. One way is to add a use lib statement in the header of your script.[6]

```
use lib "/complete/path/to/your/code/directory/lib";
```

The disadvantage of this is that if you take your script and module to another computer your Library.pm file will probably end up in a new location, so you may have to edit the script accordingly. An alternative is to use Perl's -I command-line option that lets you specify the location of a Perl library:

```
$ perl -I /path/to/library script.pl
```

You can also use this -I option as part of the 'shebang' line:

```
#!/usr/bin/perl -I /path/to/library
```

However, this raises the same issue as before – you may have to edit your script depending on the computer you are working on. The best way to have Perl read library paths is to use the $PERL5LIB environment variable. This works very similarly to your $PATH variable (which lets you provide a colon-separated list of directories). With this strategy, you don't need to edit your scripts, just your .profile or .cshrc.[7]

```
export PERL5LIB=/path/to/library:${PERL5LIB}   # bash
setenv PERL5LIB /path/to/library:${PERL5LIB}   # csh
```

The package statement
The package statement actually declares something called a *namespace*. All of the functions and global variables (and some other things) in any Perl module (or script) are part of a specific namespace. We saw above that the gc() function was part of Library, and when we called it from the script, it was named Library::gc().

---

[6] You might be tempted to use the push() function to add your personal library directory onto the @INC array, but this doesn't work because libraries are loaded during compilation and push() occurs later, at run-time.
[7] Or .bashrc or .login, etc. See Chapter 3.25 for more details.

Suppose we used another library named Module that also had a slightly different gc() function. You can imagine that it might cause problems having more than one function named gc(), but since the two gc() functions are in different namespaces, there is no conflict at all. The namespace of your program is called main:: or :: for short. If you also have a gc() function in your program, this can be called as gc(), main::gc() or ::gc(). The following example clarifies these concepts.

Example 6.2.3

```
1.    #!/usr/bin/perl
2.    use strict; use warnings;
3.    use Library;
4.    use Module;
5.
6.    my $seq = "AACGCTTAAcgA";
7.    sub gc {return 0.5}
8.
9.    gc($seq);              # calls function on line 7
10.   main::gc($seq);        # calls function on line 7
11.   ::gc($seq);            # calls function on line 7
12.   Library::gc($seq);     # calls function from Library.pm
13.   Module::gc($seq);      # calls function from Module.pm
```

Understanding the script

Lines 3 and 4 load two libraries: one that you created before (in Example 6.2.1), and another that you will need to create (see below). Note that the variable assignment on line 6 deliberately includes some lower-case characters. Line 7 defines a gc() function in the main program which simply returns 0.5.

Lines 9–11 call the function on line 7 using alternative syntaxes, but note that the syntax used on line 9 is the preferred form. Line 12 calls gc() from Library.pm and line 13 calls gc() from Module.pm. The two functions are not identical because Module::gc() only counts upper-case characters:

```
1.    package Module;
2.    sub gc {$_[0] =~ tr[GC][GC] / length $_[0]}
3.    1;
```

Package variables

It is often useful for libraries to contain variables for their own use. For example, several functions might want to access a translation table for the genetic code. Each function could define its own table, but it is better to store it just once.

Example 6.2.4

```
1.    package MolBio;
2.    use strict; use warnings;
3.    my %GENCODE = (
4.          'AAA' => 'K', 'AAC' => 'N', 'AAG' => 'K', 'AAT' => 'N',
5.          'ACA' => 'T', 'ACC' => 'T', 'ACG' => 'T', 'ACT' => 'T',
6.          'AGA' => 'R', 'AGC' => 'S', 'AGG' => 'R', 'AGT' => 'S',
7.          'ATA' => 'I', 'ATC' => 'I', 'ATG' => 'M', 'ATT' => 'I',
8.          'CAA' => 'Q', 'CAC' => 'H', 'CAG' => 'Q', 'CAT' => 'H',
9.          'CCA' => 'P', 'CCC' => 'P', 'CCG' => 'P', 'CCT' => 'P',
10.         'CGA' => 'R', 'CGC' => 'R', 'CGG' => 'R', 'CGT' => 'R',
11.         'CTA' => 'L', 'CTC' => 'L', 'CTG' => 'L', 'CTT' => 'L',
12.         'GAA' => 'E', 'GAC' => 'D', 'GAG' => 'E', 'GAT' => 'D',
13.         'GCA' => 'A', 'GCC' => 'A', 'GCG' => 'A', 'GCT' => 'A',
14.         'GGA' => 'G', 'GGC' => 'G', 'GGG' => 'G', 'GGT' => 'G',
15.         'GTA' => 'V', 'GTC' => 'V', 'GTG' => 'V', 'GTT' => 'V',
16.         'TAA' => '*', 'TAC' => 'Y', 'TAG' => '*', 'TAT' => 'Y',
17.         'TCA' => 'S', 'TCC' => 'S', 'TCG' => 'S', 'TCT' => 'S',
18.         'TGA' => '*', 'TGC' => 'C', 'TGG' => 'W', 'TGT' => 'C',
19.         'TTA' => 'L', 'TTC' => 'F', 'TTG' => 'L', 'TTT' => 'F'
20.    );
21.    sub translate_codon {
22.         my ($codon) = @_;
23.         if (exists $GENCODE{$codon})   {return $GENCODE{$codon}}
24.         else                          {return 'X'}
25.    }
26.    sub translate_sequence {
27.         my ($seq) = @_;
28.         my $pep;
29.         for (my $i = 0; $i < length($seq); $i += 3) {
30.             my $codon = substr($seq, $i, 3);
31.          if (exists $GENCODE{$codon})  {$pep .= $GENCODE{$codon}}
32.          else                          {$pep .= 'X'}
33.         }
34.      return $pep;
35.    }
36.    1;
```

Understanding the script

Lines 3–20 define a global variable containing the genetic code.[8] Since this variable is not designed to have its contents changed, it is declared in all upper-case characters to convey that it is a constant.[9]

Lines 21–35 contain two functions that both use the $GENCODE variable. The first simply returns a single amino acid character that corresponds to the codon that is passed to the function. The second function takes a sequence and sequentially extracts a string consisting of three nucleotides, i.e., codons. These are used as a look-up key in the %GENCODE hash to extract the corresponding amino acid. Each amino acid is then concatenated onto the end of a variable that stores the peptide sequence and this is returned by the function.

You might wonder why `translate_sequence()` does not call `translate_codon()`. This could replace lines 30–32 with a single function call:

```
$pep .= translate_codon(substr($seq, $i, 3));
```

While it is true that such a substitution would make the code more compact and abstract, nucleotide sequences can be very long. If you translated an entire chromosome, you would end up making millions of calls to `translate_codon()`. Generally, we advocate choosing abstraction over performance, but there are situations, like this, where the gain from abstraction is minor compared to the obvious overhead.

Nested namespaces

Namespaces can contain other namespaces separated by the double-colon scope operator. This helps organize related libraries. The concept is similar to folders that contain other folders. In fact, this is the way Perl stores nested libraries. For example, suppose you need a library that provides some descriptive statistics.[10] You would include a statement in your script such as the following:

```
use Statistics::Descriptive;
```

This instructs Perl to look for a file called 'Descriptive.pm' inside a directory called 'Statistics'. Note that you can have both a directory named 'Statistics' and a library called 'Statistics.pm' in the same directory. While nested libraries are a useful feature, we do not recommend that you make nested libraries at this point. They become necessary with some kinds of object-oriented programming that we do not cover in this book.[11]

---

[8] Most of the organisms we think about on a daily basis use this genetic code, but there are many species (especially microorganisms) that do not.

[9] It is possible to declare true `constants` in Perl using the constants pragma.

[10] See Chapter 6.9 for how to find additional libraries.

[11] In Chapter 6.8 we introduce object-oriented programming (OOP). However, we do not cover inheritance with *isa* relationships, which usually employs nested libraries. While OOP is powerful and beautiful, it can be horrible if done incorrectly. We therefore suggest you *use* libraries for a year or two before you start to write your own.

Got spreadsheets?

So far, all of the arrays we have used have been one-dimensional. For example, the following array holds three values in linear order.

```perl
my @array = ('Mouse', 'Mus musculus', 'Rodent');
```

If you placed this data in a spreadsheet, you could put all of these values in a single row or a single column.

Mouse
Mus musculus
Rodent

But what if you wanted to store the content of an entire spreadsheet in an array? How could you represent data that looked like this:

Mouse	Mus musculus	Rodent
Human	Homo sapiens	Primate
Cow	Bos taurus	Ungulate

A natural way to represent such data is a two-dimensional array. But before we show you how to do that, we must first discuss some things that are known in Perl as *references*.

One way to think about references is that they behave a little bit like a Windows shortcut or a Mac alias, which act as clickable 'bookmarks' for frequently accessed applications and files. When you create a shortcut/alias, you get an icon that mimics the original file, but does not duplicate it. You can make multiple shortcuts or aliases for a single application and place these in different folders across your filesystem. When you open a shortcut/alias of a document, you can make some edits and the changes appear in the original file. However, if you delete a shortcut/alias, it does not delete the original file. References in Perl behave a lot like shortcuts/aliases. The analogy is imperfect, but it is a useful place to begin.

Array references

Here is a simple array assignment:

```perl
my @author = ('Keith', 'Ian');
```

If we want to create a reference to this array, then we need to use the backslash operator:

```perl
my $author_ref = \@author;
```

Here we create $author_ref, which is a scalar variable that 'points to' the @author array. You may, or may not, find this confusing! Without the backslash character, we would just be assigning an array to a list and $author_ref would end up containing the size of @author. It is the backslash character that is therefore important in making $author_ref a reference.

Note that $author_ref is still a scalar in that it holds the contents of *one* thing, and yet the thing that it points to can be an array containing many things. If you try printing the value of $author_ref, you will get a strange string such as ARRAY(0x100800f40). This indicates that the variable is a reference to an array. The strange hexadecimal[12] number in parentheses is its memory address.

The Perl ref() function lets you ask if a scalar variable is a reference (as opposed to being a regular scalar variable). In this chapter, we will only be looking at array references. If you use ref() on something that is not a reference, it returns false.

```
print ref($author_ref); # prints ARRAY
print ref("something"); # prints nothing
```

Using array references

An array reference can be used very similarly to an array, but with a slightly different syntax. Since $author_ref is a scalar variable we cannot do $author_ref[0]. Instead we use the arrow operator[13] -> like this:

```
print $author_ref->[0]; # prints 'Keith'
```

You can sort of read this backwards as 'print the zero-th element of the array that is pointed to by the scalar called $author_ref'. Both $author[0] and $author_ref->[0] occupy the exact same location in the computer's memory. If you change one, you change the other:

```
$author_ref->[0] = 'Keith Bradnam';
print $author[0]; # will now print 'Keith Bradnam'
```

Conversely, if you make an array reference undefined or even turn it into a regular scalar variable, it does not affect the array that it formerly pointed to:

```
$author_ref = undef;       # make reference undefined
print $author[0];          # will still print 'Keith'
$author_ref = "cheese";    # can repurpose variable to act as a normal scalar
```

The -> operator lets you *dereference* individual elements of an array. But sometimes you need to dereference an entire array. For example, we use entire arrays when sorting. To deference the whole array, we can prefix the scalar with the array symbol. Alternatively, we can use the array symbol in front of a block containing the scalar. The following three statements are identical:

---

[12] Hexadecimal digits include 0–9 followed by a–f. For example, b is the hexadecimal equivalent of 11 and ff is the hexadecimal equivalent of 255. Hexadecimal numbers are indicated by the 0x prefix.

[13] This is actually called the infix operator, but we prefer arrow operator because it looks more like an arrow than an infix (whatever that looks like).

```
push @author,          'Nigel';
push @$author_ref,     'Nigel';
push @{$author_ref},   'Nigel';
```

The only difference here is that the first example is adding a string to an array whereas the last two examples are adding a string to a dereferenced array; but it is always the same array that is modified.

Seeing the combination of @ and $ signs together can take a while to get used to. With experience, you will glance at something like that and instinctively see that you are dealing with a reference to an entire array. Any operation you perform on an array can just as easily be performed on a dereferenced array. For example, looping through the elements of an array reference is as straightforward as:

```
foreach my $author (@$author_ref) {print "$author\n"}
```

As we will see below, there are times when the syntax of the block format @{$scalar} is preferred.

Anonymous array composer

In the above examples, $author_ref was a reference to the array @author. Creating references to arrays in this way makes sense when the array was created before you needed to create the reference. But what if you want to create the array at the *same time* as you make a reference? In this scenario we can actually skip making the @author array and go straight to the array reference using something called the *anonymous array composer*. This has a complicated name, but in practice, it just means replacing the parentheses that we would use to denote an array with square brackets:

```
my $author_ref = ['Keith', 'Ian'];
```

The reason we call this *anonymous* is because the array has no name.[14] The array reference has a name ($author_ref), but the array it points to does not. Just to be clear, the following two strategies both do the same thing.

Named array	Anonymous array

`my @array = ('Keith', 'Ian');`
`my $array_ref = \@array;`
`print $array_ref->[0];`
	`my $array_ref = ['Keith', 'Ian'];`
	`print $array_ref->[0];`

The reason we are discussing references and the anonymous array constructor is because these are the secret ingredients for making arrays of multiple dimensions. If you only need to work with one-dimensional data, then you could use an anonymous array, but it would probably make more sense to use a regular array.

---

[14] This is where the shortcut/alias analogy breaks down. An anonymous array is like an invisible file with no name that can only be accessed by its alias.

Two-dimensional arrays

Enough talk, time for some action. Let's make a two-dimensional array. Here is the data we want to represent:

Mouse	Mus musculus	Rodent
Human	Homo sapiens	Primate

We can make ordinary arrays for each row as follows:

```
my @row0 = ('Mouse', 'Mus musculus', 'Rodent');
my @row1 = ('Human', 'Homo sapiens', 'Primate');
```

Now we need to put both of these one-dimensional arrays into another array to make it two-dimensional. There are four different ways we can do this. Let's start by using array references:

```
my $row_ref0 = \@row0;
my $row_ref1 = \@row1;
my @table = (
    $row_ref0,
    $row_ref1,
);
```

Now we have a two-element array, where each element is a reference to another array. It is worth pausing to contemplate the data structure we have just created. We have a regular array (@table) that can be used just like any other array we have shown you. The only difference is that rather than containing regular scalars, it contains scalars that are references to other arrays. Also note that this is still a one-dimensional array until such a point that you need to dereference the contents of any array element. Only at that point will you be dealing with a two-dimensional data structure.

We can simplify the creation of the @table array by omitting the need to first make named references for each array – we can add references to the row arrays directly:

```
my @table = (
    \@row0,
    \@row1,
);
```

Unless you had a specific need to have a named reference to a specific row, this is a more straightforward way of creating the two-dimensional table. Now we want to show you an even simpler way of creating this table through the use of anonymous array constructors and named variables:

```
my $row0_ref = ['Mouse', 'Mus musculus', 'Rodent'];
my $row1_ref = ['Human', 'Homo sapiens', 'Primate'];
my @table = (
    $row0_ref,
    $row1_ref,
);
```

The end result is the same as before, but we now avoid having to create arrays to hold each row of data. However, this reproduces the earlier situation of making named array references that we might not need to use again. Let's move on to the fourth way of building the table of data. In the following strategy, we name only the two-dimensional array, and not the arrays or references. The relationship between data contained in the rows and columns becomes much more obvious in this syntax:

```perl
my @table = (
    ['Mouse', 'Mus musculus', 'Rodent'],
    ['Human', 'Homo sapiens', 'Primate'],
);
```

Hopefully you will agree that this is a simpler and tidier way of making our two-dimensional array. Note how the commas that occur *outside* square brackets denote each element in the first dimension, and the commas *inside* the square brackets denote elements in the second dimension. You can access an entire row as $table[1]. This holds a reference to an array. If you print $table[1] you will find it has a strange string such as ARRAY(0x1008000c0) because it is an array reference. How can you get to a single cell? From what we learned above, when you want to dereference a scalar, you use the -> operator. So $table[1]->[1] contains Homo sapiens. The arrow in this case is optional. This is a good thing, because $table[1][1] looks a lot better. To recap, if you wanted to print the contents of the first 'cell' in the table, you could do either of the following:

```perl
print $table[0]->[0];
print $table[0][0];
```

You can read both of these lines of code as 'print the zero-th element of the array that is pointed to by the reference that exists at the zero-th element of the @table array.'

Now let's put everything we have covered so far into a working script:

Example 6.3.1 Since this is longer than most scripts, you should write part of it, check that it runs, and then write some more. For example, write lines 1–9, then up to 11, then up to 17, and finally up to 23. As you write line 14, make sure you use @{$zoo[$i]} rather than @$zoo[$i].[15]

```perl
1.   #!/usr/bin/perl
2.   # zoo.pl
3.   use strict; use warnings;
4.
5.   my @zoo = (
6.        ["Mouse", "Mus musculus", "Rodent"],
```

---

[15] The reason is because @$zoo binds tighter than $zoo[$i] so Perl thinks that $zoo is a scalar variable. To force the array to be recognized, it must be in a block as {$zoo[$i]}.

```
7.          ["Human", "Homo sapiens", "Primate"],
8.          ["Cow", "Bos taurus", "Ungulate"],
9.     );
10.
11.    print "$zoo[1][2]\n"; # will print 'Primate'
12.
13.    for (my $i = 0; $i < @zoo; $i++) {
14.         for (my $j = 0; $j < @{$zoo[$i]}; $j++) {
15.              print "$zoo[$i][$j]\n";
16.         }
17.    }
18.
19.    foreach my $row (@zoo) {
20.         foreach my $column (@$row) {
21.              print "$column\n";
22.         }
23.    }
```

Understanding the script

Lines 5–9 define a two-dimensional array using anonymous array constructors for the rows.

Line 11 shows how you can access a single element of the array.

Lines 13–17 show how you can loop through all the rows and columns of the array by using a nested `for` loop.

Lines 19–23 do the same thing, but use a `foreach` loop instead.

Reading two-dimensional arrays from a file

If you have a lot of data, it would be laborious to write out all of the contents as part of your Perl code. In addition to developing sore wrists, you might make an error while typing. Therefore, it's much better to read data from files.[16] Even in the internet age, plain-text files based on the comma-separated values (CSV) and tab-separated values (TSV) formats are very common. Let's see how we can read these from Perl.

Example 6.3.2

This program will read data from a CSV file, which we also need to create, so first make a file with the following contents and name it something like species.csv:[17]

---

[16] See Chapter 6.4 for an overview of data management strategies.
[17] Or download the file from www.unixandperl.com.

Mouse,Mus musculus,Rodent
Human,Homo sapiens,Primate
Cow,Bos taurus,Ungulate

Now create the following program. When you run it, remember to specify the name of the CSV file as a command-line argument:

```
$ zoo_reader.pl species.csv
```

```
1.   #!/usr/bin/perl
2.   # zoo_reader.pl
3.   use strict; use warnings;
4.
5.   my @table;
6.   while (<>) {
7.       my ($common, $scientific, $family) = split(/,/, $_);
8.       push @table, [$common, $scientific, $family];
9.   }
```

Understanding the script

The `while` loop on line 6 reads from the file specified by the file operator (`<>`), which will be whatever files are specified on the command line or even STDIN.

Line 7 splits each input line from the file into three variables. The next line then adds these three variables to an anonymous array which is then pushed, as a single element, onto the end of the `@table` array.

While this script will work, it is a poor solution because it is hard-coded to work with only three columns of data in the input file. Additionally, we are choosing variable names that suggest a specific kind of data. To be more useful, the script should work with any number of columns. We can do that by changing lines 7 and 8:

```
7.   my @array = split(/,/, $_);
8.   push @table, \@array;
```

Now each line of the CSV file is split into an array. If there are 100 columns, the array will have 100 elements. This is pushed onto the `@table` to make a two-dimensional array.

To change from using CSV files to TSV files, all we have to edit is the pattern in the `split()` function:

```
7.   my @array = split(/\t/, $_);
```

For even more generality, you might imagine changing the string in the split to a variable that could be defined elsewhere in the program or – even better – be defined by the user as a command-line argument.

```
7.   my @array = split(/$separator/, $_);
```

Something strange?

Take another look at the code in Example 6.3.2. Does line 8 look strange to you? If not, don't worry about it. Move along. But if you're worried about @array getting deleted each time the script iterates through the while loop, then you are thinking very deeply. Lexical variables are supposed to live and die in curly braces, right? Yes. And in fact, @array is created each time we iterate through the loop. But the memory corresponding to @array does not get destroyed because it still has a reference. We discuss this somewhat confusing topic in Chapter 6.6 in the subsection about garbage collection.

Records and other hash references

We ate corned beef and potatoes last night ... yep, another hash reference

Some people read books from start to finish, whereas others like to skip around. If you're a skipper and haven't read the previous chapter, please do that first. Otherwise this part won't make much sense.

Hash references and the anonymous hash constructor

In the last chapter we showed you how to make references to arrays, so it makes sense to now show you how you can also make references to hashes. Not surprisingly, hash references work in a very similar way to array references. To make a hash reference, you also need to use the backslash operator. Let's create a simple hash and then make a reference to it:

```perl
my %sound = (dog => 'woof', cat => 'meow');
my $sound_ref = \%sound;
```

As you can see, the syntax for making hash references is exactly the same as the syntax for making array references. As you might expect, there is also an anonymous hash constructor that allows you to make a hash reference without a named hash. The syntax is a little different than before; simply substitute curly brackets for the parentheses:

```perl
my $sound_ref = {dog => 'woof', cat => 'meow'};
```

This produces a hash which has no name, but which is referenced by the variable $sound_ref. To dereference a single element of a hash, you need to use the -> operator:

```perl
print $sound_ref->{dog}; # prints 'woof'
```

You can read this as 'print the value for the hash key 'dog,' which is pointed to by the $sound_ref reference.' To dereference the entire hash, you prefix the scalar with a % symbol. In this way you can use a dereferenced hash anywhere that you can use a hash. Again, sometimes the block notation (shown below) is required:

```perl
my @animals = keys %{$sound_ref};
```

Let's put these new hash reference concepts into a script to try them out.

Example 6.4.1

```perl
1.   #!/usr/bin/perl
2.   # hash_ref.pl
3.   use strict; use warnings;
4.
5.   my $sound = {dog => 'woof', cat => 'meow'};
6.   $sound->{cow} = 'moo';
7.   foreach my $animal (keys %$sound) {
8.       print "$animal says $sound->{$animal}\n";
9.   }
```

Understanding the script

Line 5 creates a reference to an anonymous hash and line 6 adds another key–value pair to the hash. Finally, lines 7–9 loop through the entire hash, displaying the key–value pairs.

Hash references as database records

One of the most common uses of a hash reference is to represent a record from a database. For example, let's say we have a simple address book. Each *card* in the address book contains *attributes* for a person's name and email. One specific card might look like this:[18]

```perl
my $card = {
    name   => 'Ian',
    email  => 'atagcgaat@gmail.com'
};
```

An address book may contain many cards. To hold multiple cards, you use an array, of course. An array of hash references is a kind of two-dimensional data structure. Dimension 1 contains the cards; dimension 2 contains the attributes on each card. To add a hash reference to an array, we can simply push it on.

```perl
my @address_book;
push @address_book, $card;
```

To reiterate for clarity: in this example we take a reference ($card) that points to an anonymous hash. Because $card behaves like a normal scalar variable we can add it as a single item to an array. The @address_book array is a normal array, but it just happens to contain elements which are hash references.

If we wanted to, we don't even have to create a named reference to the hash. Instead, we can define the elements of the @address_book array as hash references:[19]

```perl
my @address_book = (
    {name => 'Ian',    email => 'atagcgaat@gmail.com'},
    {name => 'Keith',  email => 'whykeith@me.com'},
);
```

Let's see this in action.

Example 6.4.2

```perl
1.   #!/usr/bin/perl
2.   # address_book.pl
3.   use strict; use warnings;
```

---

[18] If you translate the email address in the genetic code, it is isoleucine, alanine, asparagine, or IAN. Yes, you can email Ian at this address. If you have learned this much Perl, Ian would be happy to hear from you.

[19] This array includes one of Keith's email addresses. Ian has no idea if Keith wants to hear from you. He didn't bother to ask. Try it and find out....

```
4.
5.   my @address_book = (
6.        {name => 'Ian',    email => 'atagcgaat@gmail.com'},
7.        {name => 'Keith',  email => 'whykeith@me.com'},
8.   );
9.
10.  foreach my $card (@address_book) {
11.       print "$card->{name} $card->{email}\n";
12.  }
```

Understanding the script

Lines 5–8 create an address book with two cards. Note how the attributes are aligned using whitespace in order to keep everything organized.

Lines 10–12 show how to loop through all the hash references in the array. Remember that the hash is anonymous, but when we use a foreach loop we have to assign a variable which will represent each item in the array. $card is just a temporary variable that is a placeholder for each element of @address_book, and each element will be a reference to a hash.

Reading records from a file

In the last chapter, we read the following CSV data into a two-dimensional data structure that used only arrays to contain the data:

```
Mouse,Mus musculus,Rodent
Human,Homo sapiens,Primate
Cow,Bos taurus,Ungulate
```

A two-dimensional array is not the best way to describe this type of data. Hashes let us give names to attributes rather than numbers. Which of the following do you find more descriptive?

```
$animal->[0];
$animal->{common_name};
```

Let's read the data into an array of hashes and navigate through the data structure by using a while loop. You may like to compare this code to Example 6.3.2 from the previous chapter:

Example 6.4.3

```
1.   #!/usr/bin/perl
2.   # zoo_reader2.pl
3.   use strict; use warnings
4.   my @database;
```

```
5.   while (<>) {
6.        my ($com, $sci, $fam) = split(/,/, $_);
7.        push @database, {
8.              common_name  => $com,
9.              scientific   => $sci,
10.             family       => $fam,
11.       };
12.  }
13.
14.  foreach my $animal (@database) {
15.       print "$animal->{common_name}\n";
16.  }
```

Understanding the script

Lines 5–12 read one line at a time from the CSV file specified on the command line. More specifically, lines 7–11 convert the data into an anonymous hash, and push a reference to the hash onto the @database array. Note how the key–value pairs on lines 8–10 are aligned with each other, and each attribute has its own line. This is a very common programming practice that makes the logic clear.

Lines 14–16 then loop through the array of hashes, printing just the common name of each animal.

Hashes of hashes and hashes of arrays

We have explored two-dimensional arrays and also arrays of hashes. In both these cases, the first dimension was an array. You can easily make this a hash instead. All you need to do is assign names instead of numbers. One reason to do this is that it makes searching much faster. Let's revisit the address book example. Suppose you have an address book containing many 'cards' and that you already have some code that retrieves this data from a file:

```
my @address_book = read_address_book("some_file");
```

Now, let's suppose that @address_book contains a few million records. We can search for any single record using the following construct, but note that this will take a relatively long time:

```
foreach my $card (@address_book) {
     if ($card->{name} eq 'Ian') {do_something($card)}
}
```

Using an unsorted list like this means that, on average, we have to search half the list to find a match. If the names are instead placed in a hash, the retrieval is much faster. For example, if you want to index your address book by name, you could use the following code:

```
my %index;
foreach my $card (@address_book) {
    $index{$card->{name}} = $card;
}
```

This code creates a new hash (%index) where each key of the hash is just the 'name' field from the hash of arrays. The value associated with each key is a reference to the array of all information about that person. If you find this syntax confusing, just remember that Perl tends to work from the inside out. The first thing that happens in the above code is that Perl resolves the $card->{name} part. This will produce a single string which can then be assigned to the hash key for the new hash (%index). Only then can Perl associate this key with a value. With this hash in place, you can retrieve any card with just a simple look-up:

```
my $card = $index{'Ian'};
```

This method assumes that names are unique, which might not be true, depending on the type of data you have. If you have redundant names, then you have to store an array of cards for each name.

```
foreach my $card (@address_book) {
    push @{ $index{$card->{name}} }, $card;
}
```

This now means that %index is a three-dimensional data structure! Dimension 1 is a hash where the key is a name, and the associated value is a reference to an array. Dimension 2 is the array of cards associated with each name. Dimension 3 is the attributes on the cards. Unsurprisingly, this can be very confusing, so let's access a single element to make the dimensions obvious. Suppose I want to get the email address for the first card whose name is 'Ian.' Here's how we can do that:

```
print $index{Ian}[0]{email}
```

In trying to decipher this, you should read from left to right. First, we have a hash key ($index{Ian}), and the value associated with that key is a reference to an array. The first element of this array ([0]) is itself a reference to a hash, and we then ask for the value associated with a specific key ({email}). We shall return to such complex data structures again in Chapter 6.5.

Don't pass up on this opportunity to learn how to pass arrays to subroutines

In the last two chapters we saw how references allow us to make useful data structures such as two-dimensional arrays and records. In this chapter we continue to explore more uses of references.

Pass arrays and hashes to subroutines as references

Up to now we have never passed multiple arrays or hashes to a subroutine. Why? Because they would get damaged by passing through the special @_ variable. Consider the following code:

```
compare_two_arrays(@a, @b);
sub compare_two_arrays {
    my (@array1, @array2) = @_;
    # etc.
}
```

The intention is to fill up @array1 and @array2 in the subroutine from the @a and @b arrays via @_. Unfortunately, in list context, Perl cannot determine the size of the arrays. So what happens is that @array1 gets all of the data from @a *and* @b, whereas @array2 ends up containing nothing. This means you can't ever pass multiple arrays to a subroutine.[20] The solution to this problem is to pass *references* to arrays:

```
compare_two_arrays(\@a, \@b);

sub compare_two_arrays {
    my ($ar1, $ar2) = @_;
    foreach my $e1 (@$ar1) {
        foreach my $e2 (@$ar2) {
            # something
        }
    }
}
```

Now the subroutine is passed two things which we can assign to a pair of variables. Those variables ($ar1 and $ar2) become references to each array, and as long as you remember to dereference them within the subroutine you can still access the original array data. The same strategy works with hashes. As a general rule, always pass arrays and hashes by reference.

Pass scalar references to minimize overhead in @_

We said earlier that you can create references to scalars, arrays, and hashes; but so far we have only dealt with array and hash references. The fact is that you probably won't

---

[20] Technically, you can pass multiple arrays … but they will always be received as a single array in the subroutine.

have to use scalar references very often. There are times, however, when they are very useful. Consider the following code:

```
my_function($scalar);
sub my_function {
    my ($thing) = @_;
}
```

The value of $scalar is copied into @_, which in turn is copied into $thing. What if $scalar contains something huge, like the DNA sequence of an entire human chromosome? Unfortunately, you will now have multiple copies of this data. This could reduce the performance of your computer if there is not much free memory left. Considering that other processes also need memory, you shouldn't be wasteful of it.

Generally, we like to focus on making code beautiful (see Chapters 7.3 and 7.4), but if a script uses an excessive amount of memory (or CPU), then some amount of ugliness may be appropriate. To minimize the amount of memory copied through @_, you can pass a scalar reference instead.

```
print gc_content(\$chromosome), "\n";

sub gc_content {
    my ($chr_ref) = @_;
    my $Gs = $$chr_ref =~ tr/G/G/;
    my $Cs = $$chr_ref =~ tr/C/C/;
    return ($Gs + $Cs) / length($$chr_ref);
}
```

So now we have a variable ($chromosome) in the main body of the script and a reference to that variable ($chr_ref) in the subroutine. Notice that to dereference a scalar variable we have to use a second $ symbol. Once again we can also use curly braces to say the same thing. The following are equivalent:

```
$$chr_ref
${$chr_ref}
```

Life beyond flatland

Some kinds of data are naturally complex and do not fit neatly into spreadsheets. For example, this book contains a title, sections, chapters, and paragraphs. Let's look at these from the bottom up. A chapter can be considered an array of paragraphs. If you wanted to store the paragraphs of Chapter 4.1 in an array, you might do something like this:

```perl
my @chapter_4_1 = (
    "paragraph 1 contents... ",
    "paragraph 2 contents... ",
    "paragraph n contents... ",
);
```

A section contains a number of named chapters. One way to store this would be as a hash, where each key is the name of the chapter and each value is a reference to the paragraphs:

```perl
my %section_4 = (
    "4.1 Hello World"        => \@chapter_4_1,
    "4.2 Scalar variables"   => \@chapter_4_2,
    "4.3 Use warnings        => \@chapter_4_3,
);
```

A book contains a title, and named subsections:

```perl
my %book = (
    title => "Unix and Perl",
    sections => {
        Introduction     => \%section_1,
        Installation     => \%section_2,
        Essential_Unix   => \%section_3,
        Essential_Perl   => \%section_4,
        Advanced_Unix    => \%section_5,
        Advanced_Perl    => \%section_6,
    },
);
```

If we wanted to get to a particular paragraph, we could do so as follows:

```perl
print $book{sections}{Essential_Perl}{"4.3 Use warnings"}[0];
```

A book is a fairly complex entity.[21] A lot of scientific data is equally complex.

---

[21] There is nothing like writing a book that makes this so apparent!

Too much nesting is confusing

When you start to use deeply nested data structures, processing the data can get very confusing. Let's consider a simplified view of a typical genome sequence and all the genes that a genome sequence can contain.[22]

Primarily, we can divide a genome into a set of named chromosomes.[23] In turn, each chromosome will contain many protein-coding genes, which will have one or more names. Each gene can encode one or more transcripts. Finally, each transcript consists of exons and each exon has a start and end coordinate. In Perl, we could represent this information by using a multi-dimensional hash/array structure. Given such an organization, you could access the beginning of an exon as follows:

```
print $genome{$chrom}{$gene}[$transcript][$exon]{begin};
```

If you find this code confusing, then that is a perfectly normal reaction. Such complex data structures can be used to solve lots of problems, but they can also create a lot of headaches. If we wanted to print out all the information for an entire genome, it would take a lot of nested foreach loops:

```
foreach my $chrom (keys %genome) {
    foreach my $gene (keys %{$genome{$chrom}}) {
        foreach my $tx (@{$genome{$chrom}{$gene}}) {
            foreach my $exon (@$tx) {
                print "$exon->{begin} $exon->{end}\n";
            }
        }
    }
}
```

It's rarely a good idea to use so many nested loops within a script. In a complex program, it can be difficult to see which parts are scoped to which loops, especially if the loop contains a lot of other code. Our own personal rule is not to descend more than two loops deep. Furthermore, we prefer to not have any set of nested loops (or a subroutine) occupy more than a single viewable page of code.

You might be wondering just how you can avoid the type of code we show above? The solution is to break the nested loop into subroutines. This also makes the code more readable. Here is a better alternative to the four loops that were used in the above code:

```
foreach my $chrom (keys %genome) {
    foreach my $gene (keys %{$genome{$chrom}}) {
        report_exons($genome{$chrom}{$gene});
    }
}
```

---

[22] The authors are biologists and sometimes we feel compelled to use biological examples.
[23] Although most chromosomes are numbered, letters (X and Y) are used to denote sex chromosomes.

```perl
sub report_exons {
    my ($gene) = @_;
    foreach my $transcript (@$gene) {
        foreach my $exon (@$transcript) {
            print "$exon->{begin} $exon->{end}\n";
        }
    }
}
```

In this new code arrangement we have two nested `foreach` loops that would loop over each gene from each chromosome. We then call a subroutine (`report_exons`) for each gene that we are inspecting, and this subroutine loops over all the transcripts and exons of each gene. In biology, a gene is often one of the more important 'units' of data and scripts may often want to report lots of statistics at the level of the gene. By reworking the code and including a subroutine, we can more easily access information for any specific gene without having to loop over all genes.

Digging yourself out of multi-dimensional hell

There are times when you want to explore a complex data structure, but you've forgotten what the structure was. Was `$data` a reference to an array of hashes or a hash of arrays? We don't advocate being disorganized enough to ask yourself this question, but we ask ourselves it often enough that we had better answer it. One way to explore a variable and all its substructures is with Perl's built-in `Data::Dumper` module. Let's give it a go.

Example 6.6.1

```perl
1.  #!/usr/bin/perl
2.  # dump.pl
3.  use strict; use warnings;
4.  use Data::Dumper;
5.
6.  my $thing = [0, 1, [2, 3, {hello => 'world'}], 6];
7.  'print Dumper($thing);'
```

This code creates a reference to an anonymous array which itself contains several tiers of higher-order structure. If you run this script you should see output as follows:

```
$VAR1 = [
          0,
          1,
          [
            2,
            3,
            {
              'hello' => 'world'
            }
```

```
              ],
              6
          ];
```

Understanding the script

The Dumper() function takes a reference and recursively descends through all levels of the data structure that are pointed to by the reference. It then prints out the data that exists at each level.

As you can see, the indentation in the Dumper() output shows the hierarchy of the data structure. The output is valid Perl, and you could paste this into another program.

We don't like the output of Dumper() very much because it is not very descriptive. It's easy to make a program that does the same type of thing with a prettier display. Let's build it! We'll call our new function display(). Since we may want to use this code in a lot of different programs, we will create a library. If you already have a library, you can add the display() function below to your library. If not, then you should create a new Perl module file called Toolbox.pm.

Example 6.6.2

```
1.    package Toolbox;
2.
3.    sub display {
4.         my ($thing, $level) = @_;
5.
6.         no warnings;
7.         my $tab = "\t" x $level;
8.         print "$thing\n";
9.         $level++;
10.        use warnings;
11.
12.        if (ref($thing) eq 'ARRAY') {
13.            for (my $i = 0; $i <@$thing; $i++) {
14.                print "\t$tab [$i] = ";
15.                display($thing->[$i], $level);
16.            }
17.        } elsif (ref($thing) eq 'HASH') {
18.            foreach my $k (sort keys %$thing) {
19.                print "\t$tab $k => ";
20.                display($thing->{$k}, $level);
21.            }
22.        }
23.    }
24.    1;
```

Understanding the code

On line 4, the display() subroutine receives two items. The first will be a reference to whatever data structure it is that we want to view; the second will keep track of how many levels deep we have descended into that data structure.

Lines 6–10 prints whatever is in $thing. Warnings are temporarily silenced so that undefined values don't cause error messages.

Lines 12–22 prints each element of an array or hash depending on whether $thing is a reference to an array or hash.

Note that lines 15 and 20 both include a call to the display() function ... from within the display() function! Such recursion can be confusing at first, but it means you can produce very powerful code. In this case we want to be able to descend through all levels of a data structure. When you reach a level which has no more references and just contains 'regular' variables, then no more recursion will occur.

Let's create a test script that will use our new Toolbox::display() function.

Example 6.6.3

```
1.    #!/usr/bin/perl
2.    # dump.pl
3.    use strict; use warnings;
4.    use Toolbox;
5.
6.    my $thing = [0, 1, [2, 3, {hello=>'world'}], 6];
7.    Toolbox::display($thing);
```

The output is as follows:

```
ARRAY(0x100863120)
        [0] = 0
        [1] = 1
        [2] = ARRAY(0x100800f00)
                [0] = 2
                [1] = 3
                [2] = HASH(0x1008001f0)
                        hello => world
        [3] = 6
```

Compared to the Dumper() output that we saw earlier, our new display() function adds more indenting to help separate out the different levels of the data structure. It also adds details to remind you whether you are exploring an array or a hash at any given level.

Garbage collection and reference counts

Consider this last part of the chapter as optional reading. You might find it helpful to know how Perl manages memory. On the other hand, you might just find this extremely confusing.

When you use the my keyword you are asking Perl to give you a piece of memory. This memory is created when you request it, and is returned to your computer once the execution reaches a closing brace in the same scope. The following endless loop allocates and frees memory continuously:

```perl
while (1) {
    my $variable = 'something';
}
```

What's happening 'under the hood' is that the memory associated with $variable is given a *reference count* of 1 when it is created on line 2. When the execution hits the closing brace at line 3, the reference count is decremented by 1 to 0. Any memory location with a reference count of 0 is given back to your computer. The act of freeing unused memory is called *garbage collection*.

Now let's consider a problem we posed in Chapter 6.3. Specifically, how can the contents of the variable @array exist in @table if @array is recreated with each pass through the loop?

```perl
1.   #!/usr/bin/perl
2.   # csv_reader.pl
3.   use strict; use warnings;
4.
5.   my @table;
6.   while (<>) {
7.       my @array = split(/,/, $_);
8.       push @table, \@array;
9.   }
```

If you make a reference to a variable with the backslash operator, you increase its reference count by one. This is why the memory associated with @array is not destroyed each time through the loop. The reference count of the memory associated with @array on line 7 is equal to 1. But on line 8, its reference count is increased to 2 due to the backslash. At line 9, the reference count is reduced by 1 when @array goes out of scope. But the memory location still has a reference count of 1, so it is not freed. Consequently, @table fills up with array references.

Reference count garbage collection is very efficient, but it has problems with circular references. The following code assigns $variable a reference to itself, so its reference count drops to 1 and never 0. Consequently, my $variable keeps allocating and never freeing memory.

```
1.   while (1) {
2.       my $variable;
3.       $variable = \$variable;
4.   }
```

Caution: this could crash your computer!

The use of options is also optional

Most of the Unix command-line programs you will use have options that control how they behave. For example, the Unix ls command lists the current directory, but if you want to see files sorted by date, you can type ls -lt. Your Perl programs can easily have this same behavior. There are two built-in modules for processing command-line options, Getopt::Std and Getopt::Long. Using these modules means you can add more functionality to your programs. Depending on the needs of your program, you can use command-line options instead of, or in conjunction with, the @ARGV array.

Getopt::Std

Getopt::Std is used for single-character options. The behavior of these options is just like other command-line Unix programs. If there are multiple options, they can be concatenated together. For example -l -t can be written as -lt or -tl. For options that take an additional parameter, such as -o for an output file, the space between the option and the parameter may be omitted: both -o file or -ofile mean the same thing. Let's see this in action. The following script can be a template for most of your programs.

Example 6.7.1

```
1.   #!/usr/bin/perl
2.   # getopt_std.pl
3.   use strict; use warnings;
4.   use Getopt::Std;
5.
6.   my $usage = "usage: getopt.pl [options] <arguments...>
7.   options:
8.      -v version
9.      -f flag
10.     -p <some parameter>
11.  ";
12.  die $usage unless @ARGV;
13.
14.  my %opt;
15.  getopts('hvfp:', \%opt);
16.
17.  if ($opt{h}) {print $usage; exit}
18.  if ($opt{v}) {print "version 1.0\n"; exit}
19.  if ($opt{f}) {print "flag is turned on\n"}
20.  if ($opt{p}) {print "Parameter is: $opt{p}\n"}
21.
22.  print "Other arguments were: @ARGV\n";
```

Understanding the script

Lines 1–4 are typical header information material with the addition of Getopt::Std.

Lines 6–11 contain a typical usage statement, which is just meant to remind the user about all of the options supported by the program. Note that the string stored in $usage is spread across several lines.

The die() function on line 12 will print the usage message only if the program is not given any arguments on the command line. Many Perl scripts will typically require the names of one or more files as input, and so you want to stop your script immediately and print the usage statement if such files are not specified. If your program does not require any arguments, then you should omit line 12.

Lines 14–15 are all that are required to remove the options from the command line and place them into the special %opt variable. The string 'hvfp:' is the crucial part. This specifies that there are four options: -h, -v, -f, and -p. The colon following p indicates that -p expects an additional argument. The other options do not take arguments.

Lines 17–20 perform different actions depending on the options present. These are simple one-liners here, but could be much more complex in a larger program.

Note that if the user specifies the -h option (for 'help'), then the script will print the usage statement and exit. This is another commonly used method of printing out a usage statement and -h will frequently, but not always, print out help information in many Unix programs. This script includes two ways of printing out the help information – this is a good thing!

Line 22 prints any other arguments that were specified on the command line.

Experiment with the following command-line options of getopt_std.pl to see what happens:

```
$ getopt_std.pl          # prints $usage
$ getopt_std.pl -h       # prints $usage
$ getopt_std.pl -v       # prints version
$ getopt_std.pl -x        # reports an error: there is no option -x
$ getopt_std.pl -fpx     # reports flag on, -p is x
```

Getopt::Long

Getopt::Long is used for longer, more descriptive options, for example --version. Long option names are preceded with two dashes. For options that take an argument, you do not omit the space between the option and the argument. You can either do --option x or --option=x. Here is the equivalent template.

Example 6.7.2

```
1.   #!/usr/bin/perl
2.   # getopt_long.pl
3.   use strict; use warnings;
4.   use Getopt::Long;
5.
6.   my $usage = "usage: getopt.pl [options] <arguments...>
7.   options:
8.    --version
9.    --help
10.   --flag
11.   --number <number>
12.   --string <string>
13.   ";
14.
15.  my $flag;     # some Boolean flag
16.  my $number;   # will contain a number
17.  my $string;   # will contain a string
18.
19.  GetOptions(
20.      "flag"       => \$flag,
21.      "number=i"   => \$number,
22.      "string=s"   => \$string,
23.      "version"    => sub {print "1.0\n"; exit},
24.      "help"       => sub {print $usage; exit},
25.  );
26.
27.  if ($flag)    {print "flag turned on\n"}
28.  if ($string)  {print "string set: $string\n"}
29.  if ($number)  {print "number set: $number\n"}
30.
31.  print "Other arguments were: @ARGV\n";
```

Understanding the script

Lines 1–4 are typical header material, but note that we are now using Getopt::Long rather than Getopt::std.

Lines 6–13 contain the familiar usage statement.

Lines 15–17 declare variables that will correspond to each of the possible command-line options that could be chosen by the user of the program.

Lines 19–25 are a call to the GetOptions() function to parse the specified command-line options. Line 20 shows how to set a simple flag option. This option will control an aspect of the program in an on-or-off manner.

Lines 21 and 22 show how to associate arguments with particular options. The argument can be checked for a numeric value as shown in line 21. The use of the equals sign tells the GetOptions() function that this option needs a mandatory argument. You can't run the script just by typing -number without additionally specifying a number.[24] The i and s signify that the options will receive an integer and a string respectively. There is also an f option for floating point numbers. If you specify that an option should contain a number and then you run the script and provide a string instead, you will get an error message from the Getopt function.

Lines 23 and 24 contain anonymous subroutines for reporting version and usage information. What is an anonymous subroutine? It is a subroutine that exists only as a reference. It's similar to an anonymous array or anonymous hash in that regard. In general, we do not suggest people use anonymous subroutines, which is why we do not specifically address subroutine references in this book. But this one example is harmless and convenient.

Lines 27–29 have been added to test which command-line options have been specified. Some programs will echo back the options that have been chosen to provide extra confirmation.

Try experimenting with the script by running the following command-line options:

```
$ getopt_long.pl                      # no help this time
$ getopt_long.pl --version            # prints version
$ getopt_long.pl --help               # prints usage
$ getopt_long.pl --flag --number=2    # flag on, number is 2
$ getopt_long.pl --string hooray      # string is hooray
```

Getopt::Std or Getopt::Long?

Both of these modules contain a great deal more functionality than is shown here. You should read the full Getopt documentation to discover what else they can do. Some people are happy to just use the standard version of the Getopt function, and others prefer the long version. As you add more and more options to your script, using the standard version can mean you run out of suitable single-letter characters for certain options – for example, if you use -h for a help option and then you want to add an option to specify the name of a valid *host* machine, you can't use -h again.[25] In contrast, if you use the long version and add lots of options then it can take a lot more typing to actually run a command. Compare:

```
$ my_script.pl -f -v -b -r
$ my_script.pl --fast --verbose --backup –recursive
```

---

[24] Sometimes it is useful to have a command-line option where the argument isn't mandatory. In this case use a colon (:) character instead of an equals sign.

[25] Though you can use -H if you so desire.

However, one advantage of using the long format for options is that they can always be reduced to the shortest unique string. All of the following would be equivalent:

```
$ my_script.pl --fast --verbose --backup --recursive
$ my_script.pl --fas --verb --back --recurs
$ my_script.pl --f --v --b --r
```

We suggest you experiment with both versions and use whichever one you prefer.[26] Just don't invent your own!

---

[26] One of the authors of this book prefers using the shorter syntax for command-line options and the other finds this maddening. Of course, the other author always uses long options, which is probably detested just as much by the first author.

OOP basics

Not oops, but object-oriented programming

Object-oriented programming (OOP) is a style of programming that focuses on data. OOP actually encompasses a large body of related programming practices that would take a long time to discuss in detail. In this chapter we will give a brief overview of OOP and focus on the parts that will help you *use* objects. We have intentionally omitted some aspects of OOP.[27]

Objects and classes

An object is some kind of thing. A person is an object. A dog is an object. Although people and dogs are both mammals, they are not the same kind of thing. People are people, dogs are dogs.[28] They look different: dogs have sharp teeth and visible tails. They behave differently: dogs bark when they are excited and greet one another by sniffing. In the language of OOP, we say that people and dogs belong to different *classes*.

All objects are instances of a class. You are an instance of the person class, as is each of the authors of this book. We are all different objects, but we all belong to the same class. In OOP, all data has to belong to a class.

The new constructor

In Perl, to create an object (an instance of a class), we use something called the new constructor. When we use new, we also have to specify the class from which we are creating a new object. The syntax looks like this:

```
my $object = new Class;
```

The new constructor is actually a function inside a file called Class.pm. Essentially this means that if you want to write the code to create a class (for people, dogs, proteins, etc.), that class has to have a function called new that will let you create a new object. As new behaves like a regular Perl function, we could also write the above code as:

```
my $object = Class::new();
```

As you can see, $object is a scalar variable. But the exact contents of that scalar are hidden. Is it a hash reference? An array reference? Don't ask. It's an object, and the internals are intentionally hidden.

Attributes

An object has certain attributes. For example, if we return to our example of a 'person' class then we can imagine that this class defines attributes for hair color, height, date of birth, etc. These attributes are shared by all objects in that class. However, each object could have attributes with very different values. To access an object's attributes you use

---

[27] In particular, we do not discuss *inheritance*, which is a wonderful bit of abstraction. In the hands of a non-expert, however, inheritance leads to poorly designed code.

[28] Some people and most dogs do not accept this as truth.

the -> operator followed by the attribute name. Let's imagine we have some object-oriented code that lets us access various fictional attributes of the 'person' class:

```
print $chris->hair_color;  # print attributes for the $chris object
print $chris->height;
print $chris->date_of_birth;

print $alex->hair_color;   # print attributes for the $alex object
print $alex->height;
print $alex->date_of_birth;
```

Is $chris->hair_color some kind of alternative syntax for a hash reference? No. $chris is an object and hair_color is an attribute. We don't ask questions about the internals of an object.

Methods

An object can perform specific actions called *methods*. A method is the OOP name for a function/subroutine. To call a method, you again use the familiar -> operator. Some methods take parameters, others do not. The syntax for calling a method is:

```
$object->method (@parameters);   # with parameters
$object->method ();              # no parameters
$object->method;                 # no parameters
```

You may have noticed that calling a method with no parameters ($object->method) appears very similar to accessing an attribute for an object ($object->attribute). The reason for this is that attributes are hidden behind accessor methods. That is, they are functions that return some attribute of the object.

Note that Perl does not interpolate functions/methods inside double quotes. The following syntax does not work as expected:

```
print "$alex->height\n"; # incorrect syntax
```

Encapsulation

One of the most important OOP concepts is *encapsulation*. This strives to separate the interface (what you interact with) from the implementation (the actual code). The only way to interact with an object is through its methods. What actually happens once a method is invoked should be irrelevant to the user. From a developer standpoint, encapsulation allows one to change the underlying code without breaking programs that depend on the object. As long as the interface to the object remains consistent, any programs that use the object will be unaffected.

OOP code reads better

Statements written in OOP code often look more readable than code which doesn't use OOP. This is because they put the object at the center of attention. Consider the clarity of the following code:

```
$ian->eat('pizza', 'salad') if $ian->is_hungry;
```

You could probably show this line of code to someone who has never programmed and they would still understand the general meaning of it. In a non-OOP style, the same functionality might be written as follows:

```
Person::eat($ian, 'pizza', 'salad') if Person::is_hungry($ian);
```

Note that in the second example we had to specify the library/class (Person) containing the eat() and is_hungry() functions. When we instead use objects, they know what class they belong to, so we don't need to tell them where to find their functions. They also pass themselves automatically to their methods, unlike in the non-OOP code example (where $ian has to be sent to each function). This is more than just syntactic sugar, it's a philosophy that data and functions are tightly bound to each other.

Your first class

The discussion so far has been mostly theoretical. Let's build a class and create objects to see how things actually work. The class will represent a circle, so we will call it Circle. The defining attribute of a circle is its radius. To construct a circle of radius 3, we will therefore want to be able to do something like this:

```
my $c = new Circle(3);
```

Before we can run this code, we first have to create the Circle class. We will do this in a step-wise fashion to explain all of the parts in detail. The class file is just like any other Perl module, so it will be named Circle.pm. It will live in the normal place where you put your libraries, and it will begin and end with the usual statements.

Example 6.8.1, Part 1: constructor

```
1.    package Circle;
2.    use strict; use warnings;
3.
4.    my $PI = 3.1415926;
5.
6.    sub new {
7.        my ($class, $radius) = @_;
8.        my $self = bless {}, $class;
9.        $self->{rad} = $radius;
10.       return $self;
11.   }
```

Understanding the code

Line 4 sets a global constant for pi. This scope of this variable is global to the entire package, i.e., it will be available to any function/subroutine we write. Internal data that can be used by several functions should always have a scope that makes them usable by all of the package. In OOP-speak, we call such variables *class variables*. We don't need $PI just yet, but we will later.

Lines 6–11 contain the new constructor, which has the syntax of a regular Perl subroutine.

Line 7 reads two arguments, the class name and the radius. Normally, we would expect a Perl subroutine that receives two parameters to be sent two parameters from the code that calls the subroutine. This is not the case when using OOP syntax. When we write code to call this subroutine, the class name will be sent automatically. This means we expect the code that calls this subroutine to pass only one parameter, the radius.

Line 8 contains the bless keyword. This turns an ordinary anonymous hash into an object of some class.[29] It is common to call an object $self or $this.

Line 9 stores the radius in the object's anonymous hash using the key rad. We could have chosen any string for this key because users never see the internals of an object.

Line 10 returns the object.

So far our class allows us to create a circle object, but not do anything with that object. We could write a program that creates a circle with the following line:

```
my $c = new Circle(3);
```

If we run this code we would create a new object ($c) which was created by sending details of a radius (3) to the Circle class. Remember, when using OOP syntax, the class name is automatically sent to the new constructor. An alternative syntax that does the same thing, but in a much uglier way, would be:

```
my $c = new Circle::new("Circle", 3); # not recommended
```

For our object to be even a little bit interesting, we will have to give it some attributes. As we discussed before, all access to an object is through a method/function interface. Therefore, we are going to need a subroutine called *radius* so we will be able to do the following:

```
print $c->radius, "\n";
```

Here is the equivalent non-OOP syntax:

```
print Circle::radius($c), "\n"; # not recommended
```

Note how the OOP syntax is much cleaner. We don't have to specify what package the radius function is in. We also don't have to pass the object to the radius function. This all happens automatically in OOP syntax. Let's see exactly how that happens by adding the necessary radius method to the Circle class:

Example 6.8.1, Part 2: accessor retrieving attribute

```
12.  sub radius {
13.      my ($self) = @_;
14.      return $self->{rad};
15.  }
```

---

[29] There is no reason you have to use an anonymous hash here. It could have been an array reference or even a scalar reference. But hash references are the most common.

Understanding the code

Lines 13 copies the object from @_. Like any other subroutine, parameters are passed to methods via @_. In OOP syntax, the object is automatically passed as the first argument.

Line 14 returns the radius from where the radius was stored. In this case we have already stored the radius using the string 'rad' (see Part 1 of the example).

As a user of this code, you would only need to know that after creating a new object for a circle, you can extract the radius for the object by using the radius method. Obviously, this might not seem that impressive, given that you would already have to know the radius in order to create the object. Time to make our object a little more useful; let's add an attribute for the area of the circle. This will be used in a simple fashion, like so:

```
print $c->area;
```

We could have computed the area and stored this when we first created the object. Instead, we are going to do something more interesting: we are going to compute the attribute. From the user's perspective, they will have no idea if the area was stored ahead of time or if it is being computed each time. More importantly, the user doesn't need to know. They just need to know that the area method will return the area of the circle object $c. Here's how we will make the area method:

Example 6.8.1, Part 3: accessor computing attribute

```
16.  sub area {
17.      my ($self) = @_;
18.      return $PI * $self->{rad} ** 2;
19.  }
```

Understanding the code

Line 17 gets the object from @_.

Line 18 computes the area of the circle by accessing the radius from the circle object and using this in conjunction with the pi constant that was defined earlier. The value of the area is then returned.

You might think it is a waste of CPU cycles to compute the area each time you need it. If you access the area attribute quite frequently, this may be true.[30] If so, then you probably should compute the area once in the constructor and store this in the object. Alternatively, if you create many circle objects, and very rarely ask about the area, it would be a waste of CPU to compute the area for every object.

Surely there must be some way to compute the attribute only once if you need it, and not at all if you don't need it? In fact, there is! All we have to do is compute and store

---

[30] Of course, such simple math is not really a burden to compute. But for the sake of the example, imagine it is.

the attribute after someone asks for it. Let's add a new method for circumference. Again, using this is straightforward:

```
print $c->circumference, "\n";
```

Example 6.8.1, Part 4: accessor memorizing attribute

```
20.  sub circumference {
21.      my ($self) = @_;
22.      if (not defined $self->{circ}) {
23.          $self->{circ} = 2 * $PI * $self->{rad};
24.      }
25.      return $self->{circ};
26.  }
```

Understanding the code

Line 21 retrieves the object from @_.

Line 22 asks if the circ key exists in the anonymous hash underlying the object. The first time this function is accessed, the circ key does not exist. Every time afterward, the key will exist because of the next line.

Line 23 computes the circumference and stores it in the hash with key circ.

Line 25 returns the circumference.

Below is a program to test-drive our new 'Circle' class. From the perspective of this script, the radius, area, and circumference accessor methods all behave in exactly the same way. As we have just seen, they are all a little different internally. Encapsulation hides these details from the user and presents a consistent interface:

```
1.  #!/usr/bin/perl
2.  # object.pl
3.  use strict; use warnings;
4.  use Circle;
5.
6.  my $c = new Circle(3);
7.  print $c->radius, "\n";
8.  print $c->area, "\n";
9.  print $c->circumference, "\n";
```

Remember, you will very likely end up using OOP code that has been written by someone else. With this example you might not care about *how* their code calculates the radius, area, and circumference of a circle, you just need to know how to use it. You might use such code and then discover that there is an update to the Circle.pm file. That update might drastically change the underlying code, but none of this would affect your script.

When to use OOP

Many software libraries use OOP syntax (see the next chapter), so all programmers should be able to use objects. However, not everyone needs to write their own classes; many expert Perl programmers have never written a class, and you don't have to either. In fact, we suggest you don't write classes just yet. Eventually you will see where OOP is useful and where it is a burden, and you will choose your programming style appropriately.

It's comprehensive

If you have come this far in the book, take a moment to celebrate! Put down the book. Sing at the top of your lungs, watch a movie, eat a delicious dessert, run around the block naked, or whatever it is you do when you celebrate.[31] You have mastered the most important parts of Unix and Perl!

Are you back? Good. Your job as a programmer is about to become a lot easier. In this chapter we are going to show you how to tap into a huge world of free Perl software: the Comprehensive Perl Archive Network (CPAN).[32]

Searching for free code

Before you start writing a complex piece of software, check CPAN first! It's possible that someone has already done some of the hard work. Good programmers write reusable code. Great programmers reuse code. Go to http://search.cpan.org and try searching for something. You may be amazed at what you can find.

Installing modules

One way to install a new module is to use one of the CPAN modules that comes pre-installed with Perl. You can use this by entering the following on the command line:

```
$ sudo perl -MCPAN -e shell
```

This opens a CPAN shell where you can type commands. The 'sudo' part is necessary to be able to have permission to copy the Perl module to somewhere on your filesystem outside of your home directory. Note that you might not have permission to do this if you are not the administrator on the computer you are using. If this works, you should see the following prompt:

```
cpan[1]>
```

Try asking for help:

```
cpan[1]> help
```

This will show you quite a bit of information.

Let's install a module for descriptive statistics:

```
cpan[2]> install Statistics::Descriptive
```

You should see some output showing that the module is being built, tested, and installed. To quit, hit 'q'; now all of your programs can use this module. For starters, try getting help about the module by using the `perldoc` command:

```
$ perldoc Statistics::Descriptive
```

---

[31] You may also do all of these things at the same time if you so desire.
[32] Not to be confused with the US cable network station that covers political news: C-SPAN.

It's also possible to download a module from the CPAN web site and install it without using the CPAN shell. However, the method for doing this is not quite as user-friendly. After downloading and unpacking the module, there will be a directory (based on the module name) which should contain a README and INSTALL file. Read and follow the instructions in these files. These will usually be along the lines of running the following commands:

```
$ cd module-name
$ perl Makefile.PL
$ make
$ make test
$ make install
```

Where do modules live?

Whatever OS you are using, there will usually be one default location where Perl modules are installed. For example, on modern Mac computers, which are based on a Unix OS, this will be the following directory:[33]

```
/System/Library/Perl/5.10.0
```

It is common to refer to a collection of modules as a 'library' and it is therefore common to also name directories that contain modules either 'library' or 'lib'. If you add new Perl modules (using the method above) they will be added to the default location. As mentioned previously, you might not have permission to add files to this directory. If you don't have permission to install them in the default location, you should always be able to add them to your home directory. You can do this (on Unix systems) by changing one of the installation steps shown above, namely:

```
$ perl Makefile.PL PREFIX=/Users/nigel/lib
```

This would work on a Unix system where /Users/nigel is the location of a home directory, and lib is a subdirectory you want to use for Perl modules (though you can name this anything you want). Obviously, this would only work if the user 'nigel' is the person installing the module.

Using modules

Some modules use OOP syntax and some do not. Some modules go so far as to offer both styles; it depends on the author. Let's try to use the Statistics::Descriptive module:

```
1.   #!/usr/bin/perl
2.   # stats.pl
3.   use strict; use warnings;
```

---

[33] This will obviously be different if you have a different version of Perl.

```
4.   use Statistics::Descriptive;
5.
6.   my $stat = new Statistics::Descriptive::Full;
7.   $stat->add_data(1, 2, 4, 8);
8.   print $stat->mean, "\n";
9.   print $stat->median, "\n";
10.  print $stat->harmonic_mean, "\n";
```

Sometimes it is relatively easy to work out what a module is doing (even when you have never used it before). In this script, we first have to make an *object* ($stat) that will handle the statistical computations (line 6) (see the previous chapter for more information about using objects). We then use the add_data() function to generate some data for the object to work with (line 7). The following lines then call various functions to calculate the mean, median, and harmonic mean of those numbers.

The documentation that accompanies each module should always provide full details of how it should be used. Do not assume you will understand everything about the module without first reading the documentation!

Using modules installed in non-standard locations

If you have installed a Perl module in a non-standard location, your script may not know where to find it. Perl will first look in the default location and also in the current working directory. If the Perl module is installed anywhere else, then Perl will not know about it and will therefore not be able to use it. One solution to this is to specify the alternative location of the module *inside* the script by adding the following line:

```
4.   use lib "/Users/nigel/lib";
```

Perl will now know to look for modules in the specified directory in addition to the default locations. This is not the best solution, however, because if you want to move your own Perl library to another directory you would have to change all of your scripts that use that library.

On Unix systems you can specify a Unix environment variable called $PERL5LIB.[34] This variable stores the location of the directory (or directories) that you want to use for any Perl library that isn't installed in the default location. On a Mac I would simply need to edit the .profile file[35] in my home directory to include the line:

```
export PERL5LIB=$PERL5LIB:/Users/nigel/lib
```

This takes the existing value of $PERL5LIB and *appends* on a new location (/Users/nigel/lib). Perl would then know to check the location of whatever directories are specified by $PERL5LIB in addition to the default library location.

---

[34] See Chapter 3.11 in order to refresh your memory about environment variables.
[35] On other Unix systems you may need to edit another file such as .bashrc or .login. See Chapter 3.25.

Programming topics 7

De bug is in de computer

In some ways, writing Perl scripts is easy. It's getting them to work properly that is the hard part. The process of removing errors and fixing problems is called *debugging* and it usually accounts for the majority of the time you will spend working on any given script. Finding and fixing bugs is one of your most important skills. Like any skill, it takes time to learn. In the beginning, you may blankly stare at an error message for what seems like an eternity, and Perl's error messages will make as much sense to you as if they were written in Klingon.[1] But with a little practice, you will learn to recognize the different types of errors and implement the appropriate debugging strategies without much thought.

It's important to realize that the problems that will require debugging will not be limited to errors in your actual code. More generally we can divide all of the issues that might require debugging into several categories:

(1) errors that are caused *before* your Perl code is even evaluated;
(2) errors in the code itself, further divided into:
    (2.1) compile-time errors;
    (2.2) run-time errors;
(3) other problems (often due to various forms of user error).

Use Perl's built-in syntax checker

You should get into the habit of checking your code *before* you run it. Perl has a built-in tool for checking the syntax of your script; this can be utilized by supplying the -c option to the Perl command. For example, if you have a script called script.pl, then simply run:

```
$ perl -c script.pl
```

Note that this does *not* run the script, it just checks the script for potential errors. If your script contains no syntax errors, you will see a message saying:

```
script.pl syntax OK
```

However, if your script contains errors you will see all of those errors just as you would if you actually run it. If possible, you should *never* run a script that produces *any* error or warning messages.

---

[1] We'd like to believe that somewhere in a parallel universe there exists a programming language which gives you error messages such as 'Hey Bob, I notice that on line 23 you forgot to put any data in variable $x, and then on line 32 you're trying to access an array element but you forgot to include the closing square bracket.'

General strategies for debugging

Programming languages like Perl have sophisticated, and therefore complicated, debugging tools. But for simple scripts, these tools can be overkill. Here is some basic advice on how to go about fixing your scripts:

(1) Stay calm and don't blame the computer. In nearly all cases, the computer is only ever doing what you have told it to do.[2]

(2) Check and re-check your code. Most errors – probably the vast majority – are due to very simple typos in your script. You will have probably looked right at the error without realizing it.

(3) Start with the *first* error message you see. Subsequent error messages are likely to all stem from the first problem in your script. Fix one, and you may fix them all.

(4) Use the ability to 'comment out' lines of code to see if 'turning off' a single line of code removes the error message (more on this later).

(5) Sometimes a program will start to work but then fail at some unknown point within your code. Consider adding simple print statements to work out where the program is failing.

(6) Use a suitable code editor. Most editors will provide *syntax highlighting*, meaning they will color various parts of code depending on their content/function. This often provides visual feedback that you have not typed a piece of code correctly.

Compile-time vs. run-time errors

In general, there are two kinds of bugs in a program, compile-time errors and run-time errors. A program that does not even begin to run has a compile-time error. A program that exhibits errors (like crashing) *after* it starts running has run-time errors.

Example 7.1.1 The line below contains two errors (unbalanced quotes and parentheses). Any program containing this line of code will not run because it produces a compile-time error.

```
print("hello world));
```

Example 7.1.2 The following line will lead to a run-time error if the value of z is zero or undefined (which Perl treats as zero in a numeric context).

```
$x = $y / $z;
```

Obviously, we want our programs to be error-free, but if you had to choose between compile-time and run-time errors, compile-time errors are better. Errors that show up only after a program has been running for minutes or hours can be laborious to debug. One of the reasons we suggest people include use strict is because it changes

---

[2] We know that some of you may never agree with us on this point. If you are the type of person who is likely to curse at their computer and believe that it is 'lying' or 'cheating' then maybe programming is not the best vocation for you.

misspelled variable names from run-time errors to compile-time errors. We also suggest including use warnings to catch run-time errors that result from undefined values. There are many more reasons to use strict and use warnings, so make sure every program has these statements.

Commenting-out code

One of the most common tools for debugging is commenting-out a line of code to determine if it is the line that is causing the error. This is as simple as placing the comment symbol # at start of a line to prevent it from compiling.[3] If the program now runs, you have isolated which line caused the error. Then it is a matter of finding out what part of the line is not syntactically correct. Most of the time, Perl error messages point you to the exact offending line, and may offer advice on how to fix it, but this is not always the case. Sometimes the error occurs in a preceding statement.

Example 7.1.3 If you run the following script, you will receive the error message shown below.

```
1.    #!/usr/bin/perl
2.    # example3.pl
3.    use strict; use warnings;
4.
5.    $x = 1;
6.    $y = 2
7.    $z = 3;
Scalar found where operator expected at example3.pl line 7, near "$z"
        (Missing semicolon on previous line?)
syntax error at example3.pl line 7, near "$z "
```

In your early days as a programmer, you will see the 'Missing semicolon on previous line' error message quite frequently. In this example the semicolon was indeed missing from line 6. If you comment-out line 6, you will find that the program runs with no errors.

Commenting-out multiple lines with pod tags

Sometimes you might want to comment-out many lines at once. While you could begin every line with a #, there is a more convenient way. Perl supports multi-line comments via its documentation system called 'pod,' which is an acronym for Plain Old Documentation. To learn more about how to use pod, see Chapter 7.6 on Documentation. For debugging purposes, we only need to know that a pod comment begins with =something and ends with =cut. The *something* can be anything. For debugging, it makes sense to use =debug.

---

[3] Though be careful. If you comment out a line which is the opening or closing part of a block of code (e.g., the first line of a for loop) you will be introducing more errors. So only comment out self-contained lines of code.

Example 7.1.4 Lines 2 through 5 are skipped by Perl. Only the first line will do anything.

```
1.  print "something\n";
2.  =debug
3.  print "for\n";
4.  print "nothing\n";
5.  =cut
```

Commenting out with __END__

Perl stops parsing the script once it sees the __END__ token. Note that this has *two* underscore characters on each side. If you want to comment-out many lines near the end of a program, this is the most convenient method. You can also store notes to yourself or even internal data after the __END__ of a program. See Chapter 7.5 for more information on storing internal data.

Example 7.1.5 The following program prints 'hello world' but nothing else.

```
1.  #!/usr/bin/perl
2.  use strict; use warnings;
3.
4.  print "hello world\n";
5.  __END__
6.  print "goodbye";
7.  I can write anything down here!
```

Debugging run-time errors

Run-time errors can be incredibly frustrating to debug, especially if they appear somewhat randomly. One of the simplest but most effective debugging strategies is using print statements at critical junctures in your code.

Example 7.1.6 The following script attempts to loop through a series of user-specified numbers to calculate a total, and then loop through them again to calculate what percentage of the total each number represents. However, the script contains a deliberate error. Can you spot it?

```
1.  #!/usr/bin/perl
2.  use strict; use warnings;
3.
4.  die "Specify at least five numbers\n" if (@ARGV < 5);
5.
6.  my @numbers = (@ARGV);
7.  my ($array_length) = @numbers;
```

```
8.
9.   my $total = 0;
10.
11.  for (my $i = 0; $i < $array_length; $i++){
12.      $total += $numbers[$i];
13.  }
14.
15.  for (my $i = 0; $i < $array_length; $i++){
16.      my $percent = ($numbers[$i] / $total) * 100;
17.      print "$numbers[$i] is $percent percent of the total\n";
18.  }
```

Understanding the script

The error is on line 7, which mistakenly tries to calculate the array length by assigning the array to a list rather than to a single scalar variable. This line should be:

```
7.   my $array_length = @numbers;
```

This error means that $array_length will just contain the first number specified on the command line. If this is less than the number of specified values, the script will run but produce misleading output. If the first number is greater than the number of specified values, then the script will produce error messages.

Debugging the script

The easiest way to debug this script would be to add some simple print statements. For example, on line 8 you could add a print statement to check that $array_length is the right length, or you could add a print statement to line 14 to check that $total is correct.

This is a highly simplistic example, but hopefully it illustrates the point that you should never assume you always know what a variable contains.

The Perl debugger

Perl includes a debugging mode that allows you to step through programs line-by-line and view the contents of variables. The debugger can be very useful as it allows you to run your script in a very controlled way; in a way, it's a bit like having a remote control for your script. You can make the debugger run your script until a specific line of code is reached, and then you can switch to executing one line of code at a time.

The debugger might seem a little intimidating if you are new to Perl and so we are not going to cover it in this book. However, if you are curious to learn more, then note that there is a Unix man page for the debugger as well as a separate man page that gives more of an introductory tutorial to how the debugger works:

```
man perldebtut
man perldebug
```

Other (non-Perl) errors

Many developers of computer code spend a lot of time trying to make their programs as fool-proof as possible. A well-designed program should hopefully reduce the ways in which the user can break it (intentionally or otherwise). However, as a species we have developed a remarkable ability to be very inventive with our own stupidity. We have all seen people hurl abuse at their computers (both verbal and physical) when they run into a problem which they think is the computer's fault, only to then realize that they forgot to do something very obvious. Here are a few situations that you will hopefully never find yourself in.

Program changes not saved.

If you make changes to your program but don't save them, then those changes will not be applied when you run the script. Always check that the script you are running is saved *before* you run it. Many code editors will offer visual feedback if a program has not been saved.[4] Repeat after us: edit, save, run; edit, save, run; edit, save, run….

The program you are editing is not the same as the program you are running.

Occasionally, you might make copies of your programs and your directory might end up with programs named things like script1.pl, script1b.pl, script2.pl, new_script2.pl. This is a bad habit to get into and you might find yourself editing one script but trying to run another. You will become very frustrated when every change you make to your script has seemingly no effect. See Chapter 7.7 on revision control to find out how to avoid this situation.

The program runs, has no errors, but doesn't print any output.

It might seem mysterious when your Perl program, which you so carefully wrote, doesn't seem to do anything. It is therefore worth asking yourself the question 'Did I ask it to do anything?' More specifically, have you made sure your program is printing any output? Making your program calculate the answer to life, the universe, and everything is one thing … but if you don't print out the answer, then it will remain a mystery.

---

[4] E.g., if you are using any form of text editor on an Apple computer, there will always be a black dot within the red 'close window' icon on the top left of any window which contains an unsaved document.

There are a number of error messages that you might see which are not really Perl errors, even though they might result from what you write (or don't write) in your Perl script. However, most error messages you see will be because of a problem, usually a typo, in your script. In this chapter, we'll first look at a few 'non-Perl' error messages before explaining some of the more common Perl error messages. Finally, we'll include a table of many of the most common error messages, along with a quick explanation as to how you can fix them.

Non-Perl error messages

Permission denied

Do you have the permission to run the script? Have you run the Unix chmod command to add executable permissions (see Chapter 3.27)?

command not found

You've either mistyped the name of your Perl program[5] *or* the program is not in a directory the Unix system knows about (technically speaking the directory is not in your *path*). If you are using a specific directory to contain all of your Perl scripts, make sure that this directory has been added to your Unix $PATH environment variable (see Chapter 3.28).

bad interpreter: No such file or directory

The first line of a Perl script should let the Unix system know where it can find a copy of the Perl program that will understand your code. This will often be in /usr/bin/perl. If you make any typo in this line (e.g., omitting a slash), then Unix will be looking for Perl in the wrong place, and things will fail.

use: command not found

Remember we told you that you can include whitespace anywhere in your program? Well, there is one exception to that rule: the first line of your script cannot be blank! This is sometimes a very hard error to spot as everything else can seem normal. Unix only looks at the first line of a file in order to determine whether it should be sent off to be interpreted by another program (e.g., Perl). If the line is blank, Unix will attempt to execute all of the following lines of Perl code as Unix commands. Because Unix also supports the use of hash characters to comment out lines, it will not report any error until it gets to the first line of Perl code, which is nearly always a use declaration.

---

[5] If you continue to type program names for yourself then you are your own worst enemy. Repeat after us: use tab-completion, use tab-completion, use tab-completion ....

Common Perl error messages

There are a lot of different Perl error messages, though in common practice a small number of messages will cover most of the problems you encounter. If you are using a Unix system you can find a description of these errors by looking at the `perldiag` man page. You can also tell Perl to produce more informative error messages by including the `diagnostics` pragma in your scripts:

```
use diagnostics;
```

What follows is a brief description of some of the most common error messages you will encounter:

`Missing right curly or square bracket at script.pl line X, or... Unmatched right curly bracket at script.pl line X`

Hopefully these two error messages are both very obvious. [These are square brackets] and {these are curly brackets}. Unless you are using them as text characters (e.g., within a print statement), then they *always* come in pairs. Make sure your ones are in pairs.

`syntax error at script.pl line X, near Y`

Syntax errors are among the most frequent errors you will see. On the plus side, they are usually very easy to fix. On the negative side, they can sometimes be very hard to spot as they frequently involve a single character that is either missing or surplus to requirements. Most commonly they might be due to one of the following:

(1) Unmatched parentheses – just like brackets, items that are in parentheses should always be a double act.
(2) A missing semicolon – if you start writing some code, then it has to end (at some point) with a semicolon. The main exceptions to this rule are for the very first line of a script (`#!/usr/bin/perl`) or when a line ends in a closing curly bracket '}'; also note that you can write one line of Perl code across several lines of your text editor, but this is still one line of code, and so needs one semicolon.
(3) A missing comma – Perl uses commas in many different ways; have you forgotten to include one in a place where Perl requires one?
(4) Inventing new Perl commands and operators – if you write `if ($a === $b)`, then you have invented a new operator (`===`), which will cause a syntax error as Perl will have no idea what you mean.

`Can't find string terminator """ anywhere before EOF at script.pl line X`

Did you make sure you have *pairs* of quotation mark characters? If you have an odd number of single or double quote characters, then you might see this error.

`use of uninitialized variable in...`

Your scripts will do many things with variables. You will add their values, calculate their lengths, and print their contents to the screen. But what if the variable doesn't actually contain any data? Maybe you were expecting to fill it with data from the command line or from processing a file, but something went wrong? If you try doing something with a variable that contains no data, you will see this error.

`Global symbol "$variable" requires explicit package name at`

You wouldn't happen to be using the `strict` package and not declaring a variable with my would you? If you definitely have included `use strict`, then maybe check that all your variable names are spelled correctly. You might have introduced a variable as my `$apple` but then later incorrectly referred to it as `$appple`.

A table of common error messages and what they mean

This table is by no means comprehensive, but it hopefully tackles most of the common error messages you might see and offers some possible solutions.[6]

Error message	Description/solution
Argument 'xyz' isn't numeric...	Perl is expecting a number, and you have given Perl something else, e.g., text or a variable containing text
Array found where operator expected...	An operator (e.g., +, ==, >, eq.) is missing and an array name has been used instead
bad interpreter: No such file or directory...	Check the first line of your script (#!/usr/bin/perl); you have probably made a typo
Bareword found where operator expected...	Most likely due to a typo in a Perl operator; e.g., typing 'eqq' rather than 'eq'
Bareword 'x' not allowed while 'strict subs' in use	You probably left off a $ @ or % on a variable name, or quotation marks off of a string
Can't find string terminator "' anywhere before EOF...	Probably a mismatched pair of quotation marks; these characters should come in pairs
Can't locate xyz.pm in...	You've added a 'use' statement, but the module name you are trying to use does not exist; possibly a typo
Can't modify constant item in scalar assignment	You may be doing an assignment with the variables swapped. $a = 5 is correct, but 5 = $a is not; 5 is a constant
Command not found	Possible typo when you typed the script name in the terminal, or the script is in a directory that is not in your Unix $PATH.
Global symbol '$xyz' requires explicit package name at...	You probably forgot to include the 'use strict' line
Name main::xyz used only once: possible typo	This generally occurs with filehandles that are misspelled; use indirect filehandles as lexical variables instead of symbol table filehandles

---

[6] The error messages in this table are simplified. The full error messages will include the script name and the line number that contains the error. They will also sometimes include variable or module names. For convenience we have just used *xyz* as a substitute for any variable/module name.

Error message	Description/solution
Permission denied...	Have you run the chmod command to give your script executable permission?
print () on unopened filehandle...	If your script is printing output to a file, you have to first open a filehandle for the output file
Scalar found where operator expected...	An operator (e.g., +, ==, >, eq.) is missing and a variable or array/hash element has been used instead
Search pattern not terminated...	When you use the matching operator (=~), there should be a pair of forward-slashes (or other matching delimiters) surrounding the search pattern
String found where operator expected...	An operator (e.g., +, ==, >, <=, eq, etc.) is missing, and some text has been added in its place
Syntax error at...	Often due to a missing semicolon/comma, or other typo (e.g., typing 'g' instead of 'gt' or 'iff' instead of 'if')
Undefined subroutine &main::xyz	You have misspelled the name of a subroutine or forgotten to save your file after making one; in either case, Perl can't find it
Unquoted string 'xyz' may clash with future reserved word	You probably left off a $ @ or % on a variable name or quotation marks off of a string
Use of uninitialized variable in...	You are working with a variable (or array/hash element) that doesn't contain any data, even though it probably should; this is more common when the data is coming from a file or is specified on the command line
xyz.pm did not return a true value	A Perl module (library) must return true after being read; the easiest way to ensure this is to have the last line of your file contain a true value (e.g., 1).

Appearance matters

Beautiful flawed programs are better than ugly correct programs.

What? How can a flawed program possibly be better than a correct one? Programs evolve over time. Features are added and removed. The data they interact with can change. Programmers also change (even if it is the same person, their programming practices change over time). While it is important for a program to complete a task now, any really useful program must complete tasks many more times in the future. Beautiful code is easier to understand, maintain, and extend. Ugly code can be nearly impossible to modify or even comprehend.

You will often be in situations where you need to write a short script which is only intended to be used once. This will tempt you to write the script quickly and you will probably not care too much about how it looks. However, you then discover that the program will actually be very useful and you promise yourself that you will tidy it all up and make it look pretty, but 'not right now.' A year later you might find yourself needing to amend that script and you open it up in your code editor only to realize that you no longer have any idea what it is doing because you never got around to tidying it up. It will be an ugly mess and you might find yourself having to rewrite the whole thing just so you can understand it. The obvious solution to this undesirable situation is to make all your scripts look beautiful … *all the time*.

Beauty in programming occurs in a variety of different contexts. There is the code itself: comments, spacing, variable names, etc. that lead to outward beauty. There is also the inner beauty of abstraction. This chapter deals with the code you see. The next chapter addresses abstraction.

This chapter has a lot of advice on how your programs should look. Don't feel you have to follow all of the advice exactly. It is important to program idiomatically. By that, we mean that your style should fit with your culture. If you are working with a team that uses a two-space indent and CamelCase,[7] by all means, do that also. But just because some particular practice is common doesn't make it correct. Use your best judgment. For example, if your programming culture doesn't include `strict` and `warnings`, you should either demand that others change or consider running away.

Indentation

As we learned all the way back in Chapter 4.6, whenever you have a block of code, you indent the block by one tab character[8] to show the logical hierarchy. This book generally uses a style of code indentation called 'one true brace.' There are several common indentation styles (see http://en.wikipedia.org/wiki/Indent_style). If you don't like one true brace indentation, feel free to choose one of the other common styles. But do not make up

---

[7] Mixing capital and lowercase letters can look like the humps of a camel.

[8] Tab characters are usually four or eight characters in width. It depends on your editor. We set our editors to four spaces. Some people prefer to indent with spaces so the indentation is consistent regardless of the tab setting. In some communities, two spaces is preferred to tabs.

your own style! Your primary job as a programmer is to write programs that can be easily understood by others, and inventing new programming paradigms defeats that goal.

Blocks of code can themselves contain other blocks, in which case the inner block is indented one extra level.

```
if (condition 1) {
    # one tab
    if (condition 2) {
        # two tabs
    }
}
```

Are these multiple levels of indentation necessary? Strictly speaking, no. But consider the following nested conditional.

```
if($x>$y){print"1\n";if($x<5){print"2\n";}}else{print"3\n";}
```

The logic is much more obvious when written with standard indentation rules.

```
if ($x > $y){
    print "1\n";
        if ($x < 5) {
            print "2\n";
        }
} else {
    print "3\n";
}
```

Line length

There is a long tradition in programming that lines should contain no more than 80 characters.[9] Although it is not necessary to follow this practice, many programmers still do. The programs in this book generally have short lines, but it is easy to make longer lines when you start nesting lots of blocks. If you find yourself writing exceedingly long lines that force your editor to scroll sideways, consider breaking the line into pieces. A long line such as this:

```
if ($my_car_is_in_the_shop == 1 and $the_buses_are_not_running == 1) {
```

can be re-written like this:

```
if ($my_car_is_in_the_shop       == 1 and
    $the_buses_are_not_running   == 1) {
```

As well as splitting the single line of code across two lines in our coding editor, we also make sure the two variables are aligned with each other. Furthermore, we insert spaces after the first variable to line up the two equality operators.

---

[9] Punch cards and terminals were often 80 columns wide.

Variable and function names

In general, variable names should be lower case. There are, of course, exceptions. Global variables should begin with upper-case characters and constants should be all capitals.

```
my $local;
my $Global;
my $CONSTANT     =   5;              # pseudo-constant
use constant PI => 3.1415926;    # true constant
```

Variables with small scopes should have small names. For example, using $i as a counter in a loop is not only okay, it is preferred. Variables with larger scope should have longer, more descriptive names. When variable names contain multiple words, there are two common conventions: underscores and CamelCase. We prefer underscores, but both styles are completely acceptable. Don't mix them in the same program, however.

```
my $long_variable_name;   # underscore
my $longVariableName;     # CamelCase
```

Variable names tend to be nouns. Function names tend to be verbs. Function names follow the same general rules as variable names. Class names begin with a capital letter. Class methods are also generally capitalized (except new). Instance methods and attributes are lower case.

```
my $obj = new Something;  # capital class name
$obj->attribute;          # lowercase
$obj->method_name;        # underscore
$obj->methodName;         # CamelCase with first letter underscore
```

Comments

One of the simplest ways to beautify your program is to employ comments to break up code into sections based on what the code does. Major sections of code can be commented with larger blocks of comments. There are a number of ways you can do this. Unlike indentation, religious wars are not being fought about this, and you could invent your own style without offending anyone:

```
###########################################################
# Major subheading
# some descriptive text about this section
###########################################################

###################
# Minor subheading #
###################          cute, but maybe too much!
                            ()_() /
#-----------------#  (o.o)
# Minor subheading #  ()_(),
#-----------------#

# minor subheading
```

Whitespace

If used properly, whitespace characters (tabs, newlines, spaces) can improve the read-ability of your code. However, if used improperly they can confuse people, especially if you are breaking with standard practices. As we saw earlier, tabs are used to show hierarchy in block structure. Spaces are used to separate operators and functions, and to align related statements on different lines. Newlines are used to separate lines and parts of the program that do different things. For a quick summary:

Bad

```
$f=1;for($i=1;$i<=$n;$i++){$f*=$i}#factorial
```

Good

```
# factorial
$f = 1;
for ($i = 1; $i <= $n; $i++){
    $f *= $i;
}
```

Spaces

Keywords like `if`, `else`, `for`, and `my`, should have whitespace on either side:

```
if($x == 1)    # bad
if ($x == 1)   # good
```

Although functions can look a little bit like keywords in some circumstance, there is no space between the name and the opening parentheses. If functions are called without parentheses, then a space is present:

```
print ("hello");   # bad
print("hello");    # good
print"hello";      # bad
print "hello";     # good
```

Most of the time, operators like `==` should have spaces on either side. The unary[10] operators such as comma and auto-increment are an exception and bind directly to variable names:

```
my ($x, $y, $z);   # trailing space after comma
$x++;              # no space
```

There should be no space between an array or hash name and its brackets:

---

[10] Operators with a single operand are called unary operators. Those with two operands are binary operators. There is only one trinary operator (see Chapter 4.6).

```
$array[0];      # good
$array [0];     # bad
```

Similarly, when using references, there should be no spaces around the arrow;

```
$ref->{something};      # good
$ref -> {something};    # bad
$ref-> {something};     # bad
$ref ->{something};     # bad
```

Aligning with spaces

When performing several assignments, align the operators:

Ugly

```
$x = 1;
$height = 4;
$precision = 7.135;
```

Pretty

```
$x          = 1;
$height     = 4;
$precision  = 7.135;
```

Similarly, when you have simple switches, break out of block structure and align the conditions and braces:

Cumbersome

```
if ($x == 1) {
    do_something;
} elsif ($x == 2 and $y == 1) {
    do_something_else;
} else {
    die;
}
```

Ugly

```
if ($x == 1)                    {do_something}
elsif ($x == 2 and $y == 1)  {do_something_else}
else {die}
```

Pretty

```
if      ($x == 1)               {do_something}
elsif   ($x == 2 and $y == 1)   {do_something_else}
else                            {die}
```

Separate thoughts with newlines

While you can add comments to break up all the different 'thoughts' in a program, sometimes a newline is more appropriate. For example, if you want to calculate the standard deviation, you first need to calculate the mean, then the variance. Using a little vertical spacing helps keep these different thoughts organized:

```perl
# descriptive statistics #

my $mean = 0;
for my $v (@v) {$mean += $v/@v}

my $var = 0;
for my $v (@v) {$var += ($v - $mean) ** 2}

my $std = sqrt($var);
```

Separate header and footer material

Separate header material from the body of the program. If you have footer material, separate that too:

```perl
#!/usr/bin/perl
# my_witty_program_name.pl
use strict; use warnings;
use SomethingElse;

# body of program

END {
  # cleanup routines
}
```

Similarly, separate header and footer material in subroutines:

```perl
sub something {
    my ($x, $y) = @_;

    # body of subroutine

    return $f, $q;
}
```

Abstraction

The inner beauty of programming

At various points in this book, we talk about the importance of abstraction. This chapter serves to organize and reinforce those ideas. One goal of this chapter is to make your programs more general and more robust. Another goal is to convey the aesthetic side of software development.

Abstraction in art is a movement away from reality and all its everyday conflicts. It is sometimes confusing, and its departure from common references means it is not always beautiful at first sight. But after some time with a piece, you may find that the abstraction makes its beauty more profound. Abstraction in programming has a similar feel. Code that solves a problem in a very general way can be confusing to read and can take a lot of effort to write. Why do all this *stuff* when all I want to do is solve one simple problem? The reason is that code which can be easily applied to new situations is not simply more useful, it also has a kind of transcendent beauty. What the heck are we talking about? Have we lost our minds? No! We just love programming! While the main goal of this book is to help you solve real-world problems with Unix and Perl, we also hope you will become a connoisseur of abstract art (well, the programming kind anyway).

Example 7.4.1 The following code computes the circumference and area of a circle that has a radius of 4.

```
1.    # circle attributes v1
2.    $circumference = 2 * 3.1415926 * 4;
3.    $area = 3.1415926 * 4 * 4;
```

On the one hand, there is nothing wrong with this code as it correctly computes the desired answers. But changing to a different-sized circle requires editing the code in several places. Not only is this laborious, but a user might make a mistake while typing.

Example 7.4.2 One obvious, and easy, improvement to Example 7.4.1 can be made simply by storing all the values in variables:

```
1.    # circle attributes v2
2.    $radius = 4;
3.    $pi = 3.1415926;
4.
5.    $circumference = 2 * $pi * $radius;
6.    $area = $pi * $radius ** 2;
```

Using variables achieves two things. Not only does it make the code easier to understand, but it is also easier to use. To change the radius, you simply change one value. Much better! Unfortunately, this still requires that the user edit the code. From a developer standpoint, you can trust users about as far as you can throw them. Even if you have super-strength, don't encourage users to edit your programs.

Example 7.4.3 It is very easy to bring user-defined parameters into your program. Simply put them on the command line.

```
1.    # circle attributes v3
2.    ($radius) = @ARGV;
3.    $pi = 3.1415926;
4.
5.    $circumference = 2 * $pi * $radius;
6.    $area = $pi * $radius ** 2;
```

In any situation where a script contains some parameter that you might want to change between separate runs of the script, you should never hard code this into your script. Use @ARGV, as in this example, or consider using the Getopt Perl module (see Chapter 6.7).

The previous examples are incredibly simple, but they illustrate two very important points: (1) *generalization* and (2) separation of *interface* and *implementation*. To make code more general, you should factor out the common parts. For example, rather than use 3.1415926 each time you need a value for pi, you can simply define a variable once and use it throughout your program. We will see greater and greater levels of generalization below.

The interface of a program defines how a user interacts with it. The interaction might be through a web page, graphical application, or just on the command line. The implementation is the code itself. Some users are developers, and their interaction is not just with programs, but also with subroutines, libraries, and classes. Regardless of who the users are, a functional unit of code should not require editing. It should have well-defined inputs and outputs and no side-effects.

Subroutines

Subroutines can make the logic of a complicated program much easier to see. We use subroutines every day in life. For example, you may think about the errands you need to do in a day as 'go grocery shopping, clean the kitchen, make dinner.' Each of these activities encapsulates a great number of specific subactivities. Defining subroutines means you can think about the big picture without getting distracted by the details.

Example 7.4.4 The following code opens two files and reports any identical lines between the files. Lines 1–3 read one file, and lines 5–7 read the other. The comparisons are in lines 9–13.

```
1.    open(my $fh1, "<", $ARGV[0]) or die "error opening $ARGV[0]";
2.    @line1 = <$fh1>;
3.    close $fh1;
4.
5.    open(my $fh2, "<", $ARGV[1]) or die "error opening $ARGV[1]";
6.    @line2 = <$fh2>;
7.    close $fh2;
8.
9.    foreach $line1 (@line1) {
10.       foreach $line2 (@line2) {
11.           print $line1 if $line1 eq $line2;
12.       }
13. }
```

This part of the book is about abstraction, but we need to make a slight digression for safety. Lines 1–3 and 5–7 are nearly identical. You might therefore write lines 1–3, copy-and-paste to get lines 5–7, and then do a little editing as necessary. This kind of copy-and-paste programming sometimes leads to horrific run-time bugs. The practice is born of laziness, and can do more damage than you can possibly imagine. Please, try not to do this! Instead, any time you need to write code twice, write a subroutine instead.

Example 7.4.5 This script is functionally identical to Example 7.4.4, but defines a single subroutine to retrieve lines from a file.

```
1.    @line1 = get_lines_from_file($ARGV[0]);
2.    @line2 = get_lines_from_file($ARGV[1]);
3.
4.    foreach $line1 (@line1) {
5.        foreach $line2 (@line2) {
6.            print $line1 if $line1 eq $line2;
7.        }
8.    }
9.
10.   sub get_lines_from_file {
11.       my ($file) = @_;
12.       my @line;
13.       open(IN, $file) or die "error opening $file";
14.       @line = <IN>;
15.       close IN;
16.       return @line;
17.   }
```

Example 7.4.5 is much easier to read and understand than Example 7.4.4 because of the abstraction provided by the function name (`get_lines_from_file`). It is also easier to modify and maintain. For example, maybe you want the script to only compare lines in the input files if those lines start with a certain character. Making this modification requires editing just the one subroutine, and not two different loops. Example 7.4.5 is longer than Example 7.4.1, and it takes more initial effort to program with abstraction. However, the reward is reaped many times over in the long run. As an added bonus, the code is more aesthetically pleasing.

You could place lines 4–8 from the above code into another subroutine called `print_identical_lines()`. Then the entire logic of the program would be stated in the first three lines. But this is going a little too far. If the code is not being used multiple times, it is a little overzealous to make it into a subroutine.

Libraries

As shown above, subroutines make parts of your code re-usable. But what if you want to share the same clever piece of code in several different programs? For example, you have written a program that reads a DNA sequence and calculates the codon usage. Later, you decide you want to make another program that reads a DNA sequence and calculates its melting temperature. Both programs must read a DNA sequence. One way to get the code from the first program to the second is to copy and paste it. STOP! Don't do that. Never copy and paste. Not only is it dangerous, there is a more general, and more beautiful solution. Subroutines can be shared between different programs. You can write code once and share it with as many programs as you like. All you have to do is place your subroutines into a *library* (also called a module).[11] This is easier than you might imagine. A library that contains various DNA analysis subroutines might be called SequenceTools. A library file ends with the suffix .pm, so the library file would then be called `SequenceTools.pm`.

The simplest form of library is just a file of subroutines. Example 7.4.6 shows how to create a library. Perl requires that libraries return a true value when read, so it is customary that the last line of a file contains the number 1:

Example 7.4.6

```
1.    # SequenceTools.pm
2.    sub read_sequence {} # actual code omitted for brevity
3.    sub codon_usage {}
4.    sub melting_temperature {}
5.    1;
```

In order for Perl to find your library, it must be in the library path (see Chapter 6.2). To use the library, you simply use it. The program to calculate melting temperature would look something like this:

---

[11] We explain libraries in more detail in Chapter 6.2.

Example 7.4.7

```
1.  #!/usr/bin/perl
2.  use strict; use warnings;
3.  use SequenceTools;
4.  my $dna   = read_sequence($ARGV[0]);
5.  my $tm    = melting_temperature($dna);
```

There is a major problem with creating libraries this way. Suppose you use two libraries and each one defines a function called read_sequence(). Which function will be called? Thankfully, Perl reports an error message, but how do you prevent this situation? In the C language, the convention is to prefix function names with the author initials or an abbreviation of the library. So read_sequence() might become st_read_sequence(). While you could do this in Perl, it would be a little strange because there is a much better alternative. The package statement allows you to associate functions with specific libraries:

Example 7.4.8

```
1.  package SequenceTools;
2.  sub read_sequence {} # actual code omitted for brevity
3.  sub codon_usage {}
4.  sub melting_temperature {}
5.  1;
```

The first line of Example 7.4.8 contains a package statement. This creates a *namespace*, here called SequenceTools. All the subroutines in SequenceTools actually have longer, more formal names. The scope operator :: separates namespaces from functions (and other namespaces). The result of this is that the name of a function such as codon_usage() can be more formally referred to as SequenceTools::codon_usage(). Namespaces can contain other namespaces. Think of a namespace as a directory, and subroutines as files.

The namespace of the main body of a Perl program is, appropriately, called main. So, the subroutine called get_lines_from_file() that we saw back in Example 7.4.5 is more formally main::get_lines_from_file(). Code inside a particular namespace does not need to use the longer name. But when you want to call a function in another namespace, you must include the namespace (which is generally the same as the library name). Let's look at an example to make this clear.

Example 7.4.9

```
1.  #!/usr/bin/perl
2.  use strict; use warnings;
3.  use SequenceTools;
4.  my $dna = SequenceTools::read_sequence($ARGV[0]);
5.  my $tm  = SequenceTools::melting_temperature($dna);
```

Libraries make code easier to share, maintain, and extend. As with other generalizing methods, there is a little initial cost in making a library. But the cost is incredibly small. Every programmer should make a library or two. Once you start using them, you won't be able to program without them.

The only thing better than your library is someone else's. There are a lot of clever programmers who have already made powerful libraries.[12] Using your imagination, a little glue code, and other people's libraries, you can create programs that do amazing things beyond your own capabilities. Want to write a program that reads graphics files and performs some image manipulations? No problem! How about connecting to a relational database and making automated queries? Trivial! Want to make a web robot that scours the web for email addresses or credit card numbers? You shouldn't, but it's easy.

Object-oriented programming

Object-oriented programming (OOP) is a level of abstraction above function libraries. Most of the Perl libraries you will want to use employ OOP principles, so we believe that all Perl programmers must be able to use objects. We covered OOP in more detail in Chapter 6.8, but we'll include a short refresher here.

The fundamental unit in OOP is the *class*. We classify things in everyday life. For example, there are a lot of vehicles we call cars. A specific car is an *instance* of a class. An instance is also called an *object*. Say it aloud, 'an object is an instance of a class.' Substitute the word *example* for *instance* if you like: 'An object is an example of a class.' You are an instance/example of the person class. You also happen to be an example of a mammal, and classes can encapsulate this kind of hierarchy. For now, let's consider a simple car class. All cars have various *attributes* such as year, make, model, and color. In Perl, you access an object's attributes with the arrow syntax as `$object->attribute` or optionally as `$object->attribute()`. Why the trailing parentheses? Because the attributes are actually parameterless functions.

```
print $car->color;
print "antique" if $car->year() < 1980;
```

Objects also have *methods*, which are actions that objects perform. For example, a car class may define a method for *drive*.

```
$car->drive("south", 10, "km");
```

This looks a little bit like a function call. In a non-OOP syntax, the code might look something like this:

```
CarLibrary::drive($car, "south", 10, "km");
```

As you can see, the OOP syntax is prettier. To create an object of a particular class, you generally call a *constructor* whose name is usually new. Constructors are *class methods*. They don't require objects. The preferred way to call the constructor is as follows:

---

[12] See Chapter 6.9 on CPAN for details of where to find many libraries written by others.

```
$car = new Car;
```

It might be more illuminating to see the more awkward syntax, where it is more apparent that the new() function is inside a file called Car.pm.

```
$car = Car::new();
```

It is common to give an object some attributes when you construct it. These are sometime passed to the object as a list, and sometime as a hash. It depends on who wrote the class and what style the author prefers.

```
$car = new Car (year => 2010, make => 'Honda', model => 'Civic');
$van = new Car (1989, 'Toyota', 'Previa');
```

An object keeps all the attributes and methods for some piece of data in a cohesive unit. With objects, you can think more about how pieces of data should interface with each other and less about the underlying implementation (code).

The persistence of memory

There are many different ways to interact with data, and no one method is best for all situations. Sometimes, a very simple solution like having your data spread across a set of CSV files on a USB flash drive is all that is necessary. At other times, an online relational database may be more appropriate. In this chapter, we will examine some of the common ways to read and write data.

Hard-coded data

The simplest kind of data is the kind that is hard-coded into your programs. For example, you might store the value for pi or e in scalar variables:

```
my $pi  = 3.1415926;
my $e   = 2.7182818;
```

Data that is frequently accessed from subroutines can be stored outside the subroutine if it is sufficiently large. For example, if you have a table for Kyte–Doolittle hydrophobicity values and a subroutine that calculates hydrophobicity for any protein sequence, the organization should be as follows:

```
my %KDH = (
    A => 1.8,   C => 2.5,   D => -3.5,   E => -3.5, F => 2.8,
    G => -0.4,  H => -3.2,  I => 4.5,    K => -3.9, L => 3.8,
    M => 1.9,   N => -3.5,  P => -1.6,   Q => -3.5, R => -4.5,
    S => -0.8,  T => -0.7,  V => 4.2,    W => -0.9, Y => -1.3,
);

sub kd_hydrophobicity {
    my ($prot) = @_;
    my $hydro = 0;
    for (my $i = 0; $i < length($prot); $i++) {
        $hydro += $KDH{substr($prot, $i, 1)};
    }
}
```

If you put the table inside the subroutine, the data structure has to be initialized every time the program runs. For small pieces of data, it is better to keep as much internal to the subroutine to make the subroutine more self-sufficient and easier to understand.

__END__ and __DATA__ tokens

If you have a large amount of data that you don't want to hard-code into variables, you can store this at the end of the program and read it with the DATA filehandle. For example, you could store the genetic code as follows:

```perl
#!/usr/bin/perl
use strict; use warnings;
my %GeneticCode;
while (<DATA>) {
    my ($codon, $amino_acid) = split;
    $GeneticCode{$codon} = $amino_acid;
}
__END__
AAA Lys
AAC Asp
AAG Lys
AAT Asp
ACA Thr
ACC Thr
:
TTT Phe
```

You can use the __END__ or __DATA__ tokens to separate data from other things at the end of the program. If you are using both __END__ and __DATA__, always put __END__ first. Otherwise everything, including the __END__ token, will be considered data.

```perl
__END__
notes to myself
__DATA__
actual data
```

CSV and TSV

Text files continue to be one of the most common interchange formats. They are easy to read, and simple to parse. They are also somewhat more future-proof compared to most other file formats. Two very common formats are comma-separated values (CSV) and tab-separated values (TSV). Both of these formats are usually record-based: each line of the file corresponds to a different unit of some kind. For example, a record-based CSV file of people might look like this:

```
Korf,Ian,F
Bradnam,Keith,R
```

Parsing such a file is very simple with the split() function. The natural way to store such records internally is as an array of hashes:

```perl
my @people;
while (<>) {
    my ($last, $first, $mi) = split(/,/, $_);
    push @people, {
        last  => $last,
```

```
        first => $first,
        mi   => $mi,
    };
}
```

It is easy to print such information by using the join() function. You might also want
to include a 'separator' variable that will easily allow you to print your data in TSV or
CSV format:

```
my $separator = $tab ? "\t" : ",";
foreach my $p (@people) {
    print join($separator,
        $p->{last}, $p->{first}, $p->{mi});
}
```

XML

One of the most convenient ways to represent complex data is XML. This text-based
format allows one to nest data to any depth. Here is an example XML file containing the
same data as the CSV file shown above:

```
<?xml version='1.0'?>
<people>
    <person>
        <last>Korf</last>
        <first>Ian</first>
        <mi>F</mi>
    </person>
    <person>
        <last>Bradnam</last>
        <first>Keith</first>
        <mi>R</mi>
    </person>
</people>
```

Comparing the XML and CSV formats, it's clear that XML is a lot more verbose.
Each value has opening and closing tags and some degree of indentation.[13] Is this extra
baggage necessary? For some kinds of data, definitely not. But there are times when data
becomes highly nested and XML becomes very convenient. Looking at the XML file,
you can imagine it would be difficult to parse. Fortunately, there is a free Perl module
called XML::Simple that makes reading and writing XML very easy.[14] For example, if
the XML information shown above was stored in a file called 'people.xml,' then here is
how you could parse this information:

---

[13] Like Perl, indentation in XML is not required but helps readability.
[14] There are several modules for reading XML. See CPAN (and see Chapter 6.9 if you don't know about CPAN).

```perl
#!/usr/bin/perl
use strict; use warnings;
use XML::Simple;
use Data::Dumper;

my $xml = new XML::Simple;
my $data = $xml->XMLin("people.xml");
print Dumper($data);
```

The output of the Dumper($data) module would look like this:

```perl
$VAR1 = {
    'person' => [
        {
            'first' => 'Ian',
            'last' => 'Korf',
            'mi' => 'F'
        },
        {
            'first' => 'Keith',
            'last' => 'Bradnam',
            'mi' => 'R'
        }
    ]
};
```

This shows that each person is a hash and the collection is stored as an array (just like how we stored the CSV records). XML::Simple hands the entire data back as a hash reference. If you wanted to loop through all of the people listed in the XML data, you could do it like this:

```perl
foreach my $p (@{$data->{person}}) {
    print "$p->{first} $p->{last}\n";
}
```

Persistent hashes with dbmopen

Imagine you have a large data file stored in some text format which can be imported by Perl as some sort of hash structure. Every time you need to access some information, your Perl script has to read all of the file. Even if you wanted to modify a single value, then you would still need to read the entire file, modify the desired value, and then write the file to disk again. Fortunately, Perl provides a more efficient way of working with large hashes. Effectively, you can store your hash as a file and treat it like a miniature database. To associate a hash with a file, you use the dbmopen() function. This takes three arguments: the hash, the name of the file, and the file permissions in octal. In this example, we create a new hash and associate it with a file called 'my_database':

```
#!/usr/bin/perl
use strict; use warnings;

my %hash;
dbmopen(%hash, "my_database", 0666);
$hash{`date`} = 1;
foreach my $date (keys %hash) {print $date}
```

After performing a dbmopen(), a new file will be created. In the example above, the file will be called my_database.db.[15] If you try to view the file with the Unix command less, you will find it is a binary file, not a text file. In addition to dbmopen() Perl also has several other ways to interact with files using the DB_File module. However, if you need to do more complex file operations, we suggest using a relational database.

Storable.pm

For reading and writing data structures that are more complex than simple hashes, Perl provides the Storable package. This allows you to read/write arbitrarily complex data structures to your filesystem with the store() and retrieve() functions. Here is an example of how to write data to a file:

```
1.   #!/usr/bin/perl
2.   # store.pl
3.   use strict; use warnings;
4.   use Storable;
5.
6.   my %table = (
7.       array    => [1, 2, 5],
8.       hash     => {dog => 'woof',cat => 'meow'},
9.       string   => "hello world",
10.  );
11.  store(\%table, "db.storable");
```

The single line of code on line 11 will both create a file ('db.storable') and write the contents of the hash to that file. This next script shows how simple it is to retrieve the data:

```
1.   #!/usr/bin/perl
2.   # retrieve.pl
3.   use strict; use warnings;
4.   use Storable;
5.
6.   my $data = retrieve('storable_file');
7.   print $data->{array}[0], "\n";
```

---

[15] Depending on your OS, the file may have a different extension.

From a retrieval point of view, the original name of the hash (%table) is not relevant, but obviously you do need to know something about the hash structure to efficiently extract information.

SQL databases

Most of the information on the internet is stored in relational database management systems (RDBMS). Common RDBMS include Oracle, MySQL, PostgreSQL, DB2, and SQLite. Some RDBMS are expensive, others are free. All are extremely useful for organizing and searching complex data. One of the great advantages of RDBMS is that they all use a common language called SQL (structured query language).[16] Therefore, if you know how to use one database, you can use them all.[17] If you are working with a lot of data, taking some time to learn SQL will be time well spent. After Perl, it's probably the most important thing a data jockey needs to know.

There are more books about SQL than there are about Perl. So how exactly are we going to cover both SQL and Perl–SQL interaction at the end of a chapter? Very briefly! To get started with SQL, the first thing you need is some RDBMS software. Even though some OSs already have an RDBMS installed, we suggest you use SQLite. One of the hassles in setting up an RDBMS is defining owners, users, and their permissions. While these security features are important to protect data, they are a distraction for the first-time user. SQLite databases are contained in a single file. If you can read from and write to this file, you can read from and write to the database. Here's what SQLite says about itself:

> SQLite is a software library that implements a self-contained, serverless, zero-configuration, transactional SQL database engine. SQLite is the most widely deployed SQL database engine in the world. The source code for SQLite is in the public domain.

SQLite may already be installed on your computer. Try typing sqlite in a terminal to find out.[18] If SQlite is not installed, download it from sqlite.org and follow the installation instructions.

Let's assume you have a file of people, addresses, and occupations as follows:

Name	Address	Occupation
Steve Jobs	Cupertino	Tycoon
Jennifer Jones	Nottingham	Sales
Willy Loman	Brooklyn	Sales
Nigel Tufnel	Squatney	Musician

---

[16] In truth, not all relational databases use SQL. There are many non-SQL databases in existence. But if you were designing a new database today, you would use a SQL RDBMS.

[17] This is also not quite true. There are subtle differences among the various databases.

[18] Remember to tab-complete. sqlite may be named sqlite3.

Let's create this in SQLite. First, you have to create a database. Enter the following command in your terminal:

```
$ sqlite3 test.db
```

This creates a file called `test.db`. It will also report a little information and enter into an interactive mode with an `sqlite>` prompt.

```
SQLite version 3.6.21
Enter . "help" for instructions
Enter SQL statements terminated with a ";"
sqlite>
```

At the prompt type the following (note that it will add indentation for you automatically):

```
create table People (
    name TEXT,
    address TEXT,
    occupation TEXT
);
```

Congratulations, you have just created your first SQL database! Now to insert some values:

```
insert into People values ("Steve Jobs", "Cupertino", "Tycoon");
insert into People values ("Jennifer Jones", "Nottingham", "Sales");
insert into People values ("Willy Loman", "Brooklyn", "Sales");
insert into People values ("Nigel Tufnel", "Squatney", "Musician");
```

You can display all the rows with an SQL `select` statement.

```
sqlite> select * from People;
```

This will show you everything in the People database. By default, fields are separated by a vertical bar:

```
Steve Jobs|Cupertino|Tycoon
Jennifer Jones|Nottingham|Sales
Willy Loman|Brooklyn|Sales
Nigel Tufnel|Squatney|Musician
```

Of course, you would never want to enter all of your data by hand and, ideally, you would be using some other application to retrieve and view the data (instead of the terminal). If you are going to be a database power-user, you should have your programs interacting directly with database.

DBI and DBD

To use Perl to connect to your SQLite database you need to use the DBI and DBD::SQLite modules. Install these as required. Now let's try a simple script that connects to the database, inserts some data, and then does some simple reporting. The following script barely scratches at the surface of what is possible with Perl and SQL. Think of this as a Hello World script for SQLite:

```perl
#!/usr/bin/perl
use strict; use warnings;
use DBI;

# connect to SQLite database
my $dbh = DBI->connect("dbi:SQLite:dbname=test.db", "", "") or die;

# insert a new row
$dbh->do('INSERT INTO People VALUES("Ian Korf", "UCD", "Professor")');

# report all people in the database
my $sth = $dbh->prepare('SELECT * FROM People');
$sth->execute;

# Loop over results from database, each result temporarily stored in $r
while (my $r = $sth->fetch row_hashref) {
    print "$r->{name} from $r->{address} works as $r->{occupation}\n";
}
```

The combination of Perl and an SQL database is incredibly powerful. But it is also a little complicated. We suggest you read the entirety of this book and also an introductory SQL book before you mix the two.

Relationships

The previous examples showed how you can use Perl to interact with a simple SQL database. However, the database we created was not actually relational. Did you notice that the 'occupation' field contained redundant data? You may have hundreds or thousands of people, yet only a handful of different occupations. We can factor out the occupations into a separate table so we have two tables that share a relationship.

Name	Address	Occupation_id
Steve Jobs	Cupertino	1
Jennifer Jones	Nottingham	2
Willy Loman	Brooklyn	2
Nigel Tufnel	Squatney	3

Occupation	Occupation_id
Tycoon	1
Sales	2
Musician	3

Factoring out all the common data into separate tables is called *normalizing*. The reverse is called *de-normalizing*. For greatest data integrity, data should be normalized as much as possible. One reason for this is that it makes it much easier to change many records at a time. For example, suppose one wanted to change the 'Tycoon' occupation to 'Business.' In a normalized database, this simply requires changing the text in the occupation table. In a non-normalized database, one would have to change each value of 'Tycoon' to 'Business.' The normalized database also uses less memory, though queries are often slower than in de-normalized databases. It is therefore a common practice to have two databases with the same information. One database is normalized and keeps live data. Periodically, a de-normalized version is created for searching. In a sense, the normalized database is the DNA and the de-normalized versions are RNA.[19]

---

[19] Probably not the best analogy, but this is what you get from biologist authors.

Self-help books for programmers

As your software becomes more complex, documentation becomes increasingly import-ant. Documentation can take the form of personal notes to yourself, instructions for other developers, or manuals for non-programming users.

Why add documentation?

When you first write a new script, it will always seem obvious and intuitive to you. The idea that you might ever forget why you wrote a particular piece of code will seem unlikely. This is why you will be tempted to not add any comments, and to omit any docu-mentation. But after you finish writing a script it becomes very easy to forget about the gory details of how it all works. More importantly, it becomes easy to forget all about *why* you chose to come up with the particular solution your code contains. Many of the scripts you write will solve particular problems and may only need to be used infrequently. It is very likely that you will sometimes need to return to a script you haven't looked at or run for many months (or even years). Let's quickly consider the four likely scenarios that will arise in which a script is needed that has remained dormant for over a year:

(1) You need to run the script again.
(2) Someone else wants to run your script.
(3) You need to edit the code (to add a new feature or fix a bug).
(4) Someone else wants to edit your code.

Without proper documentation you might only have 'vague ideas' of how to use your script. You might have to look at all the code again and spend hours, if not days, deciphering it. When someone else wants to use it, you will quickly realize that people can ask really awkward questions, like: 'How do I run it to do X?'; 'Is there an option to do X *and* Y?'; 'Can you explain to me, in detail, how I gain maximum performance from your script?' If you don't document your code, you will always need to explain your program to others.

As tedious as it can be to have to verbally explain how your program works to someone else, this is still infinitely more fun than having to edit a script that contains no comments or other documentation. When you do this with your own script, you will be shocked at how little you remember of how it worked. You might also find yourself troubled as to what the subroutine called 'process_data' is actually doing or why you ever thought that 'stuff' was a good name for an array. At least you will know that any blame lies with you and no-one else was hurt or injured by your lack of documenta-tion. Finally, consider the fourth item on the list. When you have to edit someone else's code, and when that code contains not so much as a single comment, then the whole experience can become one of the most frustrating and bitter experiences in your life.[20] Frequently, the only way you will properly understand a script that has no documenta-tion is to rewrite the whole thing from scratch. Even if you only do this once in your

---

[20] The authors can both claim this with direct personal experience.

entire career as a programmer, it will be enough to create a deep resentment toward the person who first wrote the script. Don't be that person!

> *We don't care if you lie, cheat, or steal. We do care if you don't document code.*[21]

Comments

Comments are best used for notes to yourself. You should not assume that users are programmers or that they will read your source code. So don't write anything for users in your comments. Do use comments to explain *why* you chose to solve problems in a certain way (especially if it is an unusual solution). Also use comments to explain what the point of subroutines and other large blocks of code are. Finally, use comments to explain a little bit about what type of data will be contained in certain variables. This is particularly important when you use variables whose names are not intuitive. Adding just a few words to explain a block of code can save you (and others) a lot of time in future.

Usage

A usage statement briefly tells command-line users how to run your program. Usage statements are reminders of how to use the software, and should not be the only source of documentation unless the program is very simple. There are a number of different styles of usage statements. Below we show a typical one that includes simple options processing (first covered in Chapter 6.7). This program takes two mandatory arguments and some optional parameters. The text of the usage statement is displayed if the program is run without arguments.

```
1.   #!/usr/bin/perl
2.   use strict; use warnings;
3.   use Getopt::Std;
4.
5.   my $version = "1.0b";
6.   my %opt;
7.   getopts('ab:v', \%opt);
8.
9.   my $usage = "
10.  usage: $0 [options] <file> <threshold>
11.  options:
12.    -a
13.    -b <int>
14.  ";
15.  die $usage unless @ARGV == 2;
```

---

[21] Okay, we do care a little bit when people lie, cheat, or steal ... just not as much as when people don't add any comments or write documentation.

```
16.
17.  if ($opt{h}) {die $usage}
18.  if ($opt{v}) {die $version}
19.
20.  # etc.
```

Note how the usage statement (stored in $usage) is split across multiple lines in this example (lines 10–14). You could instead write this on one line and use newline characters, but doing it this way makes it very easy to see what the final message will look like. This script also features a new variable ($0) on line 10; this special variable simply contains the name of the program that is being executed.

A good program will always prompt you with usage instructions when you fail to specify the correct set of command-line arguments. However, some programs take no arguments, and in these cases it is common to trigger usage statements with -h or -help or --help. Programs often report a version number with the -v or -version or --version flag. Try running perl -v in your terminal.[22]

__END__ and __DATA__

One convenient place to put longer internal documentation is at the end of the program. Anything after the __DATA__ keyword can be read from the DATA filehandle.[23]

```
#!/usr/bin/perl
use strict; use warnings;
# etc.
if ($opt_h) {while (<DATA>) {print}}

__END__
Private info

__DATA__
This program is designed to ...
```

README files

One common way to document your programs is to include a separate plain-text file, often called README. All by itself, a file named README might be confusing, but larger programs are generally distributed in a directory containing several files. So a file called README is not going to be confusing. Other files distributed with software often include INSTALLATION, LICENSE, and HISTORY.

The contents of a README are up to you. Ideally, you should include some general information from the outset to let people know what kind of software it is. You should also include some examples of how to use the software in addition to explicit instructions.

---

[22] At the time of this writing, Perl is at version 5.10. If you see version 6 (or later), then it may be time to find a new edition of this book because Perl is expected to undergo a number of important changes.

[23] See Chapter 7.5 on data management for more information.

Plain Old Documentation (pod)

Perl has a very powerful documentation system called pod. This is a kind of mark-up language you can use within your programs. This keeps the documentation and code in one file. One of the advantages of pod is that you can export it to a variety of formats. This gives your software documentation a professional look.

A section of pod generally begins with =head1 and ends with =cut. In the code below, lines 3–8 are not part of the code, they are part of the documentation. The blank lines at 2, 4, 7, and 9 improve readability, and may be necessary depending on the pod tool you use (see below).

```
1.   my $x = 1;
2.
3.   =head1
4.
5.   Everything in this area is documentation. The heading level
6.   determines the indentation.
7.
8.   =cut
9.
10.  my $y = 2;
```

Pod offers four levels for heading and other paragraph formatting options. There are also font tags for bold-face, italic, and such. Without getting into too much detail, let's look at some pod mark-up in a pretend module called Sequence.pm.

```
1.   package Sequence;
2.   use strict; use warnings;
3.
4.   my %GeneticCode;
5.
6.   =head1 NAME
7.
8.   Sequence
9.
10.  =head1 SYNOPSIS
11.
12.   use Sequence;
13.   my $dna = Sequence::read_fasta(file);
14.   my $pro = Sequence::translate($dna);
15.
16.  =head1 DESCRIPTION
17.
18.  Sequence is a non-OOP library containing functions for
19.  handling DNA, RNA, and protein sequences.
20.
```

```
21.  =head2 read_fasta(file)
22.
23.  Reads a fasta file and returns sequence in a single string.
24.
25.  =cut
26.
27.  sub read_fasta { }
28.
29.  =head2 translate(sequence)
30.
31.  Translates a DNA or RNA sequence into protein.
32.
33.  =cut
34.
35.  sub translate { }
```

Only five lines of this file have actual source code. Lines 1 and 2 are the familiar header. Line 4 has a variable declaration that will presumably be filled out elsewhere. Lines 27 and 35 are subroutine declarations with code omitted. Note that the documentation for each subroutine is only a few lines away. This is really convenient because if you change a line of code that affects the documentation, you don't have to open another file and edit it. Out-of-date documentation can be misleading, and keeping the code and documentation close to each other helps prevent this from happening.

To convert the embedded pod into a pretty document, we can use a number of pod translators that come with Perl (try tab-completion of pod2<tab>):

```
$ pod2text Sequence.pm
$ pod2html Sequence.pm > Sequence.html
$ pod2man Sequence.pm > Sequence.man && man ./Sequence.man
```

Additionally, there is usually a perldoc command on most Unix systems. This command will let you see just the pod documentation that is contained within a Perl script or module. Just type:

```
$ perldoc myscript.pl
```

This chapter is now at version 1.12

When you first learn to program in Perl (or any language), you are usually focused on the short-term goal of writing a program to solve a specific problem. This is obviously an important goal, but in an ideal world you should also consider some longer-term goals, namely:

(1) What is the best way of upgrading your program to meet future needs?
(2) Will you ever need to let other people contribute to a script?
(3) What can you do if you want to be able to use a script on multiple computers?

You may think that these goals do not apply to your program; you might think your script is just a small program that you might not need to use again after a week or two of use. However, it is often hard to predict what the future holds. Maybe your little program goes on to be a central tool in your research, and maybe other people's research too. Maybe you get requests to add features to your program or you forge a partnership with someone who wants to collaborate on the development of the code. You may end up having to install your program on every computer in your lab, and then you will need a way to keep them all updated with the latest version of your code.

The best approach to dealing with all of these issues is to develop your program from the outset in a way that provides flexibility to cope with all of these potential demands. This can all be done using something called *revision control software*.[24] The essence of revision control (RC) is that you use some additional software to manage the development of your scripts. Let us explore some of the concepts surrounding RC before we mention some specific software you can use.

Naming scripts and giving them version numbers

Let's imagine you have a Perl script called script.pl.[25] It works and has served you well, but now you want to experiment by adding some new functionality. However, you also want to keep the original script around in case your new code doesn't work properly. The novice programmer will copy their original script and rename the new version something like script2.pl.

You may be thinking 'where is the harm in that?' but the problem occurs when you start adding more and more functionality (as well as bug fixes, speed improvements, etc.). Over time your code directory will end up looking something like this:

```
script.pl
script2.pl
script3.pl
script_test.pl
script2b.pl
script_new.pl
script3.1.pl
```

---

[24] Also known as *version control, source control,* or *source code management* (SCM).
[25] Please don't ever give this name to any of your scripts!

Hopefully you get the picture. It can quickly become a mess and you will lose track of which script is doing what. Alternatively, you might decide you no longer need an early version of your script and delete it because you have subsequently improved it. Then you discover, a year after deleting it, that the old version actually did something right and your later versions introduced some flaws you didn't appreciate at the time.

Using RC software avoids all of this mess. Such software typically keeps one master copy of the script, which is the current, latest copy. However, it also keeps track of *all* of the changes you have ever made to your script. Every change you make increments a version number for the script. This version number is controlled by the RC software, but you can also control when you want your script to switch to a major version number change (e.g., from version 1.0 to 2.0). RC software doesn't do all of this automatically – there is some work involved on your part, but that work is worth the effort. Once you have access to the history of every change you made to a script, you can do things such as roll back to an older version, or compare two specific versions. From your point of view, you are still working with one copy of your script in your code directory, the RC software keeps track of all the changes elsewhere.

Collaborative code development

If you wanted to let someone else contribute to one of your scripts, you could maybe share a computer or email code back and forth, but this clearly isn't very practical. What if you had a really big script and there were 100 collaborators who all needed potential access to your code?

RC software allows you to do this. The exact method of how this is done varies with the particular type of software, but in general you develop your code in a way that allows you to control who else is allowed to edit your code. Most importantly, other people can edit your code independently of you. Edits that people make are not applied to the master copy of the code until people specifically sign-off on the changes they have made. This process is called *committing* changes.

In many cases, people might be editing different parts of the code, e.g., different subroutines. In these situations the change by programmer A to subroutine 1 will not affect the changes by programmer B to subroutine 2. But what if two people decide to edit the same lines of code at the same time? In these situations, the RC software will spot that this has happened and typically the person who was the last to commit their changes will be told that their changes are in conflict with the latest version of the code. The exact details of how all of this works are not important, all you need to know is that the software is smart enough to manage all of the changes.

Keeping your code in sync across multiple computers

You might develop some software that needs to be run on multiple computers. Without using RC software you could end up with different computers having different versions of your code.[26]

---

[26] If you don't plan to learn how to use any RC software, then at least try learning about the Unix rsync command.

When you use RC software you typically can run a single command, which has the effect of updating one or more scripts on that computer to whatever the latest version is. This means that if you use RC software then you can ensure your scripts are effectively backed up to multiple computers.[27]

Other benefits of using revision control

This is not the book to go into all of the details of what RC software can do for you. However, you should be aware that you can use such software to keep a production copy of your script and then also make a development version at the same time. This is called *branching*. When you have finished working on your development copy, you can *merge* changes back into the main copy of the code.

RC software also lets you comment on all of the changes you make. These comments are external to the script and allow others to clearly see *why* you made the changes you did (this is essential when developing code with others).[28] These comments can also be useful for you to review the history of your own scripts. Maybe your boss wants you to produce a list of all of the major code development you have done over the last year. With RC software, this can usually be done by typing a single command.

If you ever need to develop code with others, then you should definitely use RC software. Even if you will only ever be working on code by yourself, you should still consider learning to use it. Of course, it will initially involve some extra time and effort, but the benefits outweigh these initial costs.

Which RC software should you use?

There are many different software products you can use for revision control. Some are free and some are not. Some work across different OSs, whereas some are platform-specific. The most popular free ones are CVS, Subversion, and Git. CVS is the oldest of these, having been available since 1990. There is therefore a lot of available support in terms of online documentation, books, etc. Subversion was independently developed to be a successor to CVS and has been around since 2000. There are many similarities between the two systems, meaning that if you learn one, it will not be hard to transition to the other. Git is the newest of the three systems and was developed in 2005. It takes a very different approach to version control compared to CVS and Subversion, though all three systems achieve many of the same goals.

We suggest you compare the different RC software solutions available for your platform and choose one based on the features you desire.[29] If you work in an environment where other people may already be using some form of RC software, then it obviously makes sense to consider using the same software as them.

---

[27] This is not a substitute for having a proper back-up policy, and is no substitute at all if you fail to keep at least two computers updated with the latest version of your scripts.

[28] Commenting on code changes is only useful if you bother to write comments that describe changes in a meaningful way. 'Fixed bug' is not a useful comment, whereas 'Fixed the issue that caused a crash when trying to import files that were not in a correct format' is.

[29] Wikipedia has some excellent pages describing various RC software, including a page that compares the main features of nearly 30 different types of software: http://en.wikipedia.org/wiki/Comparison_of_revision_control_software

A chapter where we return to the end of the line

Assuming you have finished the Essential Unix and Essential Perl parts of this book, you now have a very attractive skill set. You can solve a lot of important real-world problems. You may even find yourself wanting to help other people with their data problems. Great! But before you start to unleash Unix and Perl on masses of data, learn the first rule of data wrangling:

> Rule #1: never trust anyone, not even yourself

What exactly do we mean by this? One of the greatest sources of frustration when analyzing real data is the inconsistencies caused by 'human error.' If you don't identify the errors at the beginning, it can ruin everything that follows. For example, you may be given a spreadsheet with columns containing the first name, last name, email, etc. of a large number of clients. The person who performed the data entry may have accidentally placed both the first and last names in the first column and left the second column empty. A script that processes such a file could break if the last name was absolutely required. Data that are automatically produced can also contain human errors because humans program the machines. You must get in the practice of checking the integrity of your data at every conceivable step. Sometimes these measures will seem terribly silly, but such 'sanity checks' can identify serious problems with the data and save countless hours of work.

Text vs. binary files

The very first sanity check is to ensure your data file is in a plain-text format and is not a binary file.[30] Text files often have a .txt extension to their file name, but they can have many other extensions or no extensions at all. The only thing you can really say about a file name that ends '.txt' is that whoever named it intended for it to be a text file. Binary files can have file extensions such as .exe or no extension at all. A binary file can even have a .txt extension, though it would probably be a human error if one did.

Ultimately, there is no guarantee that you can determine if a file is text or binary from the file name alone. For this reason, you should always look inside a file to determine what kind of file it is. One of the simplest ways to examine a file is to read it with the Unix less command. Locate a music or photo file on your computer and try reading it with less. If it is a binary file, less may issue a warning such as:

```
"Customized.mp3" may be a binary file. See it anyway?
```

If you continue (type 'y'), you may find some unintelligible output. That's because the file is binary. Try this with any of your Perl or Unix scripts and you will see no

---

[30] Perl can read binary files as well as text files, but in this book we only discuss text files. They are easier to process and are human-readable.

such warnings. So how does `less` know what is and is not binary? Text files have a very specific pattern due to the way they map bits to symbols. The most common encodings are UTF8 and ASCII.[31] In these eight-bit encodings, each letter (or other symbol) corresponds to exactly one byte. For example, the binary string `01100101` corresponds to the letter 'e'. Even though there are eight bits in one byte, the highest bit (farthest left) is not used. If you looked at the bits of a text file, you would therefore see a regular pattern where every byte begins with a zero. Here is an example of four bits from a text file.

`01100101 01111000 01110100 01100101`

Binary files can encode their data however the authors choose. The first bit of a byte is not always zero. This makes it fairly simple for a program such as `less` to determine if a file is binary or not: just look at the first bit of every byte.

An alternative way of determining what 'type' a file is, is to simply use the Unix `file` command. This will tell you whether a file is text or not, plus it will also provide you with a whole lot of other information:

```
$ file README.txt
REAMDE.txt: ASCII text

$ file webpage.html
webpage.html: HTML document text

$ file script.pl
script.pl: a /usr/bin/perl script text executable

$ file image.png
image.png: PNG image, 700 x 700, 8-bit/color RGB, non-interlaced

$ file song.mp3
song.mp3: Audio file with ID3 version 2.2.0, contains: MPEG ADTS, layer III,
v1, 160 kbps, 44.1 kHz, JntStereo
```

Line endings

We have now arrived at one of the most common sources of errors with OPD (other people's data). This is literally the end of the line, a subject we addressed briefly in Chapter 5.8. There are three common ways to end a line of text: Unix, Macintosh, and Windows. Whatever method is used, the goal is always the same. Any press of the return or enter key on your keyboard should move the cursor to the next line of whatever text-based program you are using.

The Unix end-of-line character is synonymously known as line feed, control-J, 10 (decimal), 0A (hexadecimal) or 00001010 (binary).

The Mac end-of-line character is the carriage return, control-M, 13 (decimal), 0D (hexadecimal) or 000010111 (binary).

---

[31] These encodings are mostly interchangeable. We always recommend people look at the Wikipedia page for ASCII.

Windows uses *two* characters, carriage return followed by line feed.

These differences can be a frequent source of confusion when working with text files that were originally produced on a different computer system to the one you are using. So what does this mean to you? If you are on a Unix system and someone gives you a file saved on Windows, your file may end up containing extra characters at the end of each line. If they give you a Mac file, it may all appear as a single line of text with no newlines.[32] If you open the file with less, the carriage returns will appear as ^M characters.[33]

Imagine someone hands you a spreadsheet saved in a text format and upon reviewing it with less you realize it has not been saved with Unix line breaks. What should you do? You can open it in your favorite text editor and then save it with the appropriate line endings (most text editors will allow you to switch to whatever line ending you like). But what if you had to convert 1000 files? That could give you an injury! So let's write our own converter.

To convert from Mac to Unix, the script needs to change every occurrence of a carriage return to a line feed (also called newline characters). Just as we have seen that Perl uses \n to represent newline characters, Perl uses \r as the special symbol for carriage return. You can imagine writing a program that looks something like this:

```
#!/usr/bin/perl
use strict; use warnings;
while (my $line = <>) {
    $line =~ s/\r/\n/g;
    print $line;
}
```

For Windows files, you would need to change the substitution part to s/\r\n/\n/g.

It turns out that Perl has some convenient command-line options that will help us accomplish this task. The -e option to the perl command is used to execute a piece of Perl code on the command line.[34] This means we can briefly revisit Chapter 4.1 and learn a new way of printing 'hello world':

```
perl -e 'print "hello world\n"'
```

There are a few other command-line options that are frequently used in conjunction with -e, and one of the most common ones is the -p option, which allows you to print each line of any input file you process (after applying some Perl code to that line). The following Perl command – and it is a command and not a script – will convert a Mac *or* Windows file to Unix:

```
perl -pe 's/\r\n?/\n/g' < mac_or_windows_file.txt > unix_file.txt
```

---

[32] Starting with OS X, line endings on the Mac have been control-J. But for backward compatibility, control-M is also allowed. Many programs still use control-M.

[33] The ^ symbol is a way of specifying what are known as 'control characters.' So you should read ^M as 'control-M.' Some people find this confusing as *two* ASCII characters are being used to represent something which internally is stored as only one character (the carriage return). Control characters often appear in inverse colors, so if your program uses white text on a black background, ^M may appear as black text on a white background.

[34] This command-line option opens up a whole new world of what are known as 'Perl one liners', which are short Perl commands you run on the command line and which often do some very powerful processing of input text files. We don't cover Perl one liners in any more detail in this book, but we encourage you to look them up on the internet.

This very simple command will loop over each line of the specified input file, apply the Perl code to that line, and then print the output, which we redirect into a new output file. Notice that we have changed the regex in the substitution so it will substitute a carriage return followed by zero or one newlines. This makes it work with both Mac and Windows files. To run this command with less typing, you can make it an alias (see Chapter 3.21):

```
alias to_unix="perl -pe 's/\r\n?/\n/g'" # sh shells
alias to_unix "perl -pe 's/\r\n?/\n/g'" # csh shells
```

Normally we don't suggest making your programs so short because they become hard to read. But this is so cute we had to share it.

Expectations about OPD

In many cases you should be able to form some firm ideas about what 'good' data should look like, even before you receive, download, or generate it. These firm ideas should form the basis of your sanity checks. For instance, if you downloaded a file containing millions of DNA or protein sequences, then you know, a priori, that there are some characters that should never appear in those sequences.[35] Equally, if you downloaded some DNA sequences that correspond to genes, then the coding portion of those sequences should have a length which is a multiple of three nucleotides.

As we have mentioned, you should always look at the data first. But there will usually be more data than you can inspect by eye. In light of that, if you have any suspicions about the data then it is entirely appropriate to spend an hour or two writing a simple 'checking' script that just ensures the data looks valid. Don't be the bad programmer who instead spends a week working on a script that will analyze your 5 GB file of genome data, only to then discover you have mistakenly downloaded a digital copy of the film 'Superbabies 2.'

Sometimes you might not know exactly what the 'rules' are to which your data (or OPD) should adhere. However, you should often be able to come up with some appropriate 'guesstimates' for what constitutes good data. For example, start coordinates should come before end coordinates; lengths of 'things' tend to be non-zero values; ages of fossils/excavations/ancient civilizations should not be older than 4.6 billion years. These types of things are all good candidates for data sanity checks. Even when all of your data can potentially take on any possible value, you might still have some expectations that a certain proportion of the data points should fall into a range between X and Y. There are a lot of bad data in the world, and at some point you *will* receive some of them. So remember:

Never trust OPD ... always examine it!

---

[35] For example, X is never a valid character by which to represent a nucleotide and J does not correspond to any amino acid.

Or how to find pearls of wisdom about the wisdom of Perl

Although it pains us to say it, there will be times when you encounter a problem and this book won't be able to help you solve it. This may be because you are learning one of the many aspects of Unix or Perl we have not covered in this book,[36] or maybe it's because we have explained something particularly badly.[37] Whatever the reason, you might have reached a point where hours – or even days – of debugging have still not fixed your problem and you now find yourself banging your head against a wall in frustration. This final chapter of the book will point you to a few places where you can go for help.

Built-in support tools

You may not need to go very far to find help for your Unix or Perl problem. There are probably a couple of support mechanisms that are already available on your computer. We have already explained, way back in Chapter 3.13, that every Unix command has its own set of documentation contained in man pages; these can be accessed by using the Unix man command. But what about Perl? Earlier in this part of the book we mentioned a perldoc command that can be used to view any pod documentation contained within a Perl script (see Chapter 7.6). This command can also help you in many other ways.

You can look up a description for any Perl function by simply providing the -f argument to the perldoc command and then adding a valid function name. Imagine you wanted to refresh your memory as to how Perl's integer function works:

```
$ perldoc -f int
int EXPR
int    Returns the integer portion of EXPR. If EXPR is omitted, uses
       $_. You should not use this function for rounding: one because
       it truncates towards 0, and two because machine representations
       of floating point numbers can sometimes produce
       counterintuitive results. For example, "int(-6.725/0.025)
       produces -268 rather than the correct -269; that's because it's
       really more like -268.99999999999994315658 instead. Usually,
       the "sprintf", "printf", or the "POSIX::floor" and
       "POSIX::ceil" functions will serve you better than will int().
```

You can also use this command with the -m option to find out about any of the built-in Perl modules that are available. Maybe you can't remember the syntax of the Getopt::std module:

---

[36] There is far, far more to Unix and Perl than we have included in this book and if you are learning new material not covered here, then this news will fill our hearts with joy.

[37] If you have found our explanations confusing or unsatisfactory then this news will fill our hearts – and potentially our wallets should you return the book – with sadness.

```
$ perldoc -m Getopt::Std
package Getopt::Std;
require 5.000;
require Exporter;

=head1 NAME

getopt, getopts - Process single-character switches with switch clustering

=head1 SYNOPSIS

  use Getopt::Std;

  getopt('oDI');  # -o, -D & -I take arg. Sets $opt_* as a side-effect.
  getopt('oDI', \%opts);  # -o, -D & -I take arg. Values in %opts
  getopts('oif:'); # -o & -i are boolean flags, -f takes an argument
          # Sets $opt_* as a side-effect.
  getopts('oif:', \%opts); # options as above. Values in %opts

=head1 DESCRIPTION

The getopt() function processes single-character switches with switch
clustering. Pass one argument which is a string containing all switches
that take an argument. For each switch found, sets $opt_x (where x is the
switch name) to the value of the argument if an argument is expected,
or 1 otherwise. Switches which take an argument don't care whether
there is a space between the switch and the argument.
```

In this example we are only showing some of the output of the `perldoc` command.
Also note that the -m option shows you any pod format documentation that is present
as well as the underlying Perl code from the module. If a Perl module doesn't have
any documentation, you will just see the actual Perl code (which still might be useful
to you).

One of the most powerful ways you can use the `perldoc` command is to search the
text of the Perl FAQ.[38] This FAQ covers an impressive list of common – and some not so
common – questions a programmer might ask. To search it, just use the -q option and
specify a word or phrase to search with. Imagine you were curious as to how you could
shuffle an array:

```
$ perldoc -q shuffle
  How do I shuffle an array randomly?

  If you either have Perl 5.8.0 or later installed, or if you have
  Scalar-List-Utils 1.03 or later installed, you can say:

    use List::Util 'shuffle';
    @shuffled = shuffle(@list);
```

Once again, we are only showing a shortened version of the actual output you would
see. In this case there is only one entry in the FAQ that mentions the word 'shuffle.' If

---

[38] Frequently Asked Questions ... asking what FAQ means is also a FAQ.

there were multiple entries they would all be listed one after another, each with its own header. If you searched the FAQ for questions relating to the word 'process' you would see the following FAQ entries:

```
How do I process/modify each element of an array?
How do I process an entire hash?
How do I process each word on each line?
How do I start a process in the background?
How can I call backticks without shell processing?
How do I close a process's filehandle without waiting for it to complete?
How do I fork a daemon process?
```

If you want to know more about how you can use the perldoc command then you will be happy to know it also has its own man page. If you are ever stumped by a Perl problem, try searching the FAQ; it has the answers to many questions.[39]

Unix and Perl support on the internet

Both Unix and Perl predate the invention of the world wide web.[40] Therefore, it is not surprising that one can find a staggering amount of Unix- and Perl-related information on the internet. There are thousands and thousands of web sites devoted to Unix and Perl that will educate and inform you at every conceivable level of complexity or functionality.[41] Given the millions of people that use these fantastic software tools, it is often easier to start any troubleshooting by assuming the following premise:

> *Whatever problem you are experiencing, someone else has had it, fixed it, and probably blogged about it.*

Realizing this means you can often expect to find dozens of web sites that will all list the solution to whatever basic problem you are having. Though, of course, you should be careful when blindly relying on other people's code or instructions as it may fix your problem but only in a flawed way.[42]

Rather than relying on what *might* be a correct solution that you find posted on a web site, it is sometimes better to ask for help. You might find code on the web that you can use to fix your problem, but you might not be able to understand such code. It can be dangerous to use code that you don't fully understand, so we recommend asking an expert. But who can you ask?

One of the great things about the Unix and Perl community is that there are lots of friendly and knowledgeable people out there who will be happy to help you with your

---

[39] Although we are nearing the very end of this book, we should point out that one important question that is answered by the FAQ is the fundamental issue of 'What is Perl?'

[40] The web dates back to 1991, whereas Unix has been around since 1969 and the first version of Perl was available in 1987.

[41] One of these sites is the official Perl site at http://perl.org.

[42] Over the years Unix and Perl have changed and they continue to change. Check the date on any advice you find on the internet. A solution to your problem that is on a web site that hasn't changed since 1998 might not be the latest and greatest solution.

problem. One of the best places to find such people is via the discussion newsgroups that make up Usenet.[43] There are over 125 Usenet discussion groups related to Perl and 40 related to Unix (not to mention more groups that focus on Linux). All of these groups can be easily viewed at Google's 'Groups' interface (http://groups.google.com). You can browse the groups, search for specific questions, or just go ahead and ask a question.[44] The most popular groups for Unix and Perl are comp.unix.shell and comp.lang.perl.

The Google Groups site also hosts another set of discussion groups that are similar to, though separate from, the Usenet groups. We have created a Google discussion group for the specific coverage of Unix and Perl issues in a biological context. The group is called 'Unix and Perl for Biologists,' which you should be able to easily find from a web search engine, or from searching at http://groups.google.com. The authors – not to mention a few hundred other group members – see every question posted on this forum and we welcome yours.

When posting questions on any discussion forum please remember to follow standard posting etiquette[45] and also make sure you add sufficient details about your problem, including:

(1) a copy of the Perl code that produces an error or problem;
(2) a description of what output you were expecting;
(3) details of the format or content of any input data;
(4) a copy of the *exact* Unix or Perl command that was run;[46]
(5) the version number of Perl;
(6) the type of Unix shell;
(7) a description of any other hardware or software details which might prove relevant (e.g., amount of RAM, OS, etc.).

The more information you provide, the more likely it is that you will get rapid and detailed help! As a final note, we will just remark that the only thing better than getting help from such a discussion group is being the person who helps someone else with their problem. As you become more proficient in the ways of Unix and Perl, it is a worthy action to help pass this knowledge on to others. If you will allow us a final indulgence, we'd like to end this book by paraphrasing a famous Chinese proverb:

> *Give a person a Perl script, and you might satisfy their needs for a day; teach a person how to write Perl scripts, and you'll satisfy their needs for a lifetime*

Happy coding!

---

[43] Usenet is a collection of thousands of different discussion forums that are available for discussing just about every subject you might care to think of (and plenty that you wouldn't). Some of these groups are over 30 years old and are still actively used, particularly those groups that relate to different aspects of computing.

[44] These groups allow posting and/or reading by email or you can subscribe to posts as an RSS feed … or you can just use a web browser (or other dedicated desktop client) to browse the groups.

[45] Demanding instant answers, insulting other posters, and generally acting in a rude manner are all behaviors that make it unlikely your question will be answered.

[46] People are often surprisingly vague in describing their computing problems and frequently there is not any problem with the Perl code, only with how it is being run.

Appendix

Solutions to problems

Remember, there's more than one way to do it!

At the end of some of the Unix and Perl chapters, we have provided one or more problems for you to tackle. For the Perl problems, it is important to realize that nearly all problems can be solved in a myriad of different ways. Some solutions will comprise elements that are clearly 'variants on a theme,' whereas others might include ideas that are completely unrelated. This can be great for creative thinkers, but at the same time it can be daunting for people who want to know *the* right way to do something. Sometimes, two different solutions may be functionally identical and will differ only in the readability of the code and in their use of Perl's many magical shortcuts.[1]

Be wary of people who claim that a specific Perl script offers the 'best' solution to a problem. The 'best' solution to any problem can be highly subjective. For example, some people want their code to run as fast as possible; others want it to be as understandable as possible; the two approaches are not always mutually compatible. In this appendix we mostly provide just one solution for each of the problems we have described. We have endeavored to choose solutions that are easy to understand but we encourage you to think of alternative solutions as well.

Unix solutions

Note that many of the Unix problems are very open ended and so in many cases the 'solutions' listed below describe very general ideas about solving the problem, or may just list a useful tip or fact about the relevant subject matter.

Solution 3.7.1
We can't provide an exact solution as it will largely depend on the layout of your filesystem. Remember, you could navigate to your root level directly by using cd /, or you could just keep navigating upwards multiple times (cd ..).

Solution 3.9.1
The only solution for this problem is to realize that simply running the cd command is enough to return you to your home directory.

Solution 3.12.1
There are several command-line options for the ls command that work very well together. Do you want to list all files (including 'hidden' files[2]) with a long-form listing that reverse-sorts them by the file modification time, shows file sizes in human readable format, *and* adds a slash after the names of directories? It's as simple as:

```
$ ls -alhtrp
```

---

[1] E.g., use of $_ or @_.
[2] Covered in Chapter 3.24.

Solution 3.13.1

Try to become familiar with the typical sections of a man page. The 'Synopsis,' 'Description,' and 'Examples' sections will probably be the most useful to you, especially the latter (if present).

Solution 3.15.1

There are times when you will have lots of files that all start with the same letter, and in those situations you might want to type 2–3 characters before reaching for the tab character to autocomplete. As your filesystem grows over time, and you find yourself having to navigate through multiple levels of directory structure, the use of tab-completion will save you incalculable amounts of time.

Solution 3.15.2

There are many other Unix tools that can be used to help navigate your history of commands, but in practice you will find yourself using the up arrow a lot just to get back to a recent command.

Solution 3.16.1

The first example should list everything in the current directory (potentially including the files you created earlier in this chapter). The second and third examples should achieve the same result, because all of the files that start with 'f' also end with 't'. The last two examples should hopefully reveal the function of the question mark character. In Unix, you use the question mark to match any single character.[3] Compared to the asterisk, it can give you a little more control and precision when matching file names.

Solution 3.17.1

Hopefully you realized that steps 2 and 3 could be combined by use of the -p option of the mkdir command:

```
$ cd /bin
$ pwd
/bin

$ mkdir ~/Tears
$ mkdir -p /tmp/123/XYZ

$ mv /tmp/123/ ~/Tears/
$ ls ~/Tears
123

$ mv ~/Tears/123/ /tmp
```

---

[3] Note that this specifically refers to when you are trying to match file names. Some Unix shells may also use the question mark for other uses.

```
$ rmdir /tmp/123/XYZ/
$ rmdir /tmp/123/
$ rmdir ~/Tears/
```

Solution 3.18.1

There are obviously many different ways you could remove these files. Assuming you have created all of the files and directories from the last few exercises, here is one possible solution:

First, navigate to the 'Temp' directory and check what you need to remove:

```
$ pwd
/Users/nigel/Unix_and_Perl
$ ls
Code   Temp
$ cd Temp
$ ls -p
Temp3/     fat    feet   heaven.txt
earth.txt  feat   fit    rags
```

We can use a wildcard to remove many files in one go, but we still ensure that we use the -i option:

```
$ rm -i f*t
remove fat? y
remove feat? y
remove feet? y
remove fit? Y
```

We can use a wildcard to remove the text (.txt) files, and then remove the 'rags' file separately:

```
$ rm -i *.txt
remove earth.txt? y
remove heaven.txt? y
$ rm -i rags
remove rags? Y
```

Notice that we could also remove all files in one go by providing them as a list to the rm command:

```
$ rm -i f*t *.txt rags
```

The 'Temp3' directory is empty so this is easy to remove with the rmdir command:

```
$ rmdir Temp3
```

Solution 3.19.1

There is no real solution for this problem, so here's a hint instead. If you use your graphical file manager to empty your trash folder in order to delete files, it sometimes doesn't work. This can happen if the OS knows that another program is still working with the file in question. In these situations you can usually use the rm and rmdir commands to remove files and directories from your trash folder. On a Mac the trash folder exists as a hidden directory in your home directory called '.Trash' (the dot character is important). You can just 'cd ~/.Trash' and then remove the file.

Solution 3.20.1

Press 'G' to jump to the end of the file, and 'g' to jump back to the beginning. You can jump 'x' percentage of the way through a file by simply typing a number followed by the percentage sign, followed by return/enter.

To search for a string, press the forward-slash character and then type your search pattern followed by return/enter. This will take you to the next match *and* highlight all matches in the file. You can jump to the next match by just pressing forward-slash and return/enter. Search backwards by using a question mark character rather than a forward-slash.

Solution 3.21.1

Here is how you would create an alias for the rm command. Note that we first illustrate the default behavior of the rm command, which is to delete a file *without* asking for confirmation:

```
$ touch aaa
$ ls
Code    Temp    aaa
$ rm aaa
$ ls
Code    Temp

$ alias rm='rm -i'
$ touch aaa
$ ls
Code    Temp    aaa
$ rm aaa
remove aaa?
```

Solution 3.22.1

To find out more about using the nano editor, press control + G to access the built-in help system.

Solution 3.24.1

Here is our solution:

```
$ pwd
/Users/nigel/Unix_and_Perl

$ mkdir .invisible

$ mv .invisible/ Temp/

$ ls -a Temp/
./          ../              .invisible/

$ rmdir Temp/.invisible/
$ ls -a Temp/
./   ../
```

Perl solutions

Solution 4.1.1

Two possible ways of printing three lines of output are shown below. First, we will use one print() statement for each line of output:

```
1.   # solution 1A - use multiple print statements
2.   print("Hello World!\n");
3.   print("How are you?\n");
4.   print("Goodbye World!\n");
```

Alternatively, we can use multiple newline characters to achieve the same output but with just one print() statement:

```
1.   # solution 1B - use one print statement
2.   print("Hello World!\nHow are you?\nGoodbye World!\n");
```

Hopefully, you can see that each time Perl sees a newline character it behaves like it is hitting the return/enter key on your keyboard. In this example, we can get three lines of output fairly easily with just one line of Perl code. However, this might not be the best solution. If each line of text was quite long then we'd have to wrap our line of code around the screen of our code editor. It would still work, but it might look a bit ugly, and we never want to write ugly code (at least not on purpose). This raises an important point: Code that is longer but easier to understand is nearly always preferable to concise code that is hard to understand.

Solution 4.1.2

```
1.   # solution 2A
2.   print("Hello World!\n);
```

Here we have removed just a single character from the script. Can you spot it? It's the missing closing quote character. A lot of things in Perl come in pairs (quotes, parentheses, etc.), and having an unmatched number will result in an error. In this case, the error should say something like:

```
Can't find string terminator '"' anywhere before EOF at helloworld.pl line 2.
```

This error message may not be the most human-friendly error message you will ever see, but it does point out that a missing ' character is part of the problem. By the way, 'EOF' means 'end of file.' If this was our error message then we would probably write 'end of file' rather than 'EOF', but as we didn't write Perl, it's a bit out of our control.

Solution 4.2.1

Without the dollar symbols Perl doesn't know that x and s are variables, and so you should see two error messages like this:

```
Can't modify constant item in scalar assignment at scalar.pl line 3, near "3;"
Substitution pattern not terminated at test.pl line 5.
```

These error messages might seem hard to understand at the moment, but you should at least recognize that both error messages correctly identify which lines of your code contain the problems.

Solution 4.2.2

The following code should hopefully show you that you can assign a variable many different values. Note the use of comments to explain various issues and note that line 9 intentionally omits a newline character.

```
1.   $x = 3;
2.   print "x is $x\n";
3.   $x = 5;
4.   print "x is now $x\n";
5.   $x = 10; # assigning $x a new value but not doing anything with it
6.
7.   $y;              # declaring a variable without assignment
8.   $y = 2;          # assigning a value
9.   print '$y is ';  # single quote printing to print the text $y
10.  print "$y\n";    # double quote printing to get the contents of $y
```

Solution 4.2.3

If you assign one variable to another, then this is like making a copy of the first variable. After you have made the copy you can then change the value of the first variable without affecting the value of the second:

```
1.   $x = 100;
2.   print "x is $x\n";
3.   $y = $x;
4.   $x = 200;
5.   print "x is $x and y is $y\n";
```

Solution 4.2.4

In this problem, the three alternatives all print the same result. You will typically assign strings to variables using single quotes or double quotes. As we will shortly learn, it is not a good idea to use no quotes at all (though it will work in this example). The difference between single quotes and double quotes is that single quotes contain exactly what is between the quotes. Single quotes do not perform variable interpolation or recognize special character codes such as \n. Let's look at an example to make this clear.

```
1.   $favorite_food = 'cheeseburgers';
2.
3.   print 'I like $favorite_food\n';
4.   print "\n";
5.
6.   print "I like $favorite_food\n";
```

   Line 3 outputs: I like $favorite_food\n
   Line 4 outputs: I like cheeseburgers
   If you want to simplify your life, you can use double quotes all the time. We use single quotes when we want to convey that a string is not going to change.

Solution 4.3.1

You should notice that if you move use warnings; so it is after the $dna = actgag-cac; line then you should no longer see the error about an 'unquoted string.' This is because Perl only turns on warnings at the point in the script where it sees use warnings; and this is why it should always be among the first few lines of code in your scripts.

Solution 4.4.1

Simply nest all three functions inside one another:

```
1.   # 3_lines_to_1.pl
2.   use warnings;
3.
4.   $answer = sqrt(int(rand(100)));
5.   print "The answer is $answer\n";
```

You could also have omitted all the parentheses from line 4:

```
$answer = sqrt int rand 100;
```

Solution 4.4.2

You should have seen that Perl handles quoted and unquoted numbers in the same way, and all three variables ($x, $y, and $z) should print the value 3. Behind the scenes, Perl has two basic data types: numbers and strings. If we include the warnings pragma in a script, then we have learned that strings should be quoted, but this doesn't apply to numbers.

Perl is generally smart enough to know that sometimes you will use numbers as strings, and conversely if you have created a string which only contains numbers, then Perl will treat it as a number instead. Actually, this in only true in those situations where you are trying to use a string as a number. For example, in line 8 of the script, we add $x, $y, and $z and put their sum in a new variable $sum. When we print $sum we get the expected answer 9. However, internally Perl noticed that $y and $z are actually strings not numbers (because of the quotes). They stay as strings until we try adding them together at which point Perl sensibly realizes that you probably want to treat them as numbers instead of strings.

If this all sounds terribly confusing, then don't worry. The take-home message is that Perl will try to make sense of your numbers and strings for you. This is in contrast to most other programming languages, where you cannot interchange numbers and strings like this. Just remember that you don't need to quote numbers when you first declare them.

Solution 4.4.3

You may (or may not) be surprised to see that this script prints:

```
int_x = 1, int_y = 1
```

If you were expecting the second integer to be rounded up to 2, you should understand that the int() function simply discards the fractional part of the number (i.e., everything after the decimal point); it doesn't do any rounding. Perl does provide a way of rounding numbers and we will see that later on.

Solution 4.8.1

There are lots of ways to test for zero. Here is one way to do it. You should try running this script several times to check that it works with different random numbers:

```perl
1.    #!/usr/bin/perl
2.    # bonus.pl
3.    use strict; use warnings;
4.
5.    my $bonus = int rand (4);
6.    print "Bonus is $bonus thousand dollars\n";
7.
8.    if ($bonus == 0){
9.        die "Boo hoo. No bonus\n";
10.  }
```

Note that we have to use rand(4) to get a random integer between 0 and 3. This is because the int() function rounds down. Here are two more ways you could test the value of $bonus:

```perl
unless ($bonus < 0 or $bonus > 0) {
    die "Boo hoo. No bonus\n";
}
```

```perl
die "Boo hoo. No bonus\n" if (!$bonus);
```

The first of these methods works but is overly complex. The second example more simply asks whether $bonus evaluates as *not true*, which it will if it is zero. Note that this last example also uses the postfix notation; it is very common to see die() statements used in this way.

Solution 4.11.1

This script will print *both* statements from lines 8 and 13 because both of the if statements will evaluate as true. This might seem strange because no pattern was specified for either matching operator. However, if you don't specify a pattern, then the matching operator matches everything, including the empty string.

Solution 4.12.1

To solve this script we need to utilize a mixture of different things that we have learned from the last few sections. Here is one possible solution.

```perl
1.   #!/usr/bin/perl
2.   # url_handler.pl
3.   use strict; use warnings;
4.
5.   my $url = "http://unixandperlforever.com";
6.
7.   # first trim the 'http://' part of the URL
8.   $url =~ s#http://##;
9.
10.  # now check length and die if too long
11.  if(length($url) > 25){
12.      die "$url contains more than 25 characters\n";
13.  }
14.
15.  # make name all upper case
16.  $url = uc($url);
17.
18.  # now count each A, B, or C
19.  my $a = $url =~ tr/A/A/;
20.  print "There are $a As in $url\n";
```

```
21.
22.  my $b = $url =~ tr/B/B/;
23.  print "There are $b Bs in $url\n";
24.
25.  my $c = $url =~ tr/C/C/;
26.  print "There are $c Cs in $url\n";
```

Note the use of comments and whitespace in this script. Also note how we declare three variables to store the counts of A, B, and C and assign them a value with the tr operator all in one go. If the tr operator makes no changes, then it will return a value of zero. Finally, we should mention that as soon as you include if statements in your code, then you should always try to test your code with example data that will trigger either possibility. If you wrote a similar program to ours but only used a web site name that was 25 characters or less, then you would never be entirely sure whether the if statement on lines 11–13 was working properly.

Solution 4.13.1

The simple solution to this is just to introduce a third variable to act as a temporary 'holding cell':

```
1.   #!/usr/bin/perl
2.   # swap.pl
3.   use strict; use warnings;
4.
5.   my ($x, $y) = ("case", "book");
6.   print "$x $y\n";
7.
8.   my $z = $x;     # make a copy of $x
9.   $x = $y;        # copy $y to $x
10.  $y = $z;        # and now copy what was $x to $y
11.
12.  print "$x $y\n";
```

Solution 4.14.1

We need to reverse the array, but as the array only has three elements, then we just need to swap the first and last elements with each other. If you remember from the section on lists, swapping items is very easy in Perl.

```
1.   #!/usr/bin/perl
2.   # abc.pl
3.   use strict; use warnings;
4.
5.   my ($one, $two, $three) = qw(alpha bravo charlie);
6.
7.   my @list = ($one, $two, $three);
```

```
8.   print "BEFORE: @list\n";
9.
10.  ($list[0], $list[2]) = ($list[2], $list[0]); # the swap
11.  print "AFTER: @list\n";
```

Solution 4.15.1

Apart from being able to use the splice() function properly, the other issue to solve for this problem is how to calculate what the middle position of an odd-length array is. You can do this simply by dividing the array length by 2, and then taking the integer of that value. If your array has seven elements, the middle position is provided by int(7/2). Because int() rounds down, this will give you the value '3,' which is the fourth (middle) position in the array.

Putting all this together, here is one solution to the problem.

```
1.   #!/usr/bin/perl
2.   # these_are_a_few_of_my_favorite_things.pl
3.   use strict; use warnings;
4.
5.   my @things_i_like = qw(cheese music spiders perl cinema);
6.   print "OLD: @things_i_like\n";
7.   my @more_things_i_like = qw(gadgets satire macs);
8.
9.   my $length_of_array = @things_i_like;
10.  my $middle_position = int($length_of_array / 2);
11.
12.  splice(@things_i_like, $middle_position, 1, @more_things_i_like);
13.
14.  print "NEW: @things_i_like\n";
```

Note that if we had wanted to, we could have avoided having to write lines 9 and 10, and instead combined this information directly into line 12:

```
splice(@things_i_like, int(@things_i_like/2), 1, @more_things_i_like);
```

As always, Perl will know that parts of this line have to be evaluated before other parts. Perl would first deduce that if we are using int() function, that this requires a number and therefore the @things_i_like array will be evaluated in scalar context to give its length. That length can then be divided by 2 and the resulting value can be passed to the int() function. Only then can Perl perform the entire splice() command. If you find this too confusing, then it's perfectly fine to do things in multiple steps. As always, it is most important that things are clear and understandable.

Solution 4.16.1

We only asked you to count the number of items in @ARGV, we don't need to actually do anything with those items. This means we can simply assign @ARGV to a single variable and that will give us the count. We then just need a simple if-else statement to stop the script if there are too few items, otherwise print out the count.

```
1.   #!/usr/bin/perl
2.   # count_things.pl
3.   use strict; use warnings;
4.
5.   my $count = @ARGV;
6.
7.   if ($count < 3){
8.       die "Please specify at least three things\n"
9.   }
10.  else {
11.      print "You specified $count things\n";
12.  }
```

Note that it would be fairly unusual to simply count the number of command-line arguments without doing anything else with them. In most scripts you would instead first assign them to another array and then count them:

```
5.   my @things = @ARGV;
6.   my $count = @things;
```

Solution 4.18.1

We can use @ARGV to store our input numbers, but we also want a suitable die() statement to check that @ARGV contains at least three elements. Then we can use the sort() function to (numerically) sort @ARGV and assign the output to a new array. Then highest and lowest values will be the first and last elements of this array. To remove the end elements, we just use the shift() and pop() functions:

```
1.   #!/usr/bin/perl
2.   # sort_input.pl
3.   use strict; use warnings;
4.
5.   # stop script if there are not enough numbers
6.   die "Specify at least 3 numbers\n" if (@ARGV < 3);
7.
8.   my @sorted_input = sort { $a <=> $b } @ARGV;
9.
10.  # discard ends of array
11.  shift(@sorted_input); # discard first array element
12.  pop(@sorted_input); # discard last array element
13.
14.  print "Final output is @sorted_input\n";
```

Solution 4.19.1

The trick to solving this script is to introduce a variable which will contain the running total. This variable needs to be declared before the for loop, but will be modified within the loop (using the += operator):

```perl
#!/usr/bin/perl
# sum.pl
use strict; use warnings;

die "usage: sum.pl x y z etc.\n" unless @ARGV;
my @input = sort { $a <=> $b } @ARGV; # numeric sort

my $total = 0;
for (my $i = 0; $i < @input; $i++) {
    $total += $input[$i];
    print "Number: $input[$i] Running total: $total\n";

}
```

Each iteration of the loop sees the current value of $input[$i] added to whatever is already in $total.

Solution 4.20.1

To solve this using a for loop, we have to realize that if we are counting backwards then we need to start with the highest array index (position 2 if there are three items in the array) and then use the decrement operator to count backwards. The validation step will therefore be true as long as the loop counter is greater or equal to zero.

For the foreach loop, we just need to first reverse the list using the reverse() function. We can then assign each item to a temporary value.

We calculate the while loop last as this loop is going to use the pop() function to remove one element at a time from the end of the array. When this loop finishes there will be no items left in the array.

```perl
1.   #!/usr/bin/perl
2.   # loops.pl
3.   use strict; use warnings;
4.
5.   my @list = qw(start middle end);
6.
7.   # for loop
8.   for (my $i = @list -1; $i >= 0; $i--) {
9.       print "$list[$i]\n";
10.  }
11.
12.  # foreach loop
13.  foreach my $item (reverse(@list)) {
14.      print "$item\n";
```

```
15.  }
16.
17.  # while loop
18.  while (@list) {
19.      my $item = pop(@list);
20.      print "$item\n";
21.  }
```

Solution 4.23.1

The key part of this problem is realizing that if you want to append to an existing file, you don't need to loop through the contents. You just open a filehandle in append mode, and then print to that filehandle.

```
1.   #!/usr/bin/perl
2.   # signature.pl
3.   use strict; use warnings;
4.
5.   my ($file) = @ARGV;
6.   die "Please specify a file to sign\n" if (@ARGV != 1);
7.
8.   my $name = "Keith Bradnam & Ian Korf";
9.
10.  open(my $out, ">>$file") or die "error appending to $file. $!";
11.  print $out "$name\n";
12.  close $out;
```

Solution 4.24.1

It is sometimes useful to create reciprocal hashes so you can effectively look up either the key or the value. You should only do this if you know that the values of the hashes are unique. Here is one possible solution to the problem:

```
1.   #!/usr/bin/perl
2.   # countries.pl
3.   use strict; use warnings;
4.
5.   die "Specify a country name or two-letter code\n" if (@ARGV != 1);
6.   my ($input) = @ARGV;
7.
8.   my %countries_to_codes = ("Australia" => "au",
9.                             "China"     => "cn",
10.                            "France"    => "fr");
11.
12.  my %codes_to_countries = ("au"  => "Australia",
13.                            "cn"  => "China",
```

```
14.                              "fr"  => "France");
15.
16.  if (exists $countries_to_codes{$input}){
17.        print "$input = $countries_to_codes{$input}\n";
18.  } elsif (exists $codes_to_countries{$input}){
19.        print "$input = $codes_to_countries{$input}\n";
20.  } else{
21.        print "$input was not found in either hash\n";
22.  }
```

This script first checks that we have only one thing specified on the command line and then proceeds to set up both hashes. Lines 16–22 then perform an if-elsif-else statement to check all possible situations. In this case the else statement captures the possibility that the specified input does not exist in either hash.

Note that the exists() function, like so many others in Perl, does not require the parentheses surrounding the name of the hash key you are testing.

In this example we know that each key *and* value is unique. It is not always common for every value in a hash to be unique, but when that is the case, you could also create the second hash more simply by using the reverse() function (which works on hashes as well as scalars and arrays[4]):

```
my %codes_to_countries = reverse %countries_to_codes;
```

Finally, note that we could also achieve the same functionality by using a single hash. This is possible because if we combine the keys from both hashes, they will still be unique. Such a hash would look like this:

```
my %countries = ("Australia" => "au",
                 "China"     => "cn",
                 "France"    => "fr",
                 "au"        => "Australia",
                 "cn"        => "China",
                 "fr"        => "France");
```

Solution 4.27.1
Here is a relatively straightforward way of solving this problem:

```
1.   #!/usr/bin/perl
2.   # hero.pl
3.   use strict; use warnings;
4.
5.   die "No file specified\n" if (@ARGV != 1);
6.   while (my $hero = <>) {
```

---

[4] Functions that do the same thing on scalars, arrays, *and* hashes are not all that common in Perl and not everyone knows that you can reverse a hash like this. Though we can't guarantee that this knowledge will impress your friends at a party, feel free to try.

```
7.        chomp($hero);
8.        if($hero =~ m/man/i){
9.            $hero =~ s/[0-9] //;
10.           print "$hero\n";
11.       }
12. }
```

We first check that a file is specified on the command line and then loop through that file with a `while` loop. Each input line is assigned to `$hero`; we then `chomp` that line to remove the trailing newline character. Although it is not necessary in this program, chomping your input lines is generally a good practice.

Line 8 establishes an `if` statement to see if the line matches the pattern 'man' (we include the ignore-case option to match 'man' and 'Man').

If there is a match we use the substitution operator to remove a single digit and space character. Remember, if you don't specify a replacement pattern, you are just removing characters.

Solution 4.27.2

Here is one possible solution:

```
1.   #!/usr/bin/perl
2.   # address_check.pl
3.   use strict; use warnings;
4.
5.   die "No file specified\n" if (@ARGV != 1);
6.
7.   while (my $address = <>) {
8.       chomp($address);
9.       print "$address: ";
10.      if ($address =~ m/^\w{2}\s\d{5}$/) {
11.          print "State and zip code present\n";
12.      } elsif ($address =~ m/^\d{5}$/) {
13.          print "state code missing\n";
14.      } elsif ($address =~ m/^\w{2}$/) {
15.          print "zip code missing\n";
16.      } else {
17.          print "state and zip code missing\n";
18.      }
19. }
```

After looping through the file we first check if each line starts with a state code (`^\w{2}`) and ends with a zip code (`\s\d{5}$`). If this does not match, then we proceed to use two more `elsif` statements to check whether either the state or zip code is present. Finally, we include an `else` statement to warn us if neither field is present.

Solution 4.28.1

The following regex includes both anchor metacharacters (^ and $) to ensure that $seq does not contain anything else. It also includes the 'ignore case' option and deals with unknown characters by adding 'N' into the character class:

$seq =~ m/^ATG[ACGTN]+(TGA|TAA|TGA)$/i;

This regular expression will match sequences that contain at least seven valid characters (e.g., 'ATGATGA' would match to this regex). Note that the pattern ends by specifying all three possible stop codons. We could also write this part like so:

T(GA|AA|GA)

On the one hand, this simplifies the regex by shortening it by two characters, but on the other hand this reduces the 'readability' of the pattern as it becomes slightly less obvious that we are dealing with stop codons.

Ensuring that the sequence must be a minimum of 100 nt can be done by swapping the + quantifier for {94,}; the remaining 6 nt will come from the start and stop codons:

$seq =~ m/^ATG[ACGTN]{94,}(TGA|TAA|TGA)$/i;

The final, and most difficult, step is to ensure the sequence is a multiple of 3 nt. One way to do this is to first define a single codon ([ACGTN]{3}), and then put that in a group with another repetition modifier to ensure that it occurs at least 32 times (this makes 96 nt, with the remainder coming from the start and stop codons):

$seq =~ m/^ATG([ACGTN]{3}){32,}(TGA|TAA|TAG)$/i);

Index